ANIMAL PEOPLE

EXPERTISE

**CULTURES AND
TECHNOLOGIES
OF KNOWLEDGE**

EDITED BY DOMINIC BOYER

A list of titles in this series is available at cornellpress.cornell.edu.

ANIMAL PEOPLE

Moral Subjects in the
Work of Animal Protection

Adam Reed

CORNELL UNIVERSITY PRESS   ITHACA AND LONDON

First published 2024 by Cornell University Press

Library of Congress Cataloging-in-Publication Data

Names: Reed, Adam, 1967– author.
Title: Animal people : moral subjects in the work of animal protection / Adam Reed.
Description: Ithaca : Cornell University Press, 2024. | Series: Expertise : cultures and technologies of knowledge | Includes bibliographical references and index.
Identifiers: LCCN 2024021682 (print) | LCCN 2024021683 (ebook) | ISBN 9781501779633 (hardcover) | ISBN 9781501779640 (paperback) | ISBN 9781501779657 (epub) | ISBN 9781501779664 (pdf)
Subjects: LCSH: Animal welfare—Moral and ethical aspects—Scotland. | Animal rights activists—Scotland—Attitudes. | Animal welfare— Scotland—Societies, etc. | Human-animal relationships—Moral and ethical aspects—Scotland.
Classification: LCC HV4807 .R34 2024 (print) | LCC HV4807 (ebook) | DDC 179/.309411—dc23/eng/20241002
LC record available at https://lccn.loc.gov/2024021682
LC ebook record available at https://lccn.loc.gov/2024021683

For Oskar and Max
And with fond memory of Lucinda Hare (1958–2020)

Contents

Preface   ix

Prologue: Making Contact   1

1. Animal Protection as a Story in the Anthropology of Ethics   7

2. Dogs in the Office   48

3. Engaging the Mainstream   79

4. The Ethical Choice   111

5. A Brief Note on the Totem of Personality   146

6. Disciplines of Investigation   156

7. Evil People (and the Bonds of Rescue)   187

8. Being Moderate in a World of Interests   217

9. Moral Subjects   252

Notes   277

References   287

Index   295

Preface

Anthropological monographs on animal activism are few and far between. But that literature can boast at least two extraordinary texts, which in some respects I regard as bookends to this one. I am speaking of Hoon Song's *Pigeon Trouble: Bestiary Biopolitics in a Deindustrialised America* (2010) and Naisargi Davé's *Indifference: On the Praxis of Interspecies Being* (2023). To be sure, there is a quite literal way in which these two books serve as a beginning and ending to the project that resulted in *Animal People*. Song's book came out the year I began working with the animal protection organization in Edinburgh, and Davé's book came out just as I was finishing up and about to send off the final draft of my manuscript. Although both written by scholars obviously invested in the forms of debate and theorization where they teach (i.e., within North American cultural anthropology), the two texts could in many ways not be more different. Song's account is driven by a growing lack of sympathy for the kinds of Anglo-American animal activism described, especially its moral tone and practice, while Davé's account provides a sometimes critical but largely empathetic treatment of the ethics of animalists in India (note that Davé prefers to not call them activists, "as it conveys both too much and too little" [2023, 14]).

The trajectory in *Pigeon Trouble* is manifested by the way the author decides to closely describe an annual pigeon shoot in Pennsylvania, annually protested by animal activists. Song is strongly committed to presenting the moral point of view and practice of the shooters as much as that of the protesters that Song first arrived at the pigeon shoot with. To me, this was a refreshing approach, particularly back in 2010 when much of my early reading for the project was informed by the dominance of approaches exemplified in multispecies ethnography and the legacy of work by Donna Haraway. For Song's text demonstrated a notable deviation from a certain ethical register that made texts like *When Species Meet* so inspiring but also sometimes risked overshadowing an understanding of how (human) subjects imagined themselves to be ethical (of which multispecies ethnography is of course also a contemporary expression). In many ways, Davé's writing travels in a very different direction. For while expressing an admiration for the relational lexicon born of the legacy of Haraway's work, Davé also insists on a divergence. However, this is a breakaway at the level of ethics. In *Indifference*, Davé claims a new basis for understanding interspecies being, this time driven by an entire reconceptualization of what the notion of indifference can be made to stand for. This is how we should now think about ethics, Davé proposes. Davé's book seeks to then recalibrate an understanding on that basis, working against the grain or queering the normative account of the grounds for an "otherwise" ethical life or stance in the context of animal protection and especially of healing and rescue. In Davé's story, the self-presentation of animalists in India is sometimes a direct inspiration but perhaps just as often a foil to the bold rearticulation of ethics advocated by the author.

In comparison, my own project is in some ways rather more modest in both scale and ambition. If Song's narrative is largely discursive in nature, grounded in a reading of activist materials and images as well as that first-person description of the pigeon shoot, and Davé's book presents an ethics couched in a panoramic vision of the movement in India, whose narrative is largely episodic and often ambulatory in nature, this book rests squarely on the description of a single animal protection organization in Scotland. Indeed, it is at one level a simple office ethnography. That modesty is reflected not just in the fact that this is a small- to medium-sized animal group but also in the nature of its activity. For unlike the direct-action activism or protest described by Song or the street rescue described by Davé, the focus of this group falls largely on parliamentary lobbying and the campaign work linked to legislative opportunity. As I go on to explore with some emphasis, this is not an organization that works *with* animals, at least not in the sense meant by Davé when developing that ethics-cum-philosophic treatment of interspecies being (Davé muses much, for instance, on the value of sensory immediacy in interspecies being and on the haptic qualities of care and attention exhibited by certain animalists). Nor is the animal group one that expresses itself in the language and tone of vast animal rights organizations such as PETA (People for the Ethical Treatment of Animals), which is clearly at the center of Song's observations. In fact, as we will also see, reservations about the expression of moral outrage and more particularly the expression of moral condemnation is a defining feature of this Scottish animal group alongside an insistence that the organization represents a form of "moderate" animal activism. That qualifier certainly reflects the values of a reformist-minded and professionalized activism, which embraces expertise as well as modeling based on fundraising, but it additionally speaks to the kinds of ethical work on behalf of animals that colleagues engage in as well as to an overriding sense of themselves as moderate kinds of moral subjects.

Putting such a qualifier in front of their activism might suggest that their commitment was a halfhearted affair. But across the book, I have also striven to portray the moral drama of working lives as moderate animal activists. For although that self-presentation emphasizes qualities such as restraint, tolerance, pragmatism, and reasonableness, it also insists that this activism is ultimately driven by the very same passions as other activists in the animal movement—that is, by the forms of individual moral experience shared as animal people. As a matter of fact, different kinds of work at the animal group are often defined considering those passions, as, for example, a discipline or sacrifice in the service of expertise, which is itself perceived as in the service of animal protection. Part of the inspiration I draw from the work of both Song and Davé comes from the sometimes-operatic tone of their narrative accounts, whether it be the polarizing melodrama or farce of the pigeon shoot or the stop-and-horror nature of street rescue, whose peculiar quality is also shaped by the remarkable calmness exhibited by certain rescuers. Relocating that drama to a field of expertise such as undercover investigation might appear obvious. However, I am equally interested in trying to communicate the moral stakes of apparently more humdrum organizational activity, in the field of policy, for instance, or communications or fundraising. Or in the context of stakeholder consultation, social media work, or organizational rebranding. Or just in the context of what colleagues eat for lunch or what nibbles the animal group decides to serve or not to serve at a public reception.

Like Davé's, this book makes strategic use of moral biography, partly because the drama I speak of is itself often expressed and reported through a biographic sense of the moral subject. But our texts also diverge at that point, since my interests additionally lie in how that subject relates or is related to the categories that typically help shape the narrative of moral biography. I am particularly thinking of pervasive categories such as character and personality but also of conscience, and of the normative assumption that these help consolidate the moral action of the biographic subject. That sense also leans into and gets articulated through the way other classic categories such as responsibility or other classic narrative arcs within moral biography such as the process of conversion get understood. As we will see, the notion of moral biography is considerably complicated by the realities of those categories' deployment and by a certain puzzlement over moral sources. Indeed, while Davé expresses confidence in how one ought to understand the origins of ethics and hence how one ought (or ought not) to frame ethics, most animal people I worked with seemed genuinely surprised by their own assured sense of moral purpose—and at the same time rather less confident not just about the origins of that purpose but about the proper basis for laying out or articulating an ethical position. As a result, colleagues often experimented with relationships to moral reasoning and with relationships to moral passions, as they did with their relationships to explanation. This included their relationship to explanations for human cruelty toward nonhuman animals and their explanations for public indifference to animal suffering as well as their explanations for why some people were good.

So, while this book can certainly be read as another story in the anthropology of ethics, it is a good deal coyer about writing in the ethical register or proposing a basis for ethics. In this regard, I take some inspiration from colleagues at the animal group, who consistently resisted the pull to philosophic reflection or first principles and who also expressed a nervousness about the impact of their own ethical register or rather about the risk of appearing "preachy." This never meant that they doubted who they were or what values they upheld. They were always confident in that respect. But it did mean that their contribution sometimes risked going unnoticed or unheard (for instance, within the wider animal movement) and that their coherence as properly moral subjects could sometimes be open to challenge. Most of the time, that was OK. As colleagues in the office would often put it, they didn't have time to stop and argue with their detractors or to worry over the finer debating points in animal ethics. For they were just too busy *getting on with it.*

* * *

I understand that one could offer all kinds of observations about the possible limitations of such an utterance, especially its apparent appeal to common sense or to a no-nonsense approach to ethics. For example, one might highlight the risk of obscuring underlying hegemonic principles or assumptions, those that might enable *getting on with it* to appear as a self-evident or unproblematic stance. Fair enough. However, in this book I aim to largely defer such a move, which of course has its own normative moral status, and instead to concentrate attention on what's involved in getting on with it, if you are a moderate animal activist. For, as I hope we will see, that process is far from straightforward; it contains its own problems and tensions as a moral project, and its own rewards.

Acknowledgments

But there is a risk of too much prevarication. To begin again, I wish to simply acknowledge the stimulus and challenge of reading *Pigeon Trouble* and *Indifference*. Although I have never communicated with the author of the former text and only quite briefly with the author of the latter, both works have been a continuing inspiration (even if, reading back my final manuscript now, I regret the comparative marginalization of an engagement with Song—the unfortunate effect of several redrafts!). Though of course there are many other perhaps more unmediated or direct debts to acknowledge.

I wish to thank my colleagues in the Department of Social Anthropology at the University of St. Andrews—first and foremost, for the ideas sparked by numerous conversations over the years, either one-on-one (especially with the late Peter Gow, Mette High, Aimee Joyce, Nigel Rapport and Huon Wardle) or in the context of debate and discussion in department seminars. But perhaps more significantly, as far as I am concerned, I want to thank them for the prompt that our particular culture of collegiality provided. For in many ways, it was my introduction to the vigorous and sometimes heated nature of department meetings (I remember with a certain affection the early days when standing up and storming out of meetings appeared quite unexceptional behavior) and to the willingness of colleagues to sometimes argue over the apparently smallest details, especially once "a matter of ethics" had been invoked, that sparked my initial curiosity for this project. It also sparked a genuine sense of feeling impressed by the moral commitment and moral fluency of others that has never completely gone away.

Naturally, there are more conventional intellectual influences and debts. Once again, I thank my first and most taxing teacher, Marilyn Strathern, both for comments provided on a couple of early chapters about fundraising (now excised) and for the ongoing challenge of the legacy of work. Though I am not entirely sure she would recognize the influence in this book, I remain grateful for the original instruction to recognize the dynamism and tension that support even the staidest of staid oppositions. I have a great deal of other readers to thank too. Most importantly, I acknowledge those who read whole draft versions or significant chunks of the book and who provided invaluable feedback. This includes the anonymous reviewers of the manuscript for their helpful and considered comments but also Paolo Heywood, with whom I have enjoyed many stimulating conversations as we steered our respective manuscripts toward completion. I must also thank Jacob Copeman, Jane Desmond, Ilana Gershon, Timothy McLellan, Rosie Jones McVey, Annelise Riles, Mette Nordahl Svendsen, and Soumhya Venkatesan for their insightful remarks at various stages. I would additionally like to express my special gratitude to Farhan Samanani, who read an earlier version of chapter 1 and provided at the time an invaluable set of queries and suggestions. For further crucial feedback at a very late stage and for the general inspiration of their work and company, I also thank Matei Candea and Tom Yarrow.

Of course, the book benefited from numerous other discussions and from the presentation of my work at various forums across the years. I would like to pick out and acknowledge a few of these, especially those that came at important moments in the project's development. Very early on, I was invited to give a department seminar at Durham, which gave me an opportunity to present reflections in and around the theme of snaring. This topic was later developed into a contribution at an American Anthropological Association conference panel (in Montreal in 2011) organized by Victoria Boydell and Katherine Dow, with Sarah Franklin as discussant, which provided early encouragement and direction. But it was a cluster of forums in 2014 that really kick-started the groundwork for the book. This included an invitation from Irus Braverman to speak at the More-Than-Human Legalities workshop in Buffalo, where I had the opportunity to meet and share work with a small group of scholars from anthropology, law, and geography, including Elan Abrell and Jamie Lorimer (both of whom went on to provide helpful feedback). I also attended the Shrinking Commons Symposium in Cambridge at the kind invitation of Ash Amin and Philip Howell and presented further work on animal activism and law at the MEGA Seminar in Sandbjerg, Denmark. More recent invitations to speak at the 2022 Natures of Europe workshop (thanks to Frederic Keck) and at the 2023 Contested Spaces: Animals, Activists and the Law workshop (thanks to Daniela Berti and Anthony Good), both in Paris, have allowed me to air late versions of other chapters.

However, it was really the opportunity to give a 2019 Munro Lecture at the University of Edinburgh that provided the spark to redraft the manuscript as a whole and particularly to begin to reorient its argument around the literature on moral patient and moral patiency. I thank the organizers for that invitation too. Indeed, a series of dialogues with anthropological colleagues at Edinburgh have remained important to this project. I am reminded of occasional but valuable conversations with Alexander Edmonds, Toby Kelly, Lotte Hoek, Rebecca Marsland and Alice Street but especially with fellow members of the now defunct Candlelit Seminar series, often generously hosted by Magnus Course and Maya Mayblin, where draft work was shared in an always convivial atmosphere. Additional thanks must also go to colleagues elsewhere, such as Samantha Hurn, Robin Irvine, James Laidlaw, Tom Rice, and Benjamin Sachs-Cobbe. I would also like to thank Naisargi Davé, William Mazzarella, and Iván Sandoval-Cervantes for various timely reading suggestions. In the case of William Mazzarella, I give additional thanks for kindly sharing a draft version of the author's thought-provoking paper on the ethics of patiency back in 2020.

In terms of the process at Cornell, special thanks go to Dominic Boyer, the series editor, and to Jim Lance, its acquiring editor, both of whom provided invaluable advice and patiently responded to all my inquiries. Among the production team, I would also like to extend my gratitude to Dina Dineva, Allison Gudenau, Alfredo Gutierrez Rios, and Susan Specter. Elsewhere, my thanks go to Fiona Menzies at the University of Edinburgh library, for assisting me with accessing images from the archive, and to Rhona Rutherford at the University of St. Andrews's Print and Design unit. I wish to acknowledge the kind permissions granted by Iain McIntosh and Alexander McCall Smith and the commissioned artwork provided by Ryan Hamill. Staying in Edinburgh, I would additionally like to thank Andrew Gardiner at the Royal (Dick) School of Veterinary Studies, and Andrew Voas and Beverley Williams, both at the Animal Health and Welfare Division of the Scottish government.

Finally, my thanks go to Shari Sabeti, not just for being a sounding board for ideas as the book developed and as I faced and largely overcame various hurdles but also for the consistent love and support shown across the project's span. Thanks too to Oskar and Max. For significant early parts of this project, Oskar committed himself to a vegetarian diet out of concern for the welfare of animals but also, as he once told me, "because it makes me feel free." Although you now claim that you only said that as you thought it the kind of thing I wanted to hear, I suspect that at the time you meant it! Anyway. Special thanks to Max for helping me with that drawing app and for being such a friend to Ruby. Cheers, Nessie!

* * *

But of course, the emergence of this book is ultimately the result of my time with colleagues at what was until relatively recently the animal group's chamber of rooms in the West End of Edinburgh. Although never directly named (early on, we agreed on a light anonymization that I have sustained across all publications), the organization has been incredibly supportive of my project from the start. This has remained the case over the years since 2010, when I began the project. And that support has been consistent, regardless of who led the organization (up to the present, there have been two CEOs as well as a current incumbent who holds the office of director) or what role they occupied within it. This includes permanent and temporary members of staff and volunteers but also various members of the board of trustees, supporters, and official patrons of the animal group. While it is true that the organization has faced considerable challenges over this period, both financially and in terms of campaign delivery (the group has had to adapt its mission to the administrative upheavals of Brexit and subsequently to economic and welfare pressures of the lockdowns during the coronavirus pandemic), I am pleased to report that today it is in good health.

Fewer in number than when I first joined them and now relocated away from its historic headquarters, the group has redefined itself as principally focused on animal protection issues in Scotland. It no longer employs an investigator, but its trimmed-down approach has brought regulative successes. As well as being involved in the lobbying process that led to a ban on wild animals in circuses, in recent years the group has led a campaign to give mountain hares in Scotland greater protections. It has also celebrated the Scottish government's announcement of an effective ban on seal shooting and welcomed the mandatory introduction of CCTV cameras in Scottish slaughterhouses (both campaign targets). Perhaps most impressively, given its centrality as a campaign goal across this book, the organization is now confident that the Scottish government will soon announce a ban on the use of snares.

In terms of colleagues, past and present, I have so many people to thank. And while it is frustrating not to be able to name them in person, I take comfort in the fact that they will recognize the names I have given them in the book, as well as the associated moral biographies. While very few of them remain at the organization thirteen or so years on, I know that they follow its progress keenly. So, where to begin. Well, I suppose it makes the most sense to start by thanking Eilidh and Craig, who first welcomed me and agreed to let me begin the project with such open permission to participate in office activities. I have very fond memories of working together at Craig's desk, listening to him run through the latest scientific reports on animal sentience. And likewise of entering Eilidh's office for one of our regular update chats. Or of the three of us sitting in a nearby café trying to work out how the hell the evaluative tool of Weaver's Triangles worked, and what the substantive difference really was between "Aim," "Outcomes," and "Outputs." Thinking organizationally, in terms of associated roles, I would next like to thank Elaine. She was always ready with a friendly smile, and my memories are of lively conversations constantly interrupted by the office doorbell or of quickly grabbed catch-ups and chuckles over a cup of tea or between phone calls.

Next, Iain and David come immediately to mind. Great friends and coworkers in the IT/social media team, the two of them both took time to explain what they were doing or to provide answers to the many rooky questions I asked (for instance, about search engine optimization or Twitter bombs). After David left the animal group, I would sometimes grab a beer with Iain in a local pub, where we would discuss what was going on in the office. Famous for his dry humor, Iain taught me much and made me laugh. He also provided useful feedback on a few early papers.

Others on the team were also welcoming and very helpful. I learned a lot from my time with the fundraisers, especially Sarah and Fraser, and from numerous chats with the communications officer, Euan. Likewise, I had frequent conversations with Mairi, who worked part time on supporter services and with the ever-present Shelia, the office manager and longest-serving member of staff. I thank Shelia for all the assistance, especially in recent times when I needed various approvals and help with accessing images and old campaign materials. There were many other colleagues who worked with me over the years but who didn't make it in pseudonymized form onto these pages. I thank all of them, especially other members of the fundraising team and the several members of the board of trustees who agreed to be interviewed. The current director of the animal group has been immensely helpful in these last stages, giving permissions for image reproduction as well as reading and providing comments on final versions of several chapters.

Finally, though, I must end with heartfelt thanks to three colleagues. As they worked away together in the back office, I tended to think of Maggie and Barry as a double bill. Although mostly out of the office, Barry's focus was always tailored to the legislative or regulative interventions that he knew Maggie to be making.

And Maggie's policy work was consistently backed up and anticipated by the evidence and materials collected by Barry. Despite their heavy schedule, both made time for me and provided close access to what their respective roles involved. I treasure the memory of occasional fieldwork trips made with Barry and the education that he provided in how to be on the land with an investigative outlook. Maggie, I don't know how to fully thank. Taking me under her wing quite early on, the policy director gave me a crash course in parliamentary lobbying and the consultative process. I especially recall the hours spent watching back recordings of committee stage hearings of the WANE bill, with Maggie providing a running commentary on the ins and outs of what was being said (or not said). Perpetually busy and overworked, Maggie somehow fitted me in. But more than that, she allowed me to regularly shadow her across a range of activities. Now very much retired, Maggie still gives time and counsel to the current staff of the animal group. In fact, it is wonderful to think that the next time we meet it will be for an award reception in recognition of Maggie's long and impressive services to the cause of animal protection.

But I choose to close these acknowledgments with my thanks to Cassie. One of the new patrons of the animal group after its relaunch in 2010, Cassie was very active in rescue circles despite or maybe partly because of ongoing health problems. Indeed, when we met, our discussions usually centered on the logistics of rescue and the labor of care for the rescued cats and dogs Cassie's circle rehomed from parts of eastern Europe. Those meetings opened another side of animal protection to me (the side that usually receives most scholarly attention) but also provided a first-person vantage point for better understanding the voluntary care and sanctuary work that many colleagues performed in their spare time. Cassie's commitment to protecting or saving these cats and dogs led her to articulate that moral concern in ways that I sometimes struggled with. But at the same time, Cassie had an unmatched capacity to capture or express what defined or drove animal people and a raw honesty about both the burdens and joys of that commitment, which included a willingness to reflect on its paradoxes. I count those conversations as a real blessing, and though I know that she never really believed it possible, I still like to think of her now happily reunited with Smudge somewhere over the Rainbow Bridge.

MAKING CONTACT

"Humankind has spent millenia wondering whether we are alone in the universe . . . whether there is other intelligent life out there, watching us. Waiting to make contact." So run the opening lines of a six-minute film, commissioned by the Scottish animal protection group that I worked with to promote the launch of a rebranded organization back in 2010. Scripted by Iain, the group's IT officer, this film was preceded by a viral video and social media campaign that took the form of a sci-fi hoax. In the weeks before its release, viewers online were fed teaser film clips, which suggested the imminent announcement of a new sci-fi movie, while a fictional social networking profile on Twitter and Facebook published promotional materials, including photographs of people recognizably in London and Edinburgh wearing billboards with a date stamped on them and looking or pointing to the skies.

The final six-minute film, shot and produced by a film company also based in the Scottish capital, continues to play with that sci-fi lexicon. After the opening lines, delivered by a suited reporter doing an outside broadcast from what we are told is East Yorkshire (a county in Northern England), the scene switches to a newsroom and the close-up face of a news anchor with a distinctively Scottish voice. "Reports are flooding into news stations across the globe," the woman announces into the camera, "of sightings of life forms with what appear to be *super*powers. . . . They appear to communicate in a wide range of languages, yet we have been unable to decipher [them]." From reading the news, the scene switches again, this time to a bearded figure in a hoodie, who sits on a sofa eating a slice of pizza, surrounded by empty wine bottles (typed white text at the bot-

FIGURE 1.   "They're Here!" human billboards in London. Image taken from 2010 website of animal group. Used by permission.

tom of the screen tells us that this is "London UK"). Apparently dumbstruck, the twentysomething watches the news story unfold on a television set. Quickly the scene changes to show a middle-aged woman wearing a waist-apron and cutting vegetables in the kitchen of a suburban house. Like the previous character, this one also pauses to watch the news and is clearly shocked (typed text now informs us that we are in "Whitelaw USA"). Another scene showing a white-whiskered face framed by a crumpled baseball cap cuts in. The character sits in what looks like a run-down outbuilding (the typed text reads "Stonebrook CA"), clutching a pitchfork. We watch as the old man tenses and then bangs a crackling radio set in frustration before bending an ear to try and listen more closely.

In these three scenes, the characters become increasingly skittish as they watch or listen to the news. Each one jumps at the sound of thuds and other noises emanating from locations just out of shot. The voice of the news anchor next interrupts to inform the audience that we are returning to our reporter on the ground. "Well, this is amazing," the outside reporter stutters. "It's true; they really are here. Just five minutes ago one of them approached us." At this point the sound becomes muffled, and the picture goes temporarily grainy and gray to indicate some kind of unexplained interference before the reporter's flickering face returns to the screen. "It was a matter of inches from our faces," the reporter excitedly recounts, "separated from us by just a thin wire fence. We could feel its breath as it quietly studied us with its huge saucer eyes. Then, without a word, it turned and moved away."

The face of the news anchor returns. "We have more information coming in," the anchor rather breathlessly informs the audience. "Eyewitnesses from across the globe have seen them flying at staggering speeds and traveling across the land far faster than any human. We're just receiving stories in live now of sightings of incredible strength." The anchor adjusts a hand on the side of their head, as if touching an earpiece. "Scientists from across the world have formed an emergency working group," the newsreader discloses. "They are convinced that these beings have been living among us and around us all this time. . . . [Scientists] think that these aliens have been performing the role as caretakers of this planet before humankind even arrived." As the tension builds to an evident reveal, the film cuts away again to the three characters. We watch as the hooded character rises from the sofa, opens a door, and bends down to pick up and stroke a cat. The anchor's voice speaks over the scene as the background music reaches a crescendo. "Film and pictures appear to show them eating, playing, and looking after their young," the anchor reports. The waist-aproned character at the chopping board kneels to welcome an obviously familiar tail-wagging dog. "We learn that just like us they have feelings of pain, distress, joy, and happiness." The anchor's face now returns to deliver a final line, half-smiling to the camera, "Clearly, we are not *that* different." As the music plays out, the film returns to the scene of the old man, this time outside in a farmyard. For the character has discovered that a tethered cow made the off-stage noises that at first unsettled him. With an affectionate pat, the farmer leads the cow away. After this last shot, the film credits announce the name of the animal protection group, urge the support of viewers, and conclude with the simple message: "Make Contact."

* * *

This book is about the professional and personal ethical work associated with the cause of animal protection. It examines the activities of a small yet important not-for-profit organization of long standing, headquartered in Edinburgh (initial fieldwork took place from 2010 to 2015 and continues intermittently until the present). In fact, now over a hundred years old, that organization has an important place in the history of such campaigning groups and in the broader story of the rise of animal protection as a moral concern, not just within Scotland but also in the United Kingdom as a whole (see chapter 2). The organization's interest in specifically Scottish animal welfare issues, for instance, has always been combined with the reality of a membership or supporter base spread more widely across Britain and, albeit to a much lesser extent, across the former settler colonies of the British Empire. Today, its staff view themselves as part of a national network of small, medium, and large campaigning or rescue groups. Within the UK, the organization's list of coalition partners include groups with titles typical of the wider sector, such as Animal Aid, Animal Concern, Animal Defenders International, Born Free Foundation, Ethical Voice for Animals, British Divers Marine Life Rescue, Captive Animals' Protection Society, Compassion in World Farming, Fish Count, Four Paws, the League against Cruel Sports, People's Dispensary for Sick Animals, Save Me, Scotland for Animals, the Society for the Prevention of Cruelty to Animals, and World Animal Protection.

However, colleagues at the animal protection group consider that they also belong to a global movement. This is partly due to the historical orientation of the organization. For at certain moments, it has had ambitions to scale its mission up. It was founded in a period of empire when animal protection was itself part of the missionizing or colonizing project (see Davis 2016),[1] and those ambitions have since manifest in a desire to participate in a transnational network of animal protection groups centered in Europe. This development was in large part due to Britain joining the European Community in 1973. Although those links have come under threat since the 2016 Brexit referendum result and subsequent departure of the UK from the European Union, the idea of cementing the organization's place in that global movement remains. Indeed, as the sponsored film I began with highlights, this has in recent years led to tentative overtures being made across the Atlantic. To an extent, the organization has been concerned to establish more formal links with animal protection groups in the United States, though the overtures expressed in that film were far more about a flirtation with the notion of attracting supporters and in particular donors from that part of the world. As far as the animal protection group was concerned, that possibility became imaginable only with the arrival of the internet and in particular the ensuing emergence of popular forms of new social media. Perhaps less positively, at least in the minds of some colleagues, that opportunity has further arisen because of the huge success of certain North American animal protection groups in establishing a reverse influence—the most well publicized example being of course the rapid spread of People for the Ethical Treatment of Animals (PETA), which in the last twenty years or so has developed a powerful voice and very strong campaigning foothold in numerous animal welfare debates in the UK.

As the film also shows, those ambitions go together with a desire to engage what the organization regards as "mainstream" publics. This includes the hope of making contact with people who are perhaps apathetic or who claim to be "animal lovers" but who nevertheless do not currently identify with the animal protection movement or actively participate in its campaigns. One of the reasons the sci-fi hoax device was selected, and considerable charitable funds sunk into the film's production, was precisely the impression that those who appreciated that genre, which includes sci-fi fandom, might thereby be drawn into the animal protection orbit of moral concerns. But the three characters depicted in the film—a couch potato or pizza-eating twentysomething in London, a middle-aged suburban householder from somewhere in Middle America, and an old Californian farmer—were intended to depict a wider cross section of that envisaged Anglo-American mainstream (perhaps tellingly, the evident whiteness and cisgender identity of the mainstream, as represented by those three characters, at the time went largely unnoticed).[2] Indeed, the characters were also chosen to communicate the message that anyone could become animal-friendly and commit to welfare values, and in turn that the animal protection group would welcome anyone who accepted their invitation to make contact. In this regard, the choice of a farmer, albeit clearly a smallholder, was particularly pertinent (dairy farming and in particular industrialized animal husbandry being classic targets of critique within the animal protection movement), a nod to the fact that colleagues at the organization wanted to reach out across old lines of division.

But the film's opening lines make clear that this message is far more than an invitation to just join an animal protection group, pledge donations, or participate in its campaigns. It is also a direct appeal for viewers to make contact themselves with animals, to notice and express concern for the individual extrahuman sentient beings that surround them. Indeed, for Iain and the rest of staff, the sci-fi metaphor worked especially well precisely because it captures the alien status of nonhuman animals in several modalities. First, it references our refusal to grant them equal sentient status, a move that can legitimate abuse and human domination. Second, and perhaps more important, it opens a positive claim that might celebrate the capacities or "superpowers" that these species possess. Why look for aliens to come from outer space, the inference runs, when there are more than enough remarkable extrahuman life forms already living among us? Central to that message is this idea of a multiform sentient world simply waiting to be acknowledged and the accompanying promise of a massive expansion of encounterable and relatable beings, whose recognition also promises unprecedented self-expansion. This moral obligation to make contact may fall on the human, but it additionally opens a space to consider individual members of other species or extrahuman "forms of intelligent life" as subjects who in turn are watching us and waiting for this contact to take place.

The film then suggests that becoming aware of the aliens among us has practical moral and ontological implications. If we are not or have never been all alone, then that means there are suddenly a lot more kinds of beings to be mindful of and if necessary to care for and protect. It further suggests that moral enlightenment itself rests on a direct experience and reflection on those encounters, rather than just on a set of philosophical axioms or the evidential status of scientific knowledge. Although, as we will see, both animal science and the moral principles derived from the classic treatise of animal liberation, including the axioms of animal rights, operate at the horizons of the animal protection group's practice and cannot be ignored, they are rarely taken to underpin a sense of ethical formation. Indeed, I argue that it is more accurate to present those bodies of knowledge as surplus to the practical knowledge *already* learned by making contact. This is an observation that informs not just my description of colleagues' working relationship to those bodies of knowledge but also my account of how, at least in this case, an ethics of animal protection may in fact play out.

ANIMAL PROTECTION AS A STORY IN THE ANTHROPOLOGY OF ETHICS

In practice, the global animal protection movement is often simply referred to as the "animal movement." Likewise, organizations working on behalf of animals typically refer to each other as simply "animal groups." This designation is also often used by government in Scotland and the rest of the UK and by other kinds of organizations who work either with those groups or against them in the world of public campaigning, in the policy sphere, or within the lawmaking process. As a shorthand, "animal groups" may be a convenient descriptor, but as everyone knows, it disguises as much as it reveals. On the one hand, it collects a very diverse range of animal activism, including more outspoken organizations focused on the arguments and protest tactics of animal liberation and those by tradition more conservatively settled on petitioning for the incremental reform of animal welfare legislation. On the other hand, it obscures significant distinctions of organizational scale and profile—for example, those between advocate groups encompassing all issues, sanctuary-based organizations, and those either specializing in a particular cause (say, antivivisection or against "cruel" field sports) or focused on the welfare of types of animals (there are all sorts of British groups just focused on the welfare of horses, for instance). Finally, it blurs the distinction between professionalized organizations and what are sometimes termed "grassroots" or nonprofessionalized activist groups. As we will go on to explore, the Edinburgh organization that I worked with currently operates as a generalist advocate for animals, running campaigns and petitioning across a broad range of welfare concerns. It is a small-scale but professionalized animal group that has a cautious attitude toward the articulation of the case for animal rights and that historically self-identifies with the "moderate" end of animal activism.[1]

Both staff and supporters of the animal group also identify strongly as "animal people."[2] Like the abbreviated terms "animal movement" and "animal group," this label can hide a good deal of difference in individual ethical practice and outlook. But it is perhaps most consistently applied when colleagues self-narrate accounts of moral awakenings or when, in casual conversation, they might wish to distinguish someone who commits their life to the protection of animals' interests, either professionally or through personal conduct. Just as importantly, colleagues regularly invoke the term to describe a type of person who claims a powerful connection to animals and in that regard feels driven or compelled to help them.

Let us take, for example, the moral biography offered by Eilidh, the CEO of the Edinburgh group during much of my fieldwork. Like the narratives presented to me by other members of staff, this one stressed that the origins of moral development derived from powerful feelings, typically an intense love of animals, sometimes accompanied by a strong aversion to eating meat and by an impulse to rescue. It further lay in the assumption that for most animal people those feelings first arose and took hold at a young age:

> It was probably as early as five or six years old that I knew I didn't want to eat meat. I can remember just refusing to eat it and my mum getting quite concerned, thinking that there's something wrong [Eilidh chuckles]. But it was literally because I'd seen blood with the meat and somehow I must have made that connection. . . . And I can remember there was a bird trapped in the stairwell at school. The janitor came to get me out of class [to help rescue it], I mean I would only have been ten, because even then I was kind of known as the girl that loved animals [more chuckles]. . . . Thinking about it now I remember doing a wee petition when I was eight, going round all the neighbors' houses with a friend because we had a donkey in the local field and they were going to build houses there. The petition was to stop Dingle the donkey from getting moved [Eilidh laughs]! We have always had pets as well, cats initially, and then later on we got dogs, which I was always close to.

In quite exemplary fashion, Eilidh then went on to narrate the gradual evolution of a recognizable moral stance. It began as something that operated without much articulation, as a kind of automatic response to the welfare needs of those animals she happened to encounter, such as the wild bird in her primary school stairwell or the donkey at risk of removal from the neighborhood just out of Edinburgh where Eilidh mostly grew up (please note, where a colleague's gender pronoun preference is known, I use it). However, over time such encounters became prompts for more considered reflection—for instance, by dwelling on the systemic causes of animal suffering—which in turn led to a series of staggered developments presented as part of an increasingly more conscious moral commitment:

> I remember that when I was around twenty there was a woman collecting for animals outside the supermarket. I stopped and chatted to her and gave her some money. "Do you ever need any help?" I asked. . . . Around that time I actually set up a local animal group for the Dian Fossey Gorilla Fund. We used to run events and do collections and petitions. So that's when I started to get more active, on a sort of grassroots level. . . . Later on, I do remember a really profound moment [in her midtwenties]. It was when I went to a big event in London, where all the different animal charities had stands and gave presentations. I went along and it was just so distressing, because they had loads of films and pamphlets, you know, especially on animal experiments, but also bear farming, on animals used for food, all the journey times and things. I remember coming out of there and thinking, "It's not enough to be a volunteer; I want to do more," and that's when I determined I would set out to get a job doing this.

Soon afterward, Eilidh did find her first professional position with the National Anti-Vivisection Society in London. In fact, animal experimentation went on to become a touchstone issue, the welfare concern that captured her moral imagination the most.

Such narratives of moral progression were commonplace. Yet the notional shift from a childhood stage of emotional or instinctive sympathy to a stage of considered reflection grounded in becoming aware of the facts or more conversant with sets of moral arguments—as well as visiting exhibitions hosted by animal groups, Eilidh started to read texts such as Peter Singer's *Animal Liberation*, often described as the founding manifesto or bible of the animal movement—could also mislead. For the experience of making contact remained an essential component of moral concern in the minds of animal people. Indeed, historicized accounts of defining encounters with individual nonhuman animals were regularly interspersed with references to the continuing importance of such encounters in their everyday lives.

By way of illustration, let me introduce another voice, this time belonging to Cassie, a patron of the Edinburgh animal group and close friend of Eilidh. Her musings were delivered through a series of at-home conversations we had, where the two of us usually sat together in her conservatory. On most occasions we were also surrounded by a small gang of dogs and by several blind cats that over the years Cassie had helped rescue from Romania and Egypt. In these reflections, accounts of early formative encounters were regularly interspersed with references to present ones:

> Well, my parents were arable farmers, and I grew up on a big estate down in East Lothian [the rural area directly south of the Scottish capital], eleven hundred acres of wood, parkland, and beaches. I spent most of my childhood wandering around there and watching the animals. You know, listening to the seals at night and the ducks, glimpsing the deer and looking out for hares, stoats, and weasels. It was bliss! . . . We had dogs and we had ponies, and I loved animals for as long as I can remember. Anything, even earwigs; I love little insects. I love spiders. See [Cassie points to the garden gate just beyond the conservatory door], I build it into things like the railings there. Do you see the little spider's web? I just love anything that is alive.

Such enthusiastic observations of encounters with local wildlife (Cassie also rescued hedgehogs) were intercut with constant nods to the presence and personalities of the various cats and dogs that shared her home.

Indeed, in the moral biographies of other staff members at the animal group, that daily attentiveness to more-than-human sentient life consistently drove a commitment to campaign and to rededicate one's life to animals. Staying with those colleagues most closely associated with Eilidh, let us consider the example of Elaine, personal assistant (PA) to the CEO. By the time I first met Elaine, she had committed much of her working life to animal protection. This had included an initial seven years at the Edinburgh group, from 1993 to 2000, which began with her working as a receptionist but eventually led to Elaine being put in charge of organizing volunteers, running street collections, and giving school talks. That period had been followed by a time at Animals Angels, a German animal group that specialized in campaigning against live-animal transportations. There, Elaine had participated in investigations, among other things tracking convoys of livestock trucks from France all the way across the continent to Greece. She had also become involved in lobbying Members of the European Parliament (MEPs). But even in those years she subsequently spent working outside of animal groups—for a while Elaine was an executive PA to the director of a pensions fund—she remained personally committed to the cause, an allegiance strongly rooted in continuing mindfulness of the needs of those animals immediately around her.

In what will prove another recurrent theme in the self-presentations of animal people I met, those encounters also fed a sustaining sense of exceptionalism. Elaine's own words stressed a contrast to an apparent public indifference or to a simple lack of awareness of sentient others in the environment:

I think I am just much more aware and much more conscious. Like, if I am outside for a walk, I will recognize a warning call, like a bird has, and I will think, Well, gosh, what is going on? Maybe there is a cat there? Whereas people just walk on by; they don't even register there is a bird there. Or a certain way a dog is howling. You think that dog needs some help or it needs something, and people just don't register that. They are not aware. They are oblivious. They are just very focused, I guess, on their own world and their iPods [chuckles]. They don't actually stop and take stock of what is happening around them.

That essential sense of animal people as subjects who have already made positive contact and the accompanying sense of a mainstream public that has not or has not *yet* made contact played out in other directions too. Perhaps most notably, one found it emphasized in the further kind of encounter that dominated their moral biographies: encounters with suffering animals, which often combined with witnessing formative scenes of human-inflicted cruelty. In this case the dominant motivating feelings were usually horror or distress. As Song (2010) observes in a study of North American animal rights groups, such reactions often tended to be garnered through mediated encounters with animal suffering—for example, by regularly consuming the magazines, street pamphlets, investigative reports, and films of such groups, and sometimes by attending the kind of public exhibitions earlier described by Eilidh. However, at least for the animal people I knew, those encounters were just as importantly directly experienced, and once again at an impressionably young age.

Returning to the moral biography of Cassie, for instance, the bliss of childhood encounters with wild animals on the arable farm where she grew up was quickly juxtaposed with other, more troubling recollections. "In fact, one of my earliest memories was seeing this rabbit dancing, leaping around in the woods," she told me, but "this time I quickly realized that he was caught in a snare; he was petrified!" Her own shock at discovering the scene prompted an immediate resolve to do something: "So I took the snare off and let him go free, and after that I started looking for other snares and began pulling them up and hiding them." The incident was, Cassie added with emphasis, "my first conscious act of rescue." Soon after that encounter, Cassie began to come across numbers of rabbits on the farm in various states of distress, many "already blind and half-stuffed." She picked them up and brought them to the local vet; as a young girl, Cassie was not yet aware of what myxomatosis was or why these rabbits were suffering such injuries. Looking back on those days, she angrily reflected, "I mean, I still think it's one of the most God-awful, disgusting things we have engineered deliberately to kill a species."

A similar kind of disgust drove the narrative of moral awakening offered by David, the organization's social media strategist and the station coworker of Iain, the IT officer (and script writer for the animal group's film). Only in his early twenties when we first met, David described growing up in the northwest English town of Lancaster and becoming involved with animal groups at school. This included participation in a campaign organized by PETA to try to abolish the use of bearskins as ceremonial fur caps in the British Army. His online advocacy continued when he went up to university in Glasgow; David volunteered to work on further social media campaigns for PETA but also on campaigns for the group Animal Aid (historical associations with a diverse range of animal groups was another common factor in the identity of most animal people). Yet once again, it was his boyhood experiences of making contact that mattered most. After telling me about a particularly close bond with a family dog as well as memories of adults in the local area regularly kicking their dogs when they took them out for walks—"I still see people doing that now; you know, it sends a shudder up my spine"—David shifted the focus of his reflections:

> I think one kind of standout moment for me was when I was about eleven. I had already quite a concern for animal welfare. Anyway, we had a rural science unit at our school. One day we were out in the garden, in the allotments planting things for rural science, and one of the kids I was with picked up a frog and threw it into the chicken coop. The chickens ripped this frog to bits! It was one of the most harrowing things I've ever seen; these chickens were just running about with the frog's legs and the rest of the body was just lying there. It was absolutely disgusting! And I remember looking at this kid and just thinking, Why on earth have you done this? You know, just not understanding his motivations for doing it. And that really kind of set the cogs in my head turning, got me thinking, trying to work out why people don't treat animals with enough respect.

Such memories crystallized the way in which for many colleagues a developing sense of their own moral stance coincided with a differentiating move from others. This included a questioning of fellow human motivations. Why don't they care? Why are they indifferent to the plight of animals? And sometimes, why are they cruel?

But those inquiries into what animal people regarded as the moralities of the mainstream also coincided with a continuing puzzlement about the extraordinary depths of their own levels of commitment. "Now that's a thing that I don't understand," Cassie once pondered. "What is in us that makes us so passionate about animals? It's not that we were educated, it's not just that we saw the ugliness [i.e., of human cruelty toward animals]." Making contact then might be essential to the ethical life of colleagues at the animal group, but, like self-education and moral reasoning, it was not always held to be a sufficient explanation for a subject's moral being. "We don't understand what makes us the way we are, and what makes them [i.e., the mainstream but also the perpetuators of cruelty]," Cassie reflected further. "Where does it come from? I don't know!" Cassie shrugged. "All I do know," she reaffirmed, "is that I love animals, that we all want to try and make life better for them, and that we recognize that feeling in each other and call ourselves animal people."

Moral Patients (and Agents)

Cassie's conviction and the conviction of other colleagues at the Edinburgh group that moral awakening was born out of and sustained by various forms of making contact centrally informs much of what I want to say about the nature of ethical lives within animal protection. However, so does Cassie's puzzlement over the force and origins of the moral drive of animal people. I am interested both in the sustaining belief that intimate embodied encounters with specific animal others can feed moral imagination and in the accompanying insistence that moral being remains a mystery. Animal people know what they know about themselves—that is, that they love animals and feel compelled to try to make their lives better—but they equally regularly claim to not fully understand what makes them who they are or, to requote Cassie, what it is in them that makes them "so passionate about animals." Further, and just as importantly, they claim to not fully understand what makes others the way they are either—why many lack that moral passion and how such people can seem so indifferent or even openly hostile toward animal welfare concerns.

As we will see, those distinctions inform much of what I also want to explore about the organizational life of this animal group. For the individual moral feelings of animal people and the wider questions thrown up by knowing and not knowing why someone is a certain kind of moral actor got reflected in the peculiar vision and expert practice of the Edinburgh group. Indeed, in many ways the identified problematics associated with making contact framed the activities of the organization as I encountered it, including its working sense of mission and the accompanying sense of where the challenges of mission implementation lay. If animals appear innately ethically demanding or viscerally powerful to some, why not to everyone else? Why has "the mainstream" Scottish or British public not made contact yet or not recognized the moral consequences of doing so? These kinds of questions animated team meetings and strategy discussions across much of the period of my initial fieldwork.

Of course, those questions sat alongside many other types of pressing question. The latter, which very often concerned working definitions and criteria for making judgments about the treatment and welfare of animals in specific policy contexts, had perhaps a more mundane scope and purpose. Often linked to the practical, technical work of being able to make a case, to the organization's role in trying to persuade government and other agencies to take the interests of animals into account, these questions nevertheless loomed large. The questions could express a narrow concern, tied for instance to specific protocols around the legal status of evidence collected through investigation or connected to what counted as admissible evidence at the committee stage of a bill going through Parliament. Or the questions could be broader in scope, closely associated, say, with argument or findings in the science of animal sentience. As we will see, colleagues could also refer to the "big" philosophical questions that they recognized as classically framing debate in the animal movement. On what basis do we recognize that a certain species has a welfare worth preserving? How do we prove that this or that measure negatively impacts quality of life or risks causing unnecessary suffering? Do animals have rights? And why do we have a duty to protect them?

Although such large definitional issues might not be regarded as ethically formative in the manner of intimate embodied encounters with animal others, they still needed to be regularly addressed. This was the case even though at work moral debate and the terms of moral reasoning were often subsumed by more pragmatic organizational concerns. Colleagues were far more likely to approach the notion of animal rights, for instance, by asking whether it made strategic sense to foreground a rights-based argument in a particular campaign or policy submission than by asking whether it was correct. Nevertheless, they had a general view about what the important questions were, which included broad consensus about how positions might be argued or how moral stances might be properly defended. Schooled as much by debate outside the workplace, most notably through the ceaseless testing of their commitment to vegetarianism or veganism experienced at home and with friends, and by secondhand as much as firsthand familiarity with classic treatises of animal liberation, colleagues operated with a shared albeit often downplayed and to a great degree assumed language of moral justification.

Instead of equivocation over why someone was a certain kind of moral actor, these arguments tended to implicitly and on occasions explicitly center on drawing a distinction between moral agents and moral patients. Indeed, it was through making that kind of distinction that colleagues' perceived duties toward animal others got typically expressed or more formally laid out. Across the book, I am interested in exploring this relationship to the emic-ish category of moral patient, both as colleagues at the animal group invoked and deployed it and as the category and its definitional distinction from moral agent resonated more widely within animal protection. As we will see, that interest also prompts a much broader inquiry into the possible utility of drawing such a distinction. This will lead me to trace an explorative and complicating genealogy for an anthropology that might properly consider the position of moral patients in tandem with the position of moral agents, including alternative readings of the distinction's value that will take us back to some of the tension points in the original equivocation offered by Cassie. However, we need to start by examining the philosophical foundation of those formal definitions of moral patient and moral agent as they would most obviously and immediately be recognizable to animal people and animal groups.

While talk about moral patients and moral agents has long been part of the vernacular of moral philosophy, in a contemporary context it is perhaps most closely associated with debates in animal ethics. Moral philosophers still tend to link an emphasis on that distinction with the work of Tom Regan and in particular the now-classic treatise *The Case for Animal Rights*. First published in 1983, that book, much like Singer's *Animal Liberation*, also remains an iconic text within the animal movement. Even if many colleagues at the Edinburgh group hadn't read it and only a few of them were truly conversant with the philosophical stakes of its argument—for instance, between a rights-based approach to animal ethics and the utilitarian or consequentialist approach embodied in the work of Singer—all had certainly heard of the title and its author (see Reed 2024). In fact, copies of both Regan's and Singer's works could be found on the shelves of the office library.

"A helpful place to begin," Regan proffers in a chapter devoted to a critique of indirect duty views, "is to distinguish between moral agents and moral patients" (2004, 151). In this first substantial example of what goes on to become a recurring series of reflections on the distinction's significance, Regan then proceeds to offer some definitions. These are provided in twofold fashion, in explanatory mode for a general reader and in a more knowing vein, addressing fellow philosophers who are assumed at some level to already understand and incorporate into their own work the purchase of such an opposition.

Regan starts with the side of the distinction that historically dominates the reflections of moral philosophers, advising the following definition: "Moral agents are individuals who have a range of sophisticated abilities, including in particular the ability to bring impartial principles to bear on the determination of what, all considered, morally ought to be done and, having made that determination, to freely choose or fail to choose to act as morality, as they conceive it, requires" (2004, 151). That statement leads on to a crucial further observation: moral agents are those fairly held morally accountable for their actions (152).

"Since it is they who ultimately decide what they do," Regan continues, "it is also they who must bear the moral responsibility of doing (or not doing) it" (152). Regan adds, by way of final qualification, "Normal adult human beings are the paradigm individuals believed to be moral agents." While admitting that this is a "large assumption," Regan states that it will hold, for all the competing moral theories that the author engages with share versions of the same supposition.

Moral patients, by contrast, "lack the prerequisites that would enable them to control their own behaviour in ways that would make them morally accountable for what they do" (2004, 152). Regan continues, "A moral patient lacks the ability to formulate, let alone bring to bear, moral principles in deliberating about which one among a number of possible acts it would be right or proper to perform. Moral patients, in a word, cannot do what is right nor can they do what is wrong" (152). Although in these definitions moral patients are largely defined by an absence of moral capacity, Regan is careful to stress that in *The Case for Animal Rights* they are depicted as full of both conscious and sentient life. Moral patients may not be *moral* agents, but they are certainly agentive. As well as being capable of experiencing pleasure and pain, some moral patients possess further "cognitive and volitional abilities" (2004, 153). Indeed, Regan is especially concerned with that subset—that is, "those who have desires and beliefs, who perceive, remember and can act intentionally, who have a sense of the future, including their own future (i.e., are self-aware or self-conscious)" and can therefore be said to also have "an emotional life," "a psychophysical identity over time," "a kind of autonomy," and what he also terms "an experiential welfare" (153). Nevertheless, Regan advises, these moral patients resemble moral agents in only one respect; like them, they "can be on the receiving end of the right and wrong acts of [other] moral agents" (154). The relationship then is marked by the essential fact that it is "not reciprocal." By their doings, moral agents may act in a fashion that "affects or involves" those distinguished as moral patients. However, moral patients, despite their general capacity to act, "can do nothing right or wrong that affects or involves moral agents" (154).

Of course, Regan is principally concerned with the category of moral patient as it pertains to (nonhuman) animals and connects to our treatment or exploitation of them and to the author's case for animal rights. But it is important to notice that this definition of that category does not entirely exclude human subjects. For Regan, "infants, young children, and the mentally deranged or enfeebled of all ages are paradigm cases of human moral patients" (2004, 153).[3] Indeed, Regan feels the need to reiterate that observation in the new preface provided for the 2004 edition, partly in response to those philosophers who critiqued the original arguments on the basis of a mistaken reading of the way the author drew that distinction. Regan stresses once again that "in my hands, 'moral patient' is not equivalent to 'animal' and 'moral agent' is not equivalent to 'human being'" (xxv). Although in the new edition the exampling continues to draw heavily on the patient status of animals, this point enables Regan to contest any charge of a false equality being made between humans and animals: that is, when those types of being are rendered absolutely incommensurate through that moral opposition.

Unlike the category of moral agent, very few of the animal people that I knew explicitly invoked the category of moral patient. Indeed, most would have struggled to define what Regan meant by the term. However, they certainly shared an instinctive appreciation both for the essential patient status of animal others and for their own obligations, as moral agents, toward them. For colleagues at the animal group, there was generally only one species held to be morally accountable for their actions (see Reed 2017b). Despite the emphasis placed on making contact and the belief that such encounters sparked and continued to affirm moral awareness, in this regard the relationship also remained essentially nonmutual. Animal people might enthusiastically celebrate the diverse sentient lives of nonhuman animals and treasure the quality of any particular connections, to the extent of insisting that such relationships were crucially intersubjective, but they otherwise operated in common with the philosophical consensus described by Regan. (Even if Regan reiterates that in his account moral agent is not equivalent to human being, the philosopher does so primarily to relocate aspects of humanity to the category of moral patient rather than to seriously render the moral agent position more than human.) In short, to reemphasize the earlier-quoted statement, animal others definitively affected the moral status of animal people, but, it was assumed, those animal others did not affect or involve animal people on the basis of any actions that could be adjudged to be right or wrong.

By way of brief illustration, let us return to the memory offered in the previous section by David, the group's social media strategist. In recalling the childhood trauma caused by witnessing a group of chickens ripping a frog to bits, he never suggested that these farmyard birds were to blame for their actions. Instead, the scene prompted David to ponder the motivations of his classmate, the boy who took it upon himself to pick up the unfortunate frog and throw it into the chicken coop. David also asked himself the wider question of "why people don't treat animals with enough respect." In this situation, then, both the chickens (i.e., those immediately doing the ripping) and the frog (i.e., the critter who was being ripped to pieces) featured as moral patients. For in David's eyes, they were each on the receiving end of wrong acts by an identifiable moral agent. Most obviously, David referred to the original act of his classmate, who at the time he held responsible for the harrowing scene. Yet by implication he additionally referred to a society or culture in which that boy grew up. Indeed, in retrospect David attributed the chief blame to the latter. Like his colleagues at the animal group, he generally accepted the principle that children had diminished moral responsibilities, that they ought in Regan's sense be viewed as far closer to moral patients.

But I want now to explore the category of moral patient in more detail, to open out its possible significance by broadening my terms of reference beyond the arguments of Regan and animal ethics. To do so, I detour to another explicit definition of the distinction between moral patient and moral agent, this time provided by someone working in the field of ethics and technology. That field has in recent years been the area of philosophical moral concern where the distinction has perhaps gained the most traction, significantly inspired by the terms of debate within animal ethics (see Gunkel 2017). More specifically, in the example I briefly explore, the focus falls not on the moral status of animals but instead on the question of whether robots can or should be assigned the status of moral entities.

Right at the beginning of an essay on artificial intelligence, Joanna Bryson (2018) offers some useful clearing remarks. In many ways, these chime with the terms of opposition set up by Regan, but they also inevitably mark a divergence. A moral agent, Bryson proposes, should be understood as "something deemed responsible by a society for its actions" and a moral patient as "something a society deems itself responsible for preserving the well-being of" (2018, 16). The stress here is immediately useful in several directions. In this account the moral status of both positions is always sociohistorically determined, but perhaps of even more relevance, Bryson insists that others crucially define the moral patient position and in doing so render it an issue of collective responsibility. There is an evident applicability in this regard to the way in which moral stances typically get expressed in animal protection. For in a very straightforward sense, colleagues at the Edinburgh group self-identify as people who express concern for the welfare or well-being of animals, who feel the responsibility for ensuring welfare improvements, and who wish to persuade others of their responsibilities and ideally to make animal welfare a wider concern in society. Whether calling on government to amend or enact legislation or petitioning public bodies to enforce existing welfare regulations, they clearly do so in the name of animals or on their behalf.

While Bryson goes on to define "moral subjects" as quite simply "all moral agents or moral patients recognised by a society" (2018, 16), the language used in the definition of those two positions suggest an alternative emphasis. Deliberately conceived as "something" rather than someone, both moral agent and moral patient in these regimes of responsibility exceed the status allotted them by Regan. In particular, Bryson is concerned to explore and consider the implications of assigning moral patient or moral agent positions to made things or artifacts. For Bryson, as an ethicist of technology, the issue chiefly falls around the question of whether such attributions are wise. So, although Bryson admits robots into the ranks of "second-order moral patients" and believes that we should find a language to consider their welfare (2018, 22), Bryson also speaks against the broader notion of making AI into moral subjects on several counts. This includes an argument that to assign moral agency to robots would risk obscuring the moral agency of their designers (21) and hence also mask the issue of human responsibility. It additionally includes an argument that to assign robots a stronger moral patient position would necessarily threaten their present nonsuffering status, one that Bryson ultimately does not believe it is in our (human) interest to alter. Indeed, for this very reason Bryson advocates against making robots look too much like familiar "empathy-deserving moral patients" (23), such as fellow humans, in particular children, or such as certain companion animals.

There are some obvious and pressing differences in the terms of this discussion, both from the kinds of argument embodied in the work of Regan and from the ways in which the animal people I knew typically assumed the moral status of animals. It is striking, for instance, that in Bryson's account one is made to consider the possibility that moral patients need not be automatically labeled empathy-deserving and further that there might even be different hierarchies or orders of both moral patients and moral agents. Most significantly, unlike the moral status of human and nonhuman sentient animals, which animal ethicists like colleagues at the animal group usually imagine as an issue of necessity, the moral status of AI can appear at least to an observer such as Bryson as a matter of choice (2018, 16). This is in part because "the nature of machines as artefacts means that the question of their morality is not simply a question of what moral status they deserve" (17); it is also a question of "what moral status we ought to build those artefacts to meet" (17). Apparently as much the outcome of the way in which the machine is constructed or designed as the outcome of any given properties—such as consciousness or intelligence—recognized in them, these moral entities are in this regard properly "second-order."

Yet those differences at the same time cloud some important similarities. While the status of animals within animal protection was never exactly presented as a choice and colleagues at the animal group would fervently resist the notion that moral attribution should be determined by human interests, the presumption that animal others were moral patients was equally not solely driven by definitional principle. In fact, colleagues' general resistance to the idea of animal others as moral agents was less philosophical in concern and more motivated by a fear of the risks perceived to accompany such a claim. Animal people worried that if animal others got publicly acknowledged as moral agents, their protection would diminish, or, to put it another way, that our sense of human responsibility for animal welfare might lessen as a result.

In offering a working definition of moral agents and moral patients, Bryson claims to restrict concerns to formal philosophical categories. Actually, the few citations given for the definition include works from social psychology (Gray and Wegner 2009) and from linguistic anthropology (Duranti 2004). But perhaps more interestingly, one of Bryson's chief references is also from the philosophy of mind. In particular, Bryson cites an essay by Mikael M. Karlsson (2002), which itself is a reappraisal of arguments by Fred Dretske. Although these discussions never really center on the *moral* dimensions of agent and patient positions, Bryson nevertheless includes the reference as a justification for the use and definition of terms. That move is helpful, I believe, in getting us to understand how we too might more radically expand an appreciation of the distinction's purchase and do so in a manner that also brings us back to some of the wider concerns animal people have about the nature of moral being.

As Karlsson points out, right at the beginning of *Explaining Behaviour: Reasons in a World of Causes*, Dretske (1988) offers the reader what is termed a familiar opposition, "the difference between things we do and things that happen to us" (1988, 1). Indeed, Dretske claims that this difference "underlies our distinction between the active and the passive—between power, agent, and action on the one hand and passion, patience, and patient . . . on the other" (1988, 1). Karlsson concurs. "Fundamental to any theory of agency," Karlsson elaborates, "is the distinction between acting and suffering, between agency and patiency, between what an agent does and what befalls him" (2002, 59). It is true that the terms of debate within the philosophy of mind are usually motivated by a concern to reinterpret the nature of action rather than the nature of its posited opposite. Dretske, for instance, is almost solely focused on redefining "agent" (rather than "patient") so as to include nonintentional action and hence allow a basis for making bold comparisons between examples of human, animal, plant, and machine behaviors; while Karlsson, by contrast, wants to highlight that in many cases "patiency" can just as easily be redefined as forms of agency, as for instance a positive or negative "power to be affected by something else" (2002, 63–64). However, the introduction of this wider set of distinctions, which draw attention to the qualitative difference between doing and being done to, has, I believe, some provocative implications for a fuller reconsideration of moral patient and agent positions. Among other things, it redirects attention to the dynamic power of redescription and hence also to the shifting claims at work or to the sheer eventfulness of consistently making that kind of distinction.

As we have seen, animal protection typically figures animal others as moral patients. But it does so in a diverse and differentiating fashion. In the context of making a case, where the designation may rest upon evidential forms and the science of animal sentience or upon abstract philosophical first principles, instinctively understood, animal others are often moral patients in a rather generic sense. They are rhetorically or statistically defined as objects of moral concern, as those things deserving of protection, to which we owe a responsibility or duty of care. When redescribed through narratives of making contact, however, those animal others appear to demonstrate a far more affective status or passive power. Imagined through intimate embodied encounters that affirm intersubjective relations, they draw out love or empathy from the moral agent. Imagined through encounters that affirm sentient talents and hence the capacity to act or do things in extraordinary manner, they remain moral patients but this time with the power to amaze or draw out admiration. And when considered through witnessed forms of human-inflicted animal suffering, they or the encounter prompted by what has happened to them can appear to exhibit powers to harrow and distress, and hence again to move the moral agent to recognize obligations and to take action.

Yet, while those I knew working in animal protection certainly regarded themselves as the only moral agents in these encounters, there were also ways in which their own suffering status or moral patient positions consistently dominated accounts of ethical life. Indeed, to repurpose the phrasing of Karlsson, I believe it may be useful to consider animal people as moral subjects closely distinguished by what they took to be the power to be affected by something else. In their accounts I was regularly struck by the fact that a discussion of moral agency could easily be reconfigured as something that looked a good deal more like a discussion of moral patiency. By this I mean an attention to a morally defining and sometimes overwhelming even if quite mysterious sensation of being done to or acted upon. Sometimes directly attributed to making contact (for instance, to the innately ethically demanding presence of animal others) but at other times not, all that my colleagues at the animal group knew for sure was that these moral feelings or passive powers persisted and that they often came unbidden, as it were to befall them.

"What is in us that makes us so passionate about animals?" Cassie asked. Having considered and valued an education in the facts about industrial farming and in the first principles of animal liberation, as well as having weighed the moral consequence of seeing "the ugliness" of abuse and mistreatment, Cassie remained unable to fully explain it. That bemusement led her to consider and reject other kinds of explanation for moral passion and conviction available to animal people. This included explanations or forms of redescription that required no reference to either the condition or the formal reasoned status or the passive powers of animal others. "It's certainly not from my parents or my sister or my brother, or anybody else I know," she told me. "There's nobody else [in her family circle] who feels the way I do." Although this time dismissed, as we will go on to see, those sorts of accounts did also inform and sometimes reconfigure a sense of ethical life within animal protection. Indeed, the broader point may be that privileging both the certainties of that moral patient position—"All I do know is that as long as I can remember I loved animals, that we all want to try and make life better for them, and that we recognize that feeling in each other and call ourselves animal people"—and its ongoing mystery was precisely what provoked or enabled such redescriptions to occur.

* * *

At this stage I think it is useful to take a step back and distinguish between at least two kinds of possible relationship to the moral patient position. For on the one hand, we clearly have in play what we might term a range of after-the-fact explanations of that position. Whether drawn from wider arguments in the animal movement or more specifically from the terms of moral reasoning in the work of Regan, these explanations focus attention on the question of whether one can or should assign a moral patient position to certain subjects. The discussion is usually concerned with debating which or what can count as moral patient or moral agent and hence be attributed moral value, and then to persuade others of that fact. As we have seen, these kinds of after-the-fact explanations can also be found elsewhere, in fields of moral inquiry influenced by but beyond animal ethics (such as the field of ethics and technology) and the immediate moral concerns of animal people.

On the other hand, we have an emerging sense of another sort of relationship to the moral patient position and more broadly to the condition of what I have identified as moral patiency. Here we might speak of a range of before-the-fact explanations characterized by an assumption that versions of this position or of the capacity for patiency are somehow innate or universal, or at least innately characteristic of moral sense and being. Within animal protection and the moral imagination of animal people, there is the dominant presumption, for instance, that animal others are innately ethically demanding in an emotion-charged way. Certainly, animal people or at least colleagues at the animal group I worked with often assume or claim this to be the case. Yet that apparently fixed possibility is also essentially problematized or complicated—for instance, by the fact that the ethically demanding or viscerally powerful presence of animals is clearly not recognized by everybody. For animal people, mainstream publics are in large part defined by an inability or failure to be moved. As well as asking why, this leads animal people to ask themselves what makes them individually and collectively morally responsive. The search for answers simultaneously sustains and recalibrates the primacy placed on moral patiency.

In the chapters that follow, I examine and explore how both these sets of explanations are produced and negotiated, in what I take to be an ongoing process, and how, from the perspective of one animal group operating in one small quarter of the transnational animal movement, that process animates the work and moral lives of those committed to the cause of animal protection. But for now I want to spend a little longer excavating an intellectual genealogy for this project and in particular for that side of it attentive to the inclination to provide before-the-fact explanations of moral stances.

Of course, there already exists literature within anthropology that might be read as concerned to foreground what I am calling the experience of moral patiency. Most obviously, one ought to acknowledge the wider influence of affect theory and, in the specific context of the study of animal activism, this time in India, the work of Naisargi Davé (2014, 2017, 2019, 2023). In part concerned to provide what Davé terms "affective histories" of "how people come to act on behalf of animals" (2014, 434), Davé zooms in on one version of what I have been describing. Indeed, the author's concern is precisely with the recurring focus in activist narratives on critical witnessed encounters with animals interpreted as suffering because of acts of (human) violence against them—and on, as in the moral biographies I began with, an accompanying sense of conversion or self-awakening. "Similar to a coming-out story," Davé tells us, "animal activists in India and perhaps elsewhere stake their commitment to a way of life based on one critical moment after which nothing can ever be the same" (2023, 14). Davé acknowledges the extent to which such accounts are a "standard trope" within the wider animal movement (animal protection in contemporary India, Davé explains, is itself significantly influenced and driven by foreign animal activists and animal groups [2014, 435]), that they are always vulnerable to the accusation of being human-centric, ultimately rendering "the animal as theoretical" or as "mere object" in the narrative of the witness "who sees and consequently acts" (2014, 434). However, Davé nevertheless believes that these moments, or at least some of them, are worth taking seriously as expansive moral experiences. Davé is particularly struck by the consistent emphasis placed on the intimacy of this event or encounter, the fact that "it occurs between two singular beings" rather than a generic animal and that it is taken to "inaugurate a bond demanding from the person a life of responsibility" (2014, 434). It is on that basis that Davé then proceeds to explore numbers of ways in which the affective dynamics of those witnessed encounters gets registered and continues to play out in the moral imagination of diverse animal activists.[4]

Although Davé makes very little explicit reference to the language of patiency—the author more often speaks of examining "moral biographies of action and inaction" (2014, 434)—there are clear resonances with its tone and emphasis. In Davé's interpretation of the expansive potential of these intimate encounters, there is occasional flirtation with borrowed concepts such as "active passivity" (2023, 114). And in recent work concerned to foreground a radically recalibrated concept of indifference, we find similar undertones of patiency. "So, I suppose another thing I mean by indifference," Davé tells us, "is getting out of our own way so that we are free to respond to what moves us" (2023, 9). In this sense, Davé continues, indifference is "an expansion and intensification of the capacity to affect and to be affected" (2023, 9). But perhaps more pertinently, Davé regularly stresses that an ethos of "surrender" figures as a guiding principle (2014, 439; 2023, 20). In some variants this attaches to a notion of humility and to a sense in which one learns that one is not free to do otherwise, that one must surrender to one's responsibility to address the inequities of animal exploitation. Elsewhere, Davé connects it to an examination of the possibility, voiced by some activists Davé knew, that one somehow "becomes an animal" through witnessing "an animal in pain" (2014, 451). Or more broadly, in Davé's final reworking, we all might prize a "surrender to the spirit of becoming" (2014, 451; 2023, 30), which is simultaneously a rupture from certain dominant moral logics grounded in the assumption of sovereign action.[5]

Perhaps unsurprisingly, it is exactly this kind of emphasis that has of late led at least one anthropologist to posit a more direct and general correlation to a notion of patiency. Writing very recently, William Mazzarella is the first anthropologist to formally propose that we ought to consider what is termed "an ethics of patiency" (2020, 5). As well as relying on affect theory, Mazzarella constructs another genealogy, this time routed through traditions of continental philosophy and a set of provocations for "thinking patiency" that Mazzarella rather inventively takes from Kant and Durkheim (32). Indeed, much like Davé, a strong current running through Mazzarella's manifesto rests exactly on what he regards to be the virtue of "surrender" or "yielding" (28), which Mazzarella interprets as an active pulling back from the sort of willful doing embodied in moral reasoning or a stress on moral choice. Here the language of affect theory conjoins with that constructed by Mazzarella for patiency to zero in on techniques or arts of "attunement" and the potency or energy tapped into by making oneself available for "resonance" and "for what resonance can bring alive between me and something else" (6).

Elsewhere, Mazzarella cites the influence of the German philosopher Peter Sloterdijk's reading of the Heideggerian concept of *Gelassenheit*, usually translated as "releasement" (2020, 9). For Mazzarella, an ethics of patiency should also be about that release from any public expectation or "commandment to act" or to be defined by intentional actions. Of special interest, given the stress in Cassie's words, is Mazzarella's further urging to embrace "the bodily ground of our preferences and our priorities," or what Mazzarella terms "the impersonal matter of our personhood" (32–33). This crucially involves paying attention to "something that is not an expression *of* us but rather something that expresses itself *in* us" (33), by implication yet another way of marking the distinction between the outlook of a notional moral agent and that of a notional moral patient.[6]

As already indicated, Mazzarella is principally concerned to make a case for an ethics of patiency. In this respect, Mazzarella offers us something closer to an anthropological version of a moral treatise. This necessarily creates some tensions when connecting the argument to ethnographic material. Nevertheless, I believe that Mazzarella's invitation to pay attention to and especially to value that which expresses itself *in us* as well as that which expresses itself as being *of us* is very useful in getting at some crucial aspects of what animal people describe and I want to identify as moral patiency. That way of marking the distinction between moral patient and moral agent positions further assists us because it allows us to hear the moments when before-the-fact explanations and indeed mystery surface within even the most formal after-the-fact explanations of moral being.

My favorite example takes us back to the animal ethics and work of Regan. For in the later preface to the 2004 edition of *The Case for Animal Rights*, something unexpected occurs. As well as reiterating the fact that the definition of moral patient includes certain human subjects, Regan takes the opportunity to deviate from the heavy emphasis on the logic of moral reasoning. Looking back at the original process of writing the book, the author chooses to highlight the "effortless work" or inspired state by which at least part of its argument developed. I can't resist quoting at length:

> However, when I began to work my way through chapter six (which is mainly devoted to a critique of utilitarianism), something happened. It was as if—and I know this will sound strange, but I'll risk it anyhow—it was as if I ceased to be the book's author. Words, sentences, paragraphs, whole pages came, from where, I did not know. What I was writing was new to me; it did not represent anything I had ever thought before. But the words took up permanent residence on the page as fast as I could write them down. This was more than enjoyable. This was exhilarating (2004, xii).

The quote continues in this fashion for the rest of the page and into the next, a passage that Regan must have known would have been received awkwardly by many fellow moral philosophers (certainly in the analytic tradition). In fact, Regan reports that this exhilaration went on for far longer than might be anticipated. "I was in this state, without interruption, for months," Regan declares. "It is no exaggeration for me to say that during this time, I had lost control over where the book was going" (xii). Even if it was ultimately attributed to the flow of logic itself, Regan insists that not until this moment of losing control did the author "begin to be transformed into an abolitionist" (xiii); up until that point, for instance, Regan had fully intended to defend biomedical research that relied on animal experimentation.

It is worth emphasizing, given my earlier point that in the animal movement experiences of moral patiency are not solely attributed to encounters with animal others, that Regan is not identifying animals themselves (whether singularly or generically imagined as leading flourishing sentient lives or in pain) as the vital source of this exhilaration. Instead, the example seems to illustrate that rational thought or moral reasoning can also sometimes be experienced through patiency or through the ways in which subjects articulate a movement back and forth between moral agent and moral patient positions. Indeed, Regan's conscience seems to be an artifact of that process. "The logic of the argument I was following had converted me to an abolitionist position" (2004, xii), Regan writes in one place, a claim that in the transitive procedures of inspiration neatly folds into reflections on the power of the book itself. Again importantly, despite the emphasis on the agentive role of a logic that acts on or through the author, Regan, like Cassie and many of the animal people I knew, stresses the incomprehensibility of what occurred. "Even as I write these words," Regan observes, "I have to shake my head in wonder, still unable to understand how it all happened" (xii).

The distinction drawn by Mazzarella, between that which expresses itself as being *of us* and that which expresses itself as being *in us*, also connects to a possible genealogy for accounts of moral patiency that one might clearly locate in debates within the anthropology of religion. Certainly, Davé acknowledges that precedent in the selected emphasis on surrender, as does Mazzarella, who admits that any appeal to the language of yielding bespeaks obvious connections to reported affective experience of the divine and to familiar exercises of "pious attunement" (2020, 7). Among the influences cited by Davé is the work of Saba Mahmood (2005) on the "passional" dimensions of ethical formation in the context of the women's mosque movement in Egypt. For the "moral-passional self" (2005, 126) that Mahmood describes is crucially empowered as moral subject through an act of submission or surrender to God. Furthermore, seemingly reinforcing the later stress laid in Mazzarella's ethics of patiency, Mahmood reports that within the mosque movement it is simultaneously held that in essential ways these passions or what we might term "passive powers" need to be trained or cultivated. The apparent contradiction is one of the many tensions that Mahmood plays with when observing that such "movements of ethical reform" can be understood to "unsettle key assumptions of the secular-liberal imaginary" (2005, 78),[7] including assumptions about the freedom of moral subjects.

Indeed, Davé's "material on animals and witnessing in urban India" can be read to unsettle in somewhat similar fashion (2014, 445). Not only can one see that "the animal subject is brought into its intimate relation with a human through its unfreedom," Davé states; but one can also see how "by entering into intimacy with an animal in pain, the activist seeks not to be more free, but to render herself even more deeply subject to unequal relations of obligation and responsibility" (2014, 445; see also 2023, 25). While, as we will see, the animal people in Scotland that I worked with were in many ways fully committed to a liberal moral imaginary explicitly and often heavily marked as secular, I also find the emphasis on the perceived unfreedom of moral subjects defined by experiences of patiency helpful. There are some important caveats, though. As the making-contact narratives offered by the Edinburgh group illustrate, colleagues were not solely concerned with encountering and responding to the animal in pain. In fact, at times the animal subject in that intimate contact was crucially envisaged as freely present to a human subject who witnesses but also feels that they participate in that encounter, and hence that witnessing might in some fashion be a two-way thing. A sense of moral patiency might still be said to flow from this encounter with an apparently free animal subject, but as we explore later, that can open up a set of quite different ramifications for the relationship between the moral patient and moral agent positions.

Likewise, from the perspective of animal people like Cassie who also don't know what makes them so passionate about animals, the unfreedom of the human moral subject can appear both more confusing and more insistent than, say, the account of yielding described by Mazzarella, which partly draws upon examples of routinized pious attunement. Certainly, there appears to be an ongoing strain between Mazzarella's invocation of yielding or surrender as a moral disposition that one ought to actively embrace and what I take to be some of the essential realities of moral patiency for animal people. Recall that in Cassie's rendering, patiency was not something one could straightforwardly *do*; instead, it was something that happened to you, often in an uncontrolled, hard-to-predict, and sometimes overwhelming manner. As we will see, that reality introduced its own set of problematics, especially into the work of the animal group and the specific practice of certain kinds of expertise within it. Although colleagues were regularly concerned to try to rein those moral passions in or to harness them to good and proper effect, they did not usually imagine that they could provoke or elicit moral passions into being. The notable exception was when the animal group envisaged the quality of engagement it might have with non–animal people or the mainstream; at these times, something close to a project of patiency was considered.

Returning to the example of debates within the anthropology of religion, the work of Amira Mittermaier (2010, 2012) might also be cited as immediately relevant. For as well as engaging with an ethics of passion, a particular target of one essay is the very assumption of self-cultivating moral subjects (see 2012). Mittermaier places some distance between this approach to what I am calling moral patiency and that of Mahmood, precisely on the basis that in Mittermaier's ethnographic story (focused on the role allotted to dreams in a Sufi community also in Egypt) and many others like it, the consistent emphasis falls on the miraculous or "elements of surprise and awe" rather than on regimes of trained self-cultivation of passive powers (2012, 250). In fact, Mittermaier's position is constructed from a series of close observations about the central valorization placed on "being acted upon," which the author identifies within certain modes of religiosity.[8] Such an attention, Mittermaier claims, pushes "the critique of the autonomous [religious-moral] subject that is central to the literature on self-cultivation" even further (2012, 251).

But what really intrigues me about Mittermaier's work is the fact that a strong part of the genealogy presented for this emphasis relies on the much earlier work of Godfrey Lienhardt and in particular *Divinity and Experience: the Religion of the Dinka* (1961). That connection draws my attention since Lienhardt is also directly cited as a source for the notion of moral patiency in a very brief examination of it offered by Michael Carrithers (2005),[9] one of the few other immediate references to the term (beyond Mazzarella) that I could find within anthropology. In the case of Mittermaier (2012, 249), though, particular attention falls on the following statement by Lienhardt, itself prompted by a reflection about the linguistic challenges of adequately conveying what Lienhardt saw as the dynamics of Dinka moral and religious concerns: "It is perhaps significant," Lienhardt advises, "that in ordinary English usage we have no word to indicate an opposite of 'actions' in relation to the human self" (1961, 151). Without access to the neologism "patiency," Lienhardt goes on to resort to an out-of-use version of "passions," or *passiones*, which Lienhardt borrows from R. G. Collingwood (not cited by Mittermaier). For in its old English meaning, Collingwood (1944, 86) advises, that word also once stood for "instances of being acted upon."

There are, for my purposes, several useful takeaway points from Lienhardt's musings. Of course, much of *Divinity and Experience* is concerned to present Dinka moral and religious life through a contrast with what Lienhardt identifies as the normative dimensions of moral being and moral action in the society of late colonial Britain. Among the Dinka, Lienhardt (1961, 150) at one point declares, "The man is the object acted upon," a statement that Lienhardt clarifies by further observing that with the people the author worked with, "we often find a reversal of European expressions which assume the human self, or mind, *as subject in relation to what happens to it* [my emphasis]." Perhaps most well-known is the way that observation feeds into Lienhardt's rendering of the stirring of conscience. In a Dinka context, Lienhardt argues, the qualities attributed to a particular fetish or "Power" work "analogously to what, for Europeans, would be the prompting of a guilty conscience" (150). But instead of being attributed to the moral subject, that sense of guilt is taken to be "a presence acting upon the self from without," and furthermore it is taken as "employed by someone to do so": that is, another person in that community who wants to instill an "experience of guilty indebtedness" toward them. "Here again," Lienhardt states, "he [i.e., the Dinka] seems to see in that which has affected him the self-determining subject of activity" and to regard "himself the object of it" (151). Indeed, on that basis Lienhardt concludes by returning to Collingwood's old English meaning for "passions" or *passiones*. "If the word 'passions' . . . were still normally current," Lienhardt advises, then "it would be possible to say that . . . [Dinka Powers or divinities] were the images of human *passiones* seen as the active source of those *passions*" (151). And furthermore, Lienhardt adds, we could therefore say that for the Dinka the identification of *passiones* is a "necessary preliminary to human action."

In considering the before-the-fact explanations of the moral patient position offered by animal people, I would want to question aspects of the assumption of reversal offered in Lienhardt's account. Likewise, I would want to do the same with aspects of the assumptions in many more recent accounts often informing reference to a "secular-liberal [moral] imaginary," especially where it is presented as a descriptive foil. It should, I hope, already be clear that animal people also sometimes regard instances of being acted on as a necessary preliminary to moral action and that they also sometimes see in that which has affected them (i.e., moral feeling or the fact that they are passionate about animals) evidence for some kind of "self-determining subject of activity" that exists in but also beyond them. While not perhaps in the radical sense described by Lienhardt (for colleagues at the animal group, moral experience such as guilt remains the authentic experience of the moral subject), animal people do concern themselves with the issue of where moral sense or conviction comes from, and they also sometimes identify a range of external agents who in an enchained fashion move or work through them. Indeed, what impresses me most about Lienhardt's work, apart from the rich and thought-provoking way in which Lienhardt explores a moral subject apparently defined as the object of activity, is that it ultimately brings this reader back to those "European" expressions of moral action. For as Lienhardt must have known, elsewhere Collingwood strives to find a way precisely to readmit the old English meaning of the passions into common usage. In other words, as far as Collingwood is concerned, what we are terming "moral patiency" is not absent from a British moral imagination; instead, its expression is merely linguistically inhibited.

To an extent, that problem of English usage clearly remains. Otherwise, what would be the need for a discussion about moral patiency or indeed for an articulation of an ethics of passion grounded in instances of being acted on? But what perhaps interests me more is the ethnographic question of whether this might also be a live problem for animal people and the wider animal movement. As Cassie's puzzlement perhaps attests, it would appear that colleagues at the animal group also sometimes noticed the absence keenly. Certainly, language often seemed to them inadequate for the task of satisfactorily conveying either to themselves or to those outside the animal movement what the experience of moral conviction truly felt like. There was a sense in which the power of those moral feelings and the drama of patiency sometimes got drowned out, that it struggled to be heard, in particular against the backdrop of often more dominant and overarticulate vocabularies of moral action, choice, and reason.

* * *

To help develop these observations, I want to end this section by briefly extending my genealogy to include a few anthropologists that have explicitly considered the dynamic interaction between patient and agent positions, even where those positions have not necessarily been marked as defining "moral" distinctions.

Apart from Lienhardt, I have found little mention of the influence of Collingwood. A notable exception, however, can be found in the work of Xin Liu (2002), which focuses less on the issue of those positions' moral categorization and more on their animating relationship to what Liu identifies as sets of dominant archetypes or moral characters (this time in a narrative about the terms of moral drama in one version of modern China at the turn of the twenty-first century). In this account, Liu cites Collingwood (1942) as the originator of the agent/ patient distinction but also reminds us that Collingwood's delineation includes a third term in the nexus, "instrument," the "what or who" that "carries out action" on behalf of others (Liu 2002, 44). "The relation between 'to act on' and 'to be acted on' is one of active versus passive," Liu at one point tells us, "whereas the relation between the end and the means is one of agents/patients versus instruments" (2002, 49). Although Liu's definition of the patient position can be said to reproduce a rather one-dimensional stress on passivity (there is no sense of patiency in the manner Mazzarella outlines), what intrigues me in Liu's usage is the emphasis on the capacity of these moral archetypes or characters to be read in dynamic relationship. Liu has them narratively interacting with each other precisely through the way they get invoked and therefore are made to shift between respective agent, patient, and sometimes instrument positions.[10]

That observation brings me to another major influence on the way in which I have envisaged relational modes of subjectivity within animal protection and the connection to drawing the vital and fluid distinction between moral patient and moral agent. I refer to the work of Alfred Gell, and in particular *Art and Agency*

(1998). For while Gell is by no means concerned with the moral status of these positions, that book does represent, I would argue, the fullest anthropological account of their terms of interaction. It is also by far the most explicit account of the animating potential of claims made from the patient position. Indeed, for some observers of Gell's project, it was exactly the addition of that patient perspective that distinguished the approach, not just within the anthropology of art and material studies but also more broadly within anthropological appreciations of agency (see Strathern 1999). "I am concerned," Gell wrote, "with agent/patient relationships in the fleeting contexts and predicaments of social life" (1998, 22). The emphasis, as with Liu, is on the dynamic capacity for those relationships to be reoriented by simple shifts in the attribution of position and especially by acts of reversal. If, Gell continues, "to be an 'agent' one must act with respect to the 'patient,'" and one understands the patient to be "the object which is causally affected by the agent's action," then it is essential to appreciate the "relational" and "transitive" implications of that fact. What this particularly means is that both agents and patients are only momentarily occupied positions; "in any given transaction in which agency is manifested, there is a 'patient' who or which is *another 'potential' agent* [original emphasis]" and vice versa (1998, 22).

Of course, when considering the after-the-fact explanations of the moral patient position offered by Regan and much of animal ethics as well as more broadly expressed across the animal movement, that assumption of reversibility can sometimes appear far less safe. Sentient animals, for instance, may well be granted the status of patients, but unlike certain human subjects, that attribution does not inevitably result in the anticipation of a potential status as moral agent. Nevertheless, I take much from Gell's stress on the eventfulness of that switching and, to borrow an insight or reading provided by Strathern (1999, 17), from the wider idea that what Gell in fact provides us through the use of the term "patient" is an analysis of a whole "field of effectual actors" in which agents are always seen to require "a relational counterpart, that which shows the effects of another's agency."[11] But what is perhaps additionally novel from my perspective about Gell's account is the stress on the fleeting but nevertheless interconnected and cascading implications of invoking alternations in agent/patient positions. It is precisely because "agent/patient relations form nested hierarchies" (Gell 1998, 23) in any given setting that they become open to description, for instance. Likewise, it is that interconnectedness that makes it possible for Gell to admit a spectrum or sliding scale of agent and patient positions, and in particular a series of active roles for the patient. "In the vicinity of art objects," Gell advises, "struggles for control are played out in which 'patients' intervene in the enchainment of intention, instrument, and result, as 'passive agents,' that is, intermediaries between ultimate agents and ultimate patients" (1998, 23). Once again, the descriptive excitement of that observation lies in the attention it allows Gell to pay to the dynamism within those relatively stable or unstable nested hierarchies.

In this book I am concerned to explore the extent to which one can extend Gell's observations to a consideration of the specifically moral distinction between patient and agent. I want to trace and describe, for example, the nested hierarchies that make up specific objects of concern in animal protection, such as the welfare considerations around wild animals caught in traps, and at an entirely different scale also make up the ethical field of animal protection as a whole. As already hinted, I believe that those nested hierarchies can also be seen operating across and within the person of animal people. Such an ambition necessarily requires me to track down the significant switches or reversals in that relationship between moral patient and moral agent, as well as the blockages to reversal. It needs me to heed the moving nature of ethical life, in both its high drama and its micro eventfulness. And of course, it requires me to foreground and examine the lived difference between "agency exercised and patient-hood suffered" (Gell 1998, 37). Indeed, in that regard I believe that Gell's work is exemplary in another fashion. For as well as exploring multiple instances of the animating consequences of "things" and "persons" being acted on, distinguishing, for instance, the diffuse and plural sorts of derivative agency and passivity assigned to patients (and agents), Gell's focus means that the author notices the differences between kinds of received action. This includes, for example, the simple difference between an impression of being acted on and an impression of being acted through, between a patient designated as the object of action and its vessel. And since in Gell's account both "things" and "persons" are continually vulnerable to reconstitution as first-, second-, third-order (etc.) agents or patients, his work can further throw a spotlight on the vigorous potential for othering within the ethical field of animal protection. This perhaps includes the othering of moral selves (see Liu 2002), as for instance it gets played out through the very claim of patiency.

In fact, for me the test of the value of such a clumsy heuristic as moral patiency lies precisely in these issues of description. Does, for instance, paying closer attention to before-the-fact explanations of the moral patient position open up or close down our capacity to better describe ethical lives? And what does that attention add to our anthropological understanding of both moral agents and moral subjects? The latter question is central, since the language of moral patiency emerges as a relevant lexicon only because of the dominance of the lexicon around moral agency; indeed, the concept of moral patiency is consciously parasitical on the obvious conventionality of that language of analysis and description (see Gunkel 2017). The genealogies I have drawn were made with that conditionality in mind. I am very much aware, for example, that to propose a new language based on putting the concept of moral patiency front stage risks introducing a new universal-izing moralism into anthropology (see Holbraad 2018). But at least in my reading of the ethical field of animal protection, there is a liveliness to the articulation of being done to and doing, in the minor and major events of moral patient/moral agent interaction, which enables a more unfolding sense of moral life to be told. Indeed, it is the fact that these interactions occur in time that provides the drama and hence the stimulus for anthropological description. And since those transformations and the invocations of moral patient and moral agent positions appear to never stop, this provides the stimulus for constant redescription too. It also provides the narrator of such an account with interesting choices about how and where to begin this story.

Animal Protection as "Modern Culture"

In much of the anthropological genealogy I have so far traced, a moral-reli-gious emphasis on instances of being acted on typically gets contrasted to a set of assumptions around the freedom or autonomy of the moral subject that are regularly ascribed to a secular-liberal imaginary. But if animal protection can be read as an expression of that moral imaginary, as one might expect, then we need also to recognize an equal centrality placed on the overriding or predetermining fact that animal activists are so passionate about animals. After all, much of their moral action is premised on a prior submission or vital surrender to these pas-sions. It is true, as we will see, that colleagues at the Edinburgh group may have also identified a self-cultivating project grounded in the professional need to sometimes strategically discipline or rein in such moral feeling. Yet, the presump-tion remained that passion precedes and continues to feed any form of expertise within animal protection as well as within the wider activism of animal people.

I want now, however, to shift our terms of reference a little. As well as continu-ing to complicate how we might understand the status of animal protection as secular-liberal moral imaginary, in this section I aim to partially redescribe what is at stake by paying closer attention to relationships to the good and in particular to the issue of moral sources. That detour will eventually lead us back to a brief reconsideration of one classic category of the secular-liberal imaginary. It also will begin to illustrate how we might find inspiration for further sets of provoca-tions and hence for a fuller excavation of the role of the moral patient position and forms of relational subjectivity both within animal protection and in that wider imaginary. This time, the insights will flow from looking again or paying a different kind of attention to some of the observations already made in those works most closely associated with anthropology's "ethical turn."

But first, let us consider another template for an exploration of what I am calling moral patiency. The example can be found in the tradition of virtue ethics, one important influence on that ethical turn (see Mattingly 2012), especially where it focuses on tracing historical shifts in the configuration of various passive powers that moral subjects have come to claim through a perceived relationship to the good. For at least in the Taylorian sense, goods and in particular "hypergoods" are those visions, "the love of which [can] move us to . . . action . . . [and] empowers us to [in turn] do and be good" (Taylor 1989, 93). Indeed, Charles Taylor defines the historical acceptance of such goods, at least in Euro-American historical contexts, through the complex manner in which subjects come to know the good as a "felt force" (73), a power that may be discerned to "rage" or "murmur" between or within them (100). Even more pertinent, for our purposes, is the spotlight Taylor throws on the shifting identification of moral sources.

For Taylor famously claimed that in a modern age of unbelief, an ethical identity is crucially defined by the subject's awareness of and sense of being moved by a diversity of moral sources (1989, 313). Alongside what is taken to be an original reference point to God or the divine, Taylor describes the emergence of forms of moral exploration that locate their source in "the agent's own powers" (314), including powers of reason, self-control, and self-expression. In addition, Taylor identifies moral sources that became located in "the depths of nature," manifested both in the world and within the human subject. "Modern culture," Taylor reiterates, "is one of multiple sources" (317), a fact that the author wants us to recognize as an existential predicament.

Operating through such broad historical terms, animal protection could be read as an exemplification of a certain kind of ethical identity within that "modern culture." Certainly, many colleagues at the Edinburgh group confidently felt and expressed a relationship to the concerns of animal protection as a hypergood: that is, the good that most centrally defined who they were (Taylor 1989, 63–64). Yet, as we have seen, they were at the same time far from assured about the location of that hypergood's moral source. This could result in moments of unresolved puzzlement but also in regular redirections or oscillations between the invocation of quite different moral sources.

On the face of it, the animal people that I knew were avowedly secular in orientation and in many ways irreligious. Indeed, one of the things that struck me most upon first joining the animal group was the general and deep-seated suspicion of organized religion, especially Christianity. As in the animal movement more widely, many held up Christian doctrine and practice as the ideological tradition most responsible for the historical justification of human dominion and hence exploitation of animals.[12] This common critique sometimes led colleagues to place an emphasis on the moral agent's powers of reason and argumentation and on the universal capacity of either animal science or forensic investigation to evidence animal welfare concerns. But it could also lead colleagues to claim that their moral feeling or passion for animals might instead be sourced to the depths of nature. In this iteration, making contact was often presented as the fulfillment of a natural connection with more-than-human sentient others or as a reconnection with an original sympathy that humanity had previously lost or become alienated from. Here, culture or the social regularly appeared as that which got in the way of animal protection or more negatively as part of those forces actively responsible for the perpetuation of cruelty and abuse.

However, suspicion or even hostility toward the influence of Christianity (in a Scottish context this usually meant Presbyterian or Roman Catholic traditions) was always conjoined with a strong desire to exhibit toleration. The group and its vision of the global animal movement might still be dominated by Anglo-American or western European perspectives on those moral concerns and, as we have seen, by sometimes unthinking assumptions about the nature and composition of its possible membership; recall that in the sci-fi teaser film produced by the animal group, all the characters were white, including those depicted as belonging to a potentially recruitable mainstream. But as usually middle class and left-leaning subjects broadly supportive of a range of progressive issues, colleagues made a conscious effort to exhibit their liberal credentials.[13] In fact, they often demonstrated those credentials through an expressed sensitivity toward what were perceived to be the customs and faith-based beliefs and practices of minorities.

For instance, although the animal group and its staff were consistently opposed to practices of pre-stun slaughter, in a British context usually associated with either Jewish or Muslim traditions of meat preparation and consumption, they were far less likely to push the campaigning point home. Indeed, the toleration shown in this regard contrasted starkly with their attitude toward majoritarian appeals to custom or heritage. An obvious example of the latter would be the sort of arguments made in defense of bullfighting, a touchstone issue in the moral biographies of many in the animal movement across the UK. In like manner, colleagues were often keen to acknowledge that the animal husbandry or hunting practices of the world's "indigenous peoples" could be endured or even condoned because in contrast to systems of industrialized agriculture the practices of these communities were needs-based and not deliberately cruel or exploitative. Some colleagues went further and embraced what they interpreted as the animal-friendly philosophy embodied in indigenous cosmologies and attitudes. For example, Cassie had always been attracted to beliefs and practices that she attributed to American Indian or indigenous American peoples and told me that she had read quite extensively on that topic. Others cited the moral standard of Amerindian peoples or spoke in more generic terms of indigenous peoples as exemplars of respectful and sustainable behaviors.

Such admiration for indigenous belief and practice got invariably linked to wider assumptions about the perceived naturalness of these systems and hence also to a perceived source in the depths of nature that it was imagined animal people might in some fashion also tap into. But that appreciation could equally be combined with what were regarded as more acceptable ways of invoking a divine or supernatural moral source. For despite suspicions about organized religion, some religious traditions could be valued precisely on the basis that they provided animal-friendly forms of "spirituality." Indeed, any appeal to a moral source based on an agent's powers of reason could easily slip into more cosmological registers typically identified as influenced by the example of Buddhism, Hinduism, or Jainism.[14] As well as a passing allusion to the concept of cycles of rebirth, this could include reference to more generic ideas about the incorporeal essence of human and other living beings.

I can recall, for instance, one colleague rehearsing a version of the commonly made point in the philosophy of animal liberation about the equivalence of historical oppressions (see Singer 2015). "As I got a little bit older and started learning a bit of history," this colleague told me, "I realized that this is what we used to do to women, to children, to black people," treat them like "they are lesser beings, like they didn't have souls." Likewise, Elaine, PA to the group's CEO, confessed to having "dabbled a bit in Buddhism and stuff like that." Elaine explained that "we [i.e., humans and other sentient animals] all have the same feelings, the same soul," but that for her that soul was "just a piece of energy that is vibrating." In fact, Elaine told me that eating an animal would be like eating a part of herself "because I'd be destroying that energy." Such notions could also surface in other ways, including via some clearly Christian-inflected beliefs. For example, after the death of a household dog or cat, some colleagues would flirt with a commitment to the notion of an inclusive or animal-friendly afterlife. This was often embodied in ideas such as the Rainbow Bridge, which envisaged an eventual reunion between that human and their close animal companion.[15]

However, any invocation of a "spiritual" dimension to the hypergood that was animal protection was usually far more vaguely conceived. Especially in the context of work, colleagues tended to be cautious, rather more hesitant in putting such beliefs forward. Eilidh, the group's CEO, once told me, for instance, that she too believed in an all-connecting energy that made "the whole planet, peoples, animals, plants and everything One." Yet almost in the same breath, she started adding caveats; of course, Eilidh advised, she was only speaking at a "personal, not organizational level." The animal group itself did not communicate its relationship to the good through direct reference to such a moral source. Other colleagues and sometimes the same colleagues at different moments actively resisted signs of spirituality or supernatural moral sources and in particular any non-materialist explanations for animal and human sentience. Or, like Eilidh, they assigned such beliefs to a realm of private opinion disconnected from the goals or ethical mission of the animal group. In such cases, moral commitment was typically reassigned to logical argument or to scientific evidence or to commonsense understandings of how other animals must feel and hence suffer at the hands of humans.

The dynamism of reattribution, which includes not just a switching between multiple moral sources for the good but also oscillations in how those sources got parsed, is another part of the ethnographic story I want to tell. For it connects us back to the wider account of how one considers the distinction between moral patient and moral agent and in particular the issue of where or under what conditions moral patiency thrives or gets acknowledged.

This is also a concern of Mazzarella. Indeed, despite consistently alluding to models of religious or pious attunement, Mazzarella ultimately cautions against the presumption that an ethics of patiency operates in a "rarefied terrain" set apart from nonmystical or nonritualized behaviors (2020, 7). "A big part of what motivated me in thinking about patiency," Mazzarella advises, "is the sense that it's a kind of wormhole concept: a concept that short-circuits any solemn distinction between the sacred and the profane or between activity and passivity" (7). We might take the point on board; it obviously makes certain sense given the thrust of the making-contact narratives we have already examined. However, at least in the context of animal protection as I knew it, we need also remember that drawing or redrawing apparently solemn distinctions (such as that between the "personal" and the "organizational") remained one of the most common ways in which both moral patiency and moral agency got registered or invoked.

Mazzarella's emphasis on the nonrarefied terrains for patiency does though nicely bring us back to the status of the secular-liberal moral imaginary. For one of the lessons that I take from a creative reading of much literature in anthropology's ethical turn is precisely the fact that one can locate the centrality of patient-thinking not just at the margins but also right at the heart of what we take to be secular-liberal traditions, including those imbued by the spirit of autonomy or freedom.

For instance, in what I regard to be a most provocative contribution, James Laidlaw (2010b) explores the dynamic and distributed status of the category of responsibility. These observations partly develop through a dialogue with a series of reflections previously made by the moral philosopher Bernard Williams (1993). In fact, what really seems to attract Laidlaw to those reflections is the insight that the components of a typical attribution of responsibility (identified by Williams as cause, intention, state of mind, and obligation of response) can actually be quite radically dispersed. They "need not coincide within the same individual, or for that matter the same collectivity" (2010b, 150). From that prompt, Laidlaw argues that moral subjects (operating within a secular-liberal imaginary) experience responsibility as a claim that only sometimes references or reinforces an image of themselves as purposefully coherent or self-aware actors. Crucially, Laidlaw chooses to illustrate this claim by examining the manifold ways that the law in a UK context can serve to simultaneously confirm and disassemble the status of the subject as moral agent. Take the example, Laidlaw points out, of the ways pleas of mitigation typically work by identifying instances of the subject being acted on either via an external cause, such as a set of extenuating circumstances, or via an internal cause, such as a psychological trait or physiological impediment. This kind of moral patient position, centered on notions of involuntariness, can be conjoined with what we might term forms of patiency enacted by legal work that expands the subject's sphere of responsibility. Laidlaw highlights the example of legal culpability for the acts of dependents or for the damage done to someone else, not directly by oneself but by one's property (152–153). "What makes any particular action, effect or state of affairs someone's responsibility," it very often then appears, "is not a matter exclusively of any characteristics that adhere in or belong to that person" (152).

This statement takes us a very long way from how the issue of responsibility tends to play out in formal animal ethics. In particular, one might make the contrast to those after-the-fact explanations that use the category of responsibility as a defining marker of what distinguishes moral agent from moral patient—that is, the former has responsibility precisely because of certain identified internal characteristics or capacities; the latter lacks those characteristics or capacities and so cannot be morally held to account. Indeed, as Laidlaw highlights, it seems that secular-liberal imaginaries are constantly innovating ways of redistributing the attributions of responsibility and hence making a wider range of individuals into the recipients or effects of actions (note: Laidlaw does not use the language of patiency). We are already familiar, Laidlaw lays out, with academic and popular acts of "statistical extenuation" that might allow an individual action like a crime to be redescribed or understood as the effect of social causes (2010b, 160). However, perhaps more interesting is the increasing degree to which statistical analysis is now being deployed "to create hitherto unimaginable responsibilities" (160). Laidlaw offers us the example of the rising concept in British institutional or organizational life of "indirect discrimination" or bias. This conceives of a singular event such as a job hiring or promotion decision that, when seen in isolation, is not the object of moral judgment but that becomes so when reassessed as part of a wider statistical pattern. Again, for our purposes what is most relevant is that the moral agents deemed accountable because of this statistical measure are not usually the subjects directly involved in making that particular decision but instead a collective agent such as organizational culture itself. Individual subjects can, though, be assigned passive powers because of that retrospective measurement.

All these fluctuating claims around responsibility of course also circulated across the field of animal protection. In part, this was because the animal movement typically regarded the law, like Christianity, as a key ideological backdrop for the justification of animal exploitation. One commonly referenced example was the power of legal definitions, including property definitions, to determine welfare obligations toward different species or different individuals within the same species. Colleagues were at times painfully aware of what to them appeared as the huge costs of arbitrary legal distinctions, such as that between pet, livestock, pest, and wild animal. But at the same time, the law was also viewed as the most viable arena for establishing protections or animal welfare reforms as well as for punishing those individuals who were held responsible for committing legally defined offenses.

Indeed, for the animal group that I worked with, law and the processes of lawmaking were perhaps the most crucial sites for organizational operations; its campaigns were nearly always tied to policy work, to lobbying, and hence also to legislative opportunity. During my time with the animal group, this included heavy investments in petitioning for the extension of principles of vicarious responsibility. For example, the animal group's policy director lobbied hard to convince lawmakers to make the owners of Scottish shooting estates also liable for the negligent or illegal actions of those they employed. Typically, these arguments centered on the example of gamekeepers laying animal traps without regard for licensing regulations or more seriously on instances of gamekeepers using banned poisons to kill predator species such as foxes or birds of prey. Other colleagues also had to regularly negotiate specific legal definitions of responsibility. In the case of the group's investigator, those negotiations were integral to the role and especially to the task of evidence gathering; as we will see, the investigator regularly participated as a cited witness in court cases linked to the prosecution of animal welfare abuses. More generally, the investigator's work depended on a prior and intimate knowledge of diverse kinds of legal culpability. To properly conduct investigations, this colleague needed to understand, for instance, when a crime or breach of welfare regulations had happened. Likewise, the investigator needed to know who or what, legally speaking, could be held to account. In terms of criminal evidence gathering, the investigator needed further to ensure that witnessing somehow appeared "accidental" or involuntary (Reed 2016a, 105), a legal dance that itself further blurred the very distinction between activity and passivity.

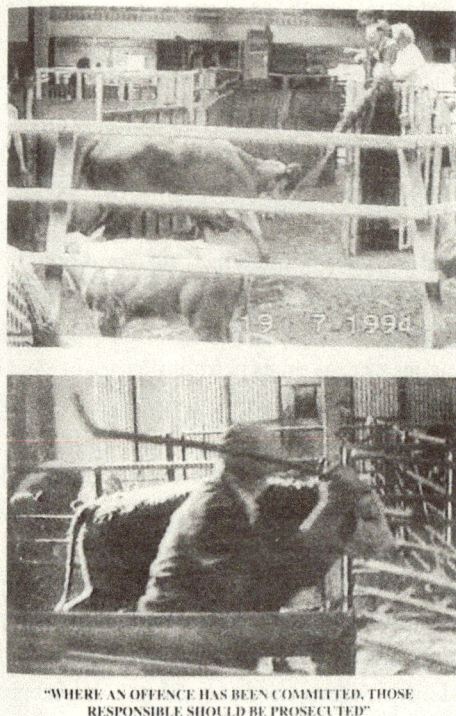

"WHERE AN OFFENCE HAS BEEN COMMITTED, THOSE
RESPONSIBLE SHOULD BE PROSECUTED"

FIGURE 2. "Where an offence has been committed, those responsible should be prosecuted." Image from 1994 annual report of animal group. Used by permission.

But colleagues also invoked assumptions of involuntariness and distributed responsibility in other ways, beyond the law. The organization often invested, for example, in the commission of surveys using forms of statistical analysis designed to reveal a range of indirect harms done to animals by humans. Indeed, in its messaging the animal group regularly stressed that mainstream publics were passive or unintentional agents of animal exploitation. Very often, it was "the system" rather than individuals that was to blame; indeed, the capacity to make such a distinction was closely tied to what colleagues saw as their professional outlook. Apparently stuck in a soup of apathy, mainstream publics were usually adjudged to be obfuscating or unthinking rather than bad, and often to be in some essential sense unfree. In this regard, members of those publics were perhaps rather closer to Regan's definition of a moral patient—like animals, somehow lacking the prerequisites that would enable them to control their own behaviors or at least to completely see the consequences of those actions and hence to be held fully morally accountable. At such moments, despite the predetermining role assigned to moral passions, colleagues and animal people more widely appeared most distinctly as moral agents in that after-the-fact sense defined and laid out by Regan.

I Am the Mainstream

Invoking problematic relationships to mainstream publics or what colleagues mostly just termed "the mainstream" is of course not unusual within movements of ethical reform. Nor is it unusual within wider animal activism. We are familiar with the presumption that activists might conceive themselves as operating at the margins of that mainstream or at some moral distance from it and with the idea that the moral stance and status of such *ethical* subjects might be clarified by drawing such a contrast. We are familiar too with the notion that their moral cause might contain an overt critique of mainstream values or lifestyles. But in the case of the animal group that I worked with, what does stand out, I believe, is the way in which colleagues instantiated or understood the terms of that relationship, at least at the historic moment that I first encountered them.

For, as already mentioned, back in 2010 the Edinburgh group had just redefined its organizational mission to be newly based around a concern to actively "engage the mainstream." In many ways, that ambition was emblematic of the long-standing moderate status of the animal group and of the kind of animal activist it represented. Within the animal movement, being moderate partly meant showing a willingness to dialogue with public bodies, including government, and to accept a route toward animal liberation based around reform rather than revolution. However, in the minds of colleagues, the turn to engage the mainstream marked something altogether more radical (by the logic of a moderate activist imagination). As we will see, the animal group envisaged that this would necessitate the organization itself in some manner becoming mainstream or, in one version of the mission, sponsoring its own mainstream animal-friendly movement. According to Eilidh, the group's CEO, this would require colleagues finding a way to embrace aspects of the values and outlook of mainstream publics, as it were to get inside those publics and initiate change from within. To do so, Eilidh regularly advised, there would need to be a complete transformation in organizational practice but also in the mindset of colleagues. If the animal group's new public messaging was to be "we are the mainstream," then everyone working for the animal group needed to start thinking as if "I am the mainstream" too. As well as a professional challenge, the CEO anticipated that for many colleagues this would be a personal, ethical struggle, as indeed it was. Over the following years, the notion of engaging the mainstream would shift in and out of organizational focus, and its interpretation would certainly vary a lot (between, say, a more radical endeavor to become mainstream and a more traditional model of dialogue with the mainstream), but it nevertheless remained a constant.

All of this points to another important theme across the book: the question of what it means to *work* on behalf of animals. In introducing readers to the professionalized world of animal protection, as it is lived and perceived by those who consider themselves to be moderate animal activists, I necessarily explore the interrelationships between cause and organizational culture. This includes an examination of the ways in which roles or recognized forms of expertise intersect with the moral experience and priorities of animal people. As such, I also make a contribution to the wider call to personalize or "humanise the expert" (Boyer 2008). On one level, this involves a straightforward consideration of how expertise, professionalization, and what we might call everyday ethics interact. But on another level, it leads me to consider how those professional roles or recognized forms of expertise can themselves become ethical exercises for animal people.[16] And in addition, it prompts reflections on the degree to which those working to improve the welfare of animals can be said to possess an expertise in a certain kind of ethical practice, or, as in the issue of engaging the mainstream, an aspirational expertise in the moralities of human others.

When I first became aware of the animal group back in 2009 and began working with the organization a year later, I had only the sketchiest sort of understanding of the priorities and concerns of animal protection. In fact, my original motivation for getting in touch—I had found the group through an internet search of campaigning organizations based in Edinburgh—was not informed by an interest in animal protection at all. Rather, I was concerned to find an organization with an overt ethical mission, one that tapped into wider moral debate within the public domain in Scotland and the rest of the UK and that supported a popularly recognizable ethical stance. I contacted the animal group because it seemed to fit that bill and started working with colleagues there because they seemed open to hosting my project. The timing was fortuitous precisely because the organization was undergoing a reappraisal of its rationale and strategic purpose. As well as envisaging a turn to engage the mainstream, this involved a new determination to reach out to the scientific community and the broader academy. Although that ambition was largely focused on making connections with animal scientists, especially those working on the latest research into the sentience levels of diverse species, my inquiry was received and welcomed in that spirit. Indeed, the first member of staff I met and discussed my project with was Craig, the science and research manager tasked with establishing the group's status as a mediating body with academic researchers. He introduced me to Eilidh, the organization's relatively new CEO, and pretty soon afterward I began my placement in the office.

At first assigned a free desk next to the small staff of fundraisers, I started to participate in team meetings, in planning discussions on campaign goals and policy opportunities, and in social media and general communications strategizing sessions. As time went on, I also began to shadow key role-holders in the organization with quite specific forms of expertise. In addition to Craig and Eilidh, this included accompanying Maggie, the animal group's long-standing policy director, on her visits to committee hearings and general sessions at the Scottish Parliament and on her trips to the annual conferences of key political parties. It involved working closely with Barry, the animal group's investigator, which included joining him on exploratory trips to Scottish shooting estates. I attended large numbers of campaigning and fundraising events, some of which brought me into contact with supporters and patrons of the animal group; these events also began longer conversations with senior members on the group's board of trustees. Over the years my participation has widened to include attendance at the Cross-Party Group for Animal Welfare in the Scottish Parliament and dialogues with a range of other animal groups or individual advocates. However, the core ethnographic focus has remained centered on the working (and ethical) lives of these colleagues.

As well as being part of the academy, I always had a secondary status at the animal group as myself a member of the mainstream. Without any background in animal protection or animal ethics and with no conspicuous signs of commitment to the cause, such as a consistent history of vegetarianism or veganism, my presence in the office was marked by a shared understanding that I was not really animal people. In fact, this was self-evident to colleagues from the many stupid or ill-informed questions I asked! However, given the revised mission of the organization, that status was itself fortuitous. Colleagues generally tolerated or even welcomed my mainstream credentials; those credentials often added a useful and informative specificity to our interactions. As well as providing a firsthand sense of what it meant to be positioned by animal people as part of the mainstream, the terms of those interactions enabled a stronger understanding of how someone like me could become less apathetic and more animal friendly.

So, the chapters of this book aim to explore the multiple entanglements within animal protection as an ethical field, which includes a shifting relationship to moralities of the mainstream but also to animal protection as a form of work or expertise. Various essential instances of the invoked alternation or reversal between moral patient and moral agent positions are examined along with specific role-holding perspectives that draw out certain expressions of that interaction. The emphasis throughout is on the fluctuating stabilities and instabilities of that ethical field and on what it means to be animal people and to work on behalf of animals in a professional yet also a moderate activist fashion.

Chapter 2 begins this exploration by providing a brief organizational history of the Edinburgh animal group. This includes a look at the connections between the group's shifts in orientation over time and those shifts reported for the broader history of the animal movement. It also considers the challenges of working for a generalist animal protection organization. For unlike many other animal groups such as those involved in rescue or running a sanctuary or those specializing in the protection of specific kinds of animal or types of harm done to animals, this organization worked on behalf of animals from a conceptual and material distance (i.e., there were no or very few immediate human-animal interactions in the workplace). As animal people, colleagues sometimes struggled with that reality. However, it also positively defined who they were both as moral subjects and as certain sorts of experts operating within the wider field of animal protection. Their positionality further defines aspects of my own, especially with regard to existent anthropological debate (and moral concern) linked to the study of human-animal and other multispecies relations. This includes that subset of the literature that directly considers cultures of expertise as expressed through those human-animal relations.

Chapter 3 introduces a significant aspect of the contemporary animal group's focus as an organization with an ethical mission. As the title suggests, attention principally falls on the new moral project to address mainstream publics and on the consequences of that move for organizational self-presentation and for the group's wider identity within the animal movement. In addition to the active promotion of a new ethos of positivity, which placed a radical constraint on how animal people and animal groups normatively express their moral concerns, this development led to the conscious foregrounding of specific virtues. First and foremost, as the chapter explores, it resulted in a new, reinvigorated emphasis on acts of kindness and hence a recalibrated role for moral agents. But the project to mobilize the mainstream also contained a new implied sociality for animal people (and those members of mainstream publics that the group successfully converted toward animal-friendly ways) that drew directly on their status and the status of animals as certain kinds of moral patients.

In chapter 4, I aim squarely to examine the most dominant and still apparently clearest expression of a subject's freedom or moral agency within the animal movement: the much-narrated decision to turn vegetarian and then usually by logical extension to turn vegan. I briefly explore how both vegetarianism and veganism tapped into wider forms of consumer activism promoted by the animal group. But as well as considering how either choice typically crystallized the experience of moral agency and sense of responsibility for colleagues, I also look at how the exemplary status of those choices simultaneously appeared to winnow down the number of other acts deemed assignable to a moral agent. In certain ways, *the* ethical choice seemed to reduce other apparent moral choices to a state of patiency or at least to hang the question of their autonomous status in the balance. This discussion includes colleagues' reflections on kinship and especially on the possibility that moral stances might be inherited rather than earned or chosen. It dwells on the primacy placed on idioms such as friendship and on what colleagues, as moderate animal activists, identify as a general wariness to preach one's values. Finally, the chapter examines the interconnection between all these issues and the core moral principle of do no harm within animal protection.

Chapter 5 is comparatively short and by its very musing nature offers something of an interlude. Indeed, this chapter returns us to the motif of sentience with which the book began. But this time the emphasis falls on a formal category of measurement within the animal science of sentience that also happens to operate as a category of moral suasion within animal protection: "personality." Of course, that category can also be invoked in a rather more commonsense fashion to describe the essential nature of human persons and their effects on one another. In this chapter, I explore those reverberating connections a little further by drawing on the ways in which colleagues at the animal group understood and shifted between these different registers of personality. That story brings us back to the tension between before-the-fact and after-the-fact forms of explanation of moral status, and once again to a certain puzzlement over the issue of moral sources.

Chapter 6 offers the first of several closer examinations of the sorts of moral disposition attached to specific forms of expertise within the organization. In this case, the spotlight falls on Barry and the role of undercover investigator, a task assigned to a specific member of staff but also one that has a high profile across the animal movement. Indeed, the investigator has something of the status of a moral exemplar for colleagues and animal activists more widely. Partly due to the perceived risks involved and the sacrifices required to do this kind of work effectively, that figure has consistently attracted the admiration of others. However, in this chapter I am equally concerned to explore the ethical disciplines that the investigator highlighted at the heart of this peculiar expertise. This includes the strict duties attached to the work of detection and observation and to the task of gathering evidence. As we will see, those duties drew forth a need to closely monitor and sometimes actively restrain expressions of ethical commitment typically associated with animal people, most notably to control the influence and impact of moral feelings or passions. As the only colleague with up-close-and-personal experience not just of systems of animal exploitation but also of the lives of animal exploiters, the investigator further offered colleagues a unique perspective on what motivates abuse. In fact, part of the expertise that Barry claimed as investigator was precisely an insider's knowledge of how animal abusers morally justify their involvement in those institutional processes.

Although the organization had no direct involvement in animal rescue, as we have already seen, that activity still had an important place in the moral imagination of staff. Alongside the turn to veganism, it provided a key reference point in the ethical lives of animal people, many of whom devoted spare time outside of work to volunteering at animal sanctuaries or involved themselves in rescue networks online. In chapter 7, I examine some of the connections and tension points between these two forms of moral labor (i.e., rescue work and the labor of working on behalf of animals in general). In particular, I am concerned to measure the difference through contrasting approaches to the category of evil. For in their unpaid work of rescue, individual acts of cruelty loomed large in the moral imagination, and, consequently, so too did the idea that animal people were engaged in a struggle against evil. Here the question of the moral status of those evil people often took center stage. Were such persons (and their cultures) agents of evil or instead patients of evil forces? Was it possible to redeem evil people? Such inquiries return us to the issue of whether being good is a form of moral patiency and hence lead to a parallel set of questions. To what extent did animal people control the will to rescue or help animals? Were they agents of goodness or alternatively also patients of moral forces?

Chapter 8 looks at the moral dispositions attached to another vital form of expertise within animal protection, the role of policy director and lobbyist. It considers again the essential virtue of being a moderate animal activist, this time enacted through the attitude and working practice of Maggie. This includes a look at the value placed on being moderate within the lawmaking process as a whole and the regulatory context around the welfare of animals. As well as examining diverse ways in which that virtue gets invoked by competing lobbyists, by government, and by other political parties involved in the process, the chapter focuses on the strategies deployed by the policy director to try to persuade politicians to prioritize animal protection concerns. Once again, the question of the nature of moral stances crops up, this time because of bodies outside the animal group, such as government ministers, civil servants, or parliamentary committees, regularly insisting that all lobbying positions on a given legislative issue represent the perspective of "interested parties." That move constantly risked flattening the distinction that Maggie absolutely wanted to uphold—that is, between the animal group's arguments and wider moral stance and what she and her colleagues saw as the self-interested or commercially motivated positions of many other lobbying groups. If the lawmaking process in many ways inflated the profile of the organization as a rational or strategic actor, the response of the policy director seemed to challenge that status. In her attempts to win over politicians, Maggie consistently emphasized that all interested parties or agents in the lawmaking process were in fact capable of change. However, she stressed that such a transformation in attitude might best occur by individuals surrendering to moral passions and hence becoming something much closer to moral patients.

By way of conclusion, I introduce an alternative reading of the virtue placed on being moderate. Here, pragmatism and the willingness to compromise to achieve legislative wins or to engage the mainstream become revalued as *compromised* moral behavior or even as acts of complicity or collaboration with the enemy. While such a perspective is usually located externally (for instance, in the outlook and accusations made by other animal groups), it also sometimes surfaces within the organization, in the sometimes-conflicted thoughts and feelings of many colleagues. In this eternal dilemma for moderate animal activists and perhaps for animal protection more broadly, "purist" moral stances were often simultaneously desired and derided as naive or dangerous. Swithering back and forth in this fashion was in many ways definitive, a form of oscillation that enabled multiple further dynamic shifts between moral patient and moral agent positions.

DOGS IN THE OFFICE

Paddy was Craig's dog, and, as Craig always insisted, he was Paddy's human. Paddy was also an office dog who came to work with Craig each day. Indeed, one of the first things I noticed when I joined the animal group was that Paddy, an adopted mutt, seemed to have free range of the animal group's rooms. Often, I found Paddy under a desk in one of the back rooms that Craig shared with Euan, the communications officer. I also regularly found Paddy in the street-facing room that housed the fundraising team and their four conjoined workstations. There, Paddy tended to sit beside Fraser, the fundraising director, or beside Sarah, the assistant fundraiser, and her companion Esme, an elderly rehomed Staffordshire bull terrier. Some days Paddy spent a few hours with Iain and David, the group's IT and social media specialists. At those times Paddy usually plonked down on the floor between Iain's bicycle and the filing cabinets that stored old reports and policy documents, campaign material, and folders of newspaper cuttings (i.e., the archive of the organization's life before digital storage). But Paddy could just as easily wander into the adjoining room where Elaine sat with her back to the "office library." This modest row of bookshelves contained a mix of large-format photographic volumes on animal lives; paperbacks about zoos, the shooting industry, vivisection, and industrial farming; works on popular animal science; a few legal texts on welfare regulations; and philosophical texts on animal rights. Doubling as a waiting room for visitors, the library space additionally had a few office armchairs and a coffee table.

From that spot, Paddy had the option of entering the room occupied by Eilidh, the CEO, and the office manager Shelia. Alternatively, Paddy could head along a narrow corridor toward the office's front door, a route that Elaine had to take many times each day to answer the doorbell, to collect deliveries or receive any callers. That passage could get especially busy around lunchtime or tea breaks, since it also gave access to the small staff kitchen. Finally, at the end of the corridor Paddy could push open the door into the boardroom. Framed by small windows overlooking a series of flat roofs, this room was notoriously cold in winter, so unless someone had put the bar heater on in preparation for a team meeting, Paddy, like the office staff, tended to avoid it.

Everyone, then, knew Paddy's comings and goings. Indeed, it was expected that when doors were closed, colleagues would get up to let Paddy in. This unwritten rule had no exceptions. Even Maggie and Barry did not hesitate to do so. As the occupiers of the last back room, they often had to discuss confidential matters linked to legal cases or to Barry's covert investigations and so quite routinely shut their door, a sign to other colleagues not to enter. The only caveat to Paddy's free movement occurred when Amber, the policy director's canine companion and one of the stars of the animal group's sci-fi hoax film (see the prologue), came into the office. For, as Maggie once explained to me, the two dogs did not get on and so were best kept apart.

Among staff, having dogs in the office was a straightforward reflection of a caregiver's commitment and of the value generally placed on the mutual affection and support these and other companion animals typically provided. But to all colleagues, the inter-room wanderings of Paddy as well as the characteristic wheeze and snorts emitted by Esme (Staffordshire bull terriers are one of those breeds that commonly suffer from brachycephalic airway syndrome) further signaled something essential about the ethos of the organization. Most simply, it was assumed that their presence naturally affirmed the ordinary concerns and obligations of animal people and thus of an animal group. But the small gestures colleagues made daily toward supporting the free movement of Paddy also quietly declared in a more outward-looking fashion that here was a workplace run on quite different principles from other workplaces in the city, including those many office spaces (without the presence of companion animals) that members of staff had visited over the years or previously known firsthand as employees.

Welfare of Animals in General: A Brief History

Alongside the presence of Paddy and other dogs in the office, the other thing that struck me almost immediately was the sheer breadth of welfare concerns to which colleagues attended. For as a "generalist" organization, the animal group could potentially advocate on any protection issue. This fact was brought home to me in some of the first team meetings that I attended, where a regular agenda item featured discussion about the planned development of an in-house policy handbook. Designed as a resource largely in anticipation of organizational growth, the handbook was intended to outline the group's position on a broad range of likely policy issues, the idea being that all staff but especially volunteers or new appointees could refer to this document should they be faced with a pressing inquiry. "Now we have fundraisers going out to represent our case to potential donors, other colleagues talking to the press or talking to politicians," Maggie explained to us in one meeting, "we've got to be saying the same thing every time." As a quick perusal of any extract from the draft table of contents revealed, there was a lot of information for any incoming colleagues to absorb.

ANIMALS KEPT AS PETS

General	7
Commercial breeding	7
Keeping	8
Commercial trade	8
Sale to children	9
Pets as prizes	9
Pet vending	9
Pet fairs	10
Exotic pets	10
Spay/neutering	11
Mutilations	11
Organ donation	11
Training and training aids	12
Stray dogs and dog registration	12
Euthanasia	12
Animal sanctuaries	12
Education	13

ANIMALS USED IN FOOD PRODUCTION

General	14
Human diet and lifestyle	15
Breeding	15
Intensive rearing and production systems	16
Farm licensing	17
Food labeling	17
Transport	17
Markets	18
Mutilations	18

ORGANIC FARMING

Procurement	19
Slaughter	19
Pigs	20
Dairy cattle	20
Beef cattle	21
Poultry for meat	22
Laying hens	22
Sheep	22
Game birds	23
Farmed fish	23
Fishing industry	24
Foie gras	24

ANIMALS USED IN EXPERIMENTS

General	26
Environment and housing	27
Genetic modification	27
Human-animal hybrid embryos	28
Xenotransplantation	28
Use of primates	29
Use of animals in education	29
Data sharing and transparency	30

ANIMALS IN THE WILD

General	31
Culling of wild animals	32
Non-native species	32
Gray squirrels	32
Reintroduction of native species	32
Traps and snares	33
Deer	33
Seals	34

ANIMALS USED IN SPORT AND ENTERTAINMENT	
General	35
Hunting, shooting, and fishing	35
Horseracing	36
Greyhound racing	37
Circuses	37
Zoos	37
Bullfighting	38

ANIMALS USED IN NONFOOD PRODUCTION	
General	39
Fur-farming	39
Seal hunting	39
Karakul	39

In fact, members of staff regularly distinguished their work on the basis of that generalist brief. They particularly stressed the contrast with those animal groups that specialized in either specific kinds of species protection or specific kinds of welfare concerns, such as those solely attached to the protection of farmed animals or to the protection of wild animals targeted by field sports or to the protection of captive animals or to the protection of different kinds of household pets. Not all these diverse organizations worked directly with the sort of animals that they sought to protect—for example, through running or sponsoring an animal shelter or sanctuary. Nevertheless, colleagues often expressed the feeling that in many ways such groups had an easier task; the very nature of specialism, it was felt, enabled a more immediate and intimate expertise to develop and to be communicated. It was often much more straightforward, for instance, to demonstrate a clear link between organizational endeavor and the difference being made. However, highlighting that distinction could also be identified as a source of pride and organizational strength. Colleagues liked to point out that specialist animal groups often wanted to work with the Edinburgh group precisely because its generalist outlook was seen to offer a valuable (and noncompetitive) set of complementary skills. This was an arrangement that also often benefited the organization's campaign goals.

When colleagues began a campaign against snaring, for instance, they chose to work closely with the League against Cruel Sports, a bigger organization that exclusively campaigned against "hunting or killing animals for leisure." Similarly,

the group had at times worked in alliance with Compassion in World Farming—for example, coauthoring a series of parliamentary briefings and a guideline
document on farm assurance standards. It had additionally partnered with Captive Animals' Protection Society or CAPS, in this case to jointly petition for a ban
on the use of wild animals in circuses. Of course, the group, like all animal groups
in Scotland, also liaised with the largest and most dominant animal protection
organization, the Scottish Society for the Prevention of Cruelty to Animals or
SSPCA. This was the sister organization to the Royal Society for the Prevention
of Cruelty to Animals or RSPCA, the first animal protection group in Britain
(founded in 1824), which operated across the rest of the UK. As well as a campaigning and rescue body, the SSPCA was the only nongovernment agency with
statuary powers to report and enforce animal welfare legislation and as such an
essential partner for any animal group concerned to draw practical attention to a
specific protection issue or to a breach of welfare regulation.

The animal group I worked with, then, sat within a wider terrain of animal
protection organizations. While all of them were committed to the reduction of
unnecessary suffering, the extent to which each one invested in or promoted a
language of animal liberation or a philosophical language of animal rights varied
greatly. As already mentioned (see chapter 1), colleagues strongly identified as
moderate animal activists and saw the animal group positioned squarely in the
pragmatist camp. Tending more toward forms of petitioning and lobbying that
stressed the need for negotiated compromise, the organization usually sought to
attain a consultative status within legislative and regulatory processes and hence
exert a gradualist influence on the well-being of animals (see chapter 8). As a
result, its messaging usually avoided terminology too easily associated with the
more combative side of the animal movement. Individually and between themselves, colleagues might regularly invoke the principles of liberation or the logic
of rights, but they were also very wary of triggering still-existent government
or public impressions of animal groups as disruptive or extremist in their views
and action.

However, in terms of internally drawn comparisons, the more outspoken or
direct-action animal groups did provide an important yardstick for organizational success. Most obviously, colleagues regularly cited the example of another
generalist organization, the originally North American but now global animal
rights group People for the Ethical Treatment of Animals (PETA), whose banner
line reads "Animals Are Not Ours." Indeed, like many small or medium-sized
animal groups in Scotland and the rest of the UK, the Edinburgh group was
somewhat envious of the evident spread and influence of PETA (founded only in
1980, today its website claims that it "is the largest animal rights organization in

the world, with more than 6.5 million members and supporters"). But such comparisons tended to focus far more on strategic examination of its campaign tactics and fundraising models. Colleagues appreciated, for instance, the effectiveness with which PETA first deployed direct mailing and subsequently social media as a leading technology for animal rights activism and the high-profile nature of its campaigning stunts and celebrity endorsements. In fact, PETA's expertise in publicity, including the pioneering of the canny use of photography and video recordings of industrial-scale animal abuse and emotive stories of animal rescue and reunion (see Song 2010, 42–43; Gruen 2011), was widely acknowledged to have had a massive impact on the way all animal groups in Britain now worked. Equally important was the emphasis PETA placed on the power of the individual consumer, typically portrayed as someone who could "spare animals excruciating pain" simply "by making better choices" (https://www.peta.org/about-peta/learn-about-peta/).

But admiration for the organizational strategies of PETA was usually mixed with distancing statements. Colleagues typically expressed reservations about the aggressive nature of PETA's campaigning and in particular the demonizing language often deployed to deride opponents or to describe those members of the public identified as participating in cruel practices such as angling. For them, this did not merely mark a difference in campaigning approach; instead, it signaled a difference in temperament or moral disposition. For instance, colleagues regarded the kind of activism supported by PETA as just too loud and self-righteous in its blame attribution, a quality of behavior sometimes figured as the result of its North American roots. In these criticisms, a valorization of politeness or good manners could be identified, often by implication presented as a Scottish or British virtue. So, the very success of PETA was also what made it a problem, at least from the perspective of animal groups like the Edinburgh group that either didn't want to or couldn't afford to adopt its full-throated and more unapologetic stance. While notions such as speciesism and Singer-inflected rereadings "of human history as a history of the vicitimization of animals" (Song 2010, 54) did resonate with many colleagues, few of them wanted to turn those critiques into strident attacks on either individuals or particular sectors of society. In the same team meeting where Maggie outlined the purposes of the new policy handbook, for instance, it was stressed that the public-facing positions outlined in that document were designed to be "radical but not frightening." As this book goes on to explore, that ambition closely tied into the animal group's new mission to engage the mainstream. But the emphasis also touched on another wider question for the organization, about the degree to which it wished to continue identifying cruelty toward animals and indeed animal suffering itself as the defining theme of its activism.

For Song (2010, 41), one of the few anthropologists to provide an ethnographic description of the animal rights movement (in the US), the early history of animal protection is usually told through the prism of the shifting relationship between attributions of cruelty, ideas of animality, and practices of interhuman othering. Sometimes, especially in colonial contexts, interpreted through the lens of race and whiteness, that interhuman othering is most regularly identified as heavily class-based in origin. Certainly, scholars tend to view early anticruelty organizations (begun first in England in 1824 and then several decades later in North America) as grounded in an aristocratic or bourgeois preoccupation with the cruelty of other classes, in particular the public cruelty of working people toward working animals (Song 2010, 57–58; see also Boddice 2009). There is much written, for instance, about the dominant genre of witnessing in early anticruelty pamphleteering and literature, which typically placed an aristocratic or bourgeois male observer at the scene of animal abuse in a crowded urban space. In these narratives, "the cruel subject was most frequently portrayed as greedy," principally driven by base economic interests (Song 2010, 59). However, it is argued that the overall moral tone of such reports focused less on the end goal of reducing animal suffering and more on the opportunities for humanity's improvement (2010, 57, 63). Indeed, a mood of optimism existed, fueled by the belief that humanity's progress was assured and that cruel subjects could be in a relative sense straightforwardly enlightened.

But as Song also observes, historians tend to describe a series of changes in the first half of the twentieth century. Most notably, they identify a rise in the participation of aristocratic and bourgeois women; a defining turn toward antivivisection as the core issue of animal protection, especially in Britain; and a simultaneous shift in the movement's narrative attention from "tales concerning working animals" to "those about smaller domestic animals" (2010, 58; and see Howell 2015). It was precisely at this point that the history of the animal group that I worked with properly began.

Founded in 1911, the organization was a breakaway offshoot of the National Anti-Vivisection Society (NAVS), one of the two main campaigning groups against animal experiments in imperial Britain. At the time, NAVS was famous for initiating the first undercover investigations of "scientific torture," most notably the well-publicized "Little Brown Dog" exposé of cruel practices in University College London's scientific laboratories in 1903 (Lansbury 1985). But it also had a reputation for gradualist tactics, especially in contrast to the more abolitionist platform of the BUAV or British Union against Vivisection (Kean 1998). As the dominant group in Scotland in this period, NAVS principally focused on growing support for the cause among the "great and the good." This included running fundraising events in the Scottish capital and encouraging "at homes" organized by local middle-class and wealthy enthusiasts for the cause. It was notable too for the prominent role of women in its foundation and committee work.

When it eventually happened, the historical splintering of NAVS seemed to be generated less by philosophical or tactical differences and more by arguments about where the monies collected through fundraising should go. Either they should continue to be sent to the group's headquarters in London, thence to be distributed nationally, or they should be held at its branch office in Edinburgh to fund specifically Scottish campaigning activities. Indeed, the actions and concerns of the newly independent antivivisection organization that did emerge in 1911 appeared to fairly closely mirror those of NAVS, at least for the next four or five decades after the split. One of its very first significant acts, however, was to purchase its own headquarters office in the West End of Edinburgh. These were the same set of chamber-like rooms on the second floor of a shop-lined Georgian terrace block that I first entered nearly one hundred years later.

Across much of the twentieth century, the independent antivivisection organization continued to lobby and call for reforming bills at the Westminster Parliament, seeking to abolish or restrict the number of experiments and any associated activities linked with the breeding and supply of laboratory animals. This was to prove an uphill task, since numbers continued to rise dramatically; by the 1970s annual experiment figures for the UK were as high as six million. Despite this climb, the animal group continued to pursue a gradualist approach. The 1980s, for instance, saw it support compromise legislative measures and launch a call for working alliances with the scientific community, at the time a highly controversial move among antivivisection campaigners. That call was partly prompted by political context. The previous decade had seen the high-profile emergence of the Animal Liberation Front in the UK with occasional instances of more violent protest such as the firebombing of laboratories (Molland 2004). Beyond the issue of antivivisection, the period had also seen the rise of other more direct-action groups such as the Hunt Saboteur Association (Ryder 2000). Drawing a deliberate contrast to such tactics, the organization publicly sought to bring "moderate activists" and "moderate scientists" together in order to negotiate welfare reform.

But the move also occurred in the wider context of the rise of the contemporary animal movement, especially the turn to ideas of animal liberation and revamped ideas of animal rights. First initiated during the 1970s by the Oxford Group of postgraduate philosophy students, which included figures like Peter Singer (see Phelps 2007, 206–7), that turn had been subsequently reinvigorated by the arguments of American philosophers such as Tom Regan (see chapter 1).[1] It was further reinvigorated by the way animal groups in the United States adopted those ideas and then reimported them back to the UK with different organizational structures and innovative campaign actions. Indeed, by the last quarter of the twentieth century, the animal movement had started to attract new kinds of supporters, especially among university students, academic scholars, and the professional middle classes. The latter, often previously committed to human rights campaigning (Song 2010, 53–54), also brought animal protection into dialogue with New Age notions and newly prominent issues around environment or ecology.

One sees those changes also reflected in the composition of paid staff (and volunteers) at the Edinburgh animal group. By 2010, when I first met them, a vast majority of colleagues had a similar background. Most had been educated at university, which was where they typically accelerated their participation in formal animal activism. Many had professional qualifications, some engaged in New Age practices and beliefs often inflected by Buddhist and other alternative forms of spirituality (see chapter 1), and all of them linked their animal activism to a broader range of moral concerns, including human rights, poverty action, child protection, and environment threat or degradation. But perhaps the most significant change in the organization's late twentieth-century history occurred back in 1990, when it renamed itself to remove direct reference to antivivisection and instead to emphasize a wider advocacy role. For although this renaming did not occur until the last decade of that century, the group had in fact begun to gradually advocate on a far greater range of welfare issues from the late 1950s onward. The new name then was seen as a confirmation of what had already in effect taken place; it confirmed the breadth of campaigning activities appropriate to a generalist-inclined and professionalized animal protection organization.

The 1990s and the first decade of the twenty-first century saw other notable changes too. Crucially, it marked a period when a generation of Scottish (and English) war widows and unmarried women began to pass away in significant numbers. Because these women were some of the most active supporters of causes such as animal protection in postwar Britain, this indicated issues of financial pressure on the horizon. While the animal group did benefit from receiving a proportion of these women's legacies, the question increasingly arose of where future funds and support would come. Indeed, the CEO and board of trustees for the organization were painfully aware that its female and largely aging supporter base was diminishing and that this situation compared unfavorably with the apparently thriving, younger, and more diverse supporter base of global animal rights organizations such as PETA. These growing concerns led to reflections on alternative models for financing the work of animal protection as well as to a renewed focus on professionalization. For at least three decades, the animal group had funded a paid staff of four to five persons, working in tandem with volunteers from its membership. However, by the mid-2000s it appeared increasingly self-evident to those in senior positions that staff numbers needed to rise to enable a more expert or professionally skilled team to take the organization forward (one can see this kind of move replicated, often a decade or so earlier, across other parts of the animal movement in the UK [Wrenn 2019][2] and across the wider not-for-profit sector, especially within larger activist organizations such as Amnesty International [Hopgood 2006]). But those organizational challenges were not just perceived to be financial; the CEO and board of trustees also spoke of that transformation toward greater professionalism as a response to a range of cultural and technological pressures. Among the questions typically being asked, the following seemed some of the most pressing. What should animal protection become as its activities increasingly move online? How should the group engage its supporters in the future? Who might those supporters be, and where might they be located?

It was in this climate of genuinely existential inquiry about the organization's future that a consensus began to emerge in favor of another name change or what, by 2010, most colleagues now termed a rebranding exercise. But this time the proposal was not so much regarded as a fait accompli or simple recognition of changes already occurring (i.e., toward a generalist turn in orientation). Instead, colleagues spoke of a deliberate and self-conscious attempt to radically reconceive the group's rationale and constitution. When I arrived in their office, they perceived that they were on the cusp of this change. Indeed, the number of paid staff had by then already escalated to a previously unthinkable total of thirteen. Plans were afoot for a newly conceived organization, potentially global in its scope and reach and with the ambition to bring the concerns of animal protection into the moral mainstream.

"Putting a Kilt on It"

As Maggie, the group's policy director, taught me, in a devolved parliamentary settlement a crucial part of legislative work fell less on the construction of genuinely new and original laws and much more on the transplantation of legislation from elsewhere. In the case of the Scottish Parliament at Holyrood, that meant laws previously drafted and enacted at the UK Parliament in Westminster or, at least before Brexit, in the European Parliament. This process typically involved the implementation of surface-level, apparently symbolic or gestural amendments to acts or directives—a procedure captured by legislators and lobbyists alike through the commonly uttered phrase, "putting a kilt on it." Much more cynical in tone or, depending on whom you asked, simply a more realistic assessment of the difference sometimes indexed by Scottishness, the move implied in that phrase to me also captured something of the animal group's own ambivalent relationship to national identity. On the one hand, the organization regularly stressed Scottish credentials, including an imagined relationship to the rest of the UK, and especially to England and Englishness, grounded in notions of confirmed or historical difference. On the other hand, it also often stressed that the group's national identity was subtler in gradation, much more a matter of degree or of emphasis and style. But either way, Maggie knew that when addressing the Parliament at Holyrood she needed to ensure that the animal group's policies appeared properly dressed to recognize the perceived need for a Scottish twist or stamp. Likewise, I want to argue, the organization could occasionally be read as wanting or having to conduct a similar kind of exercise on its other activities. In short, at various times in its history, the group sought ways of putting a kilt on the cause of animal protection.

On the front cover of the newly independent organization's first annual reports in 1911, there is a rather elaborate hand-drawn design. From each of the top corners of the frame, two cherubs face one another, wrapped in descending coils of banner. "Ring out the old," reads one in small letters; "Ring in the new," reads the other. Of course, those lines customarily mark the passing of the year and the ushering in of the next, entirely appropriate to a set of annual reports. They also nod to Tennyson's popular verse, which stages that transition as a call or appeal for a wider set of moral transformations ("Ring in the love of truth and right, / Ring in the common love of Good . . . Ring in the valiant man and free, / The larger heart, the kindlier hand; / Ring out the darkness of the land," etc.). Running down and around the rest of the cover flows a large encircling ribbon that serves as a platform for drawings of various animals. At the midpoint, two dogs sit to the right side of the page and a working horse to the left. Right at the bottom of the page's frame and positioned directly below the other animals are a small group of sheep and a pair of rabbits. And at the center of the encircling ribbon is the organization's name, which includes reference to the fact that this is now an explicitly "Scottish" antivivisection society. The emphasis is rammed home by other design flourishes: for example, sets of thistles growing straight up from the frame's corners and between the sheep and rabbits and a small Celtic bow hanging just above the group's name.

In fact, that bow belongs to the bottom of another design flourish, an oval pendant that occupies the top center of the page and displays an immediately recognizable profile of the national bard, Robert (Rabbie) Burns. The image is clearly a redrawn version of the most famous and much-reproduced portrait of Burns by Alexander Nasmyth. Spreading around the upper part of the pendant, mirroring its oval shape, is a quote taken from the poem "To a Mouse," which reads, "I'm truly sorry Man's dominion / Has broken Nature's social union." The script chosen for the lines seems to stress their spoken quality, as if declaimed from the mouth of the bard himself, and, in a final animal illustration, it appears punctu-ated at beginning and end by tiny sketches of wee mice with long curling tails.

Back then, as today, invoking the words of Burns was a highly conventional way of emphasizing a distinctively Scottish interpretation of a moral concern. Since Burns wrote in a mix of Scots and English, its effectiveness partly relied on appeal to a linguistic register. But in this case, the lines also worked because they appeared to affirm a common assumption among those at the time com-mitted to antivivisection: that animal exploitation or abuse was the legacy of a biblically informed assertion of human dominion. And, furthermore, that this dominion ruptured an original connection between humans and other animals, or in the language of Burns, that it broke "nature's social union."[3] As previously mentioned (see chapter 1), such ideas still reverberated. Colleagues at the Edin-burgh group often invoked critiques of Christianity for providing an ideological justification for dominion, and they bemoaned what they saw as the resulting alienation of mainstream publics from natural connection. Indeed, the animal group made regular appeal to that Burnsian legacy. For instance, it drew on the bard's lines to articulate a Scottish heritage or genealogy for the organizational emphasis placed on making contact (see the prologue). More broadly, the group consistently outlined key virtues such as compassion through reference back to the moral imagination of Burns:

> ### Celebrating Rabbie Burns' Compassion for Animals
>
> January the 25th is Burns night, when Scotland and its friends celebrate the great bard (poet) and his contribution to Scottish culture. But did you know that Rabbie Burns delighted in nature and the animals we share this world with and was an outspoken critic of animal cruelty? [the post next quotes the two lines from Burns I cite above] . . . These pow-erful words have only got more poignant and relevant in the 220 years since his death. The poem is about a mouse whose nest is destroyed by a plough. Burns goes on to refer to the mouse that is disturbed by a plough as an "earth-born companion" and a "fellow mortal." This idea is also reflected in our name, which, like Burns' writings, reminds us that we humans have so much in common with animal kind.
>
> Another great Burns poem that reflects his compassion for animals is *The Wounded Hare*. This is an angry but touching and thoughtful response to witnessing the shooting and maiming of a hare. It starts like this:
>
> > Inhuman man! curse on thy barb'rous art,
> > And blasted be thy murder-aiming eye;
> > May never pity soothe thee with a sigh,
> > Nor ever pleasure glad thy cruel heart!

Go live, poor wand'rer of the wood and field!
The bitter little that of life remains:
No more the thickening brakes and verdant plains
To thee a home, or food, or pastime yield.

This poem is a reminder that compassion for animals and opposition to cruelty is not a modern phenomenon as it is often made out to be. In fact, it has a long tradition that we should celebrate and take encouragement from. Happy Burns night!

You can read both these poems by following these links, or why not learn more about how much we have in common with animal kind on our animal behaviour pages.

Taken from the organization's website in 2017, this sort of public messaging was typical. Beyond reference to the two lines with a historic association for the organization, citations tended to be opportunistically tailored to meet campaign goals. So, in the example above, the inclusion of lines from "The Wounded Hare" served to help highlight current work at the Scottish Parliament, where the group was petitioning for a ban on the annual mass culling of mountain hares. However, when I first joined the organization, such references were heightened further by the fact that the group was imminently due to celebrate its centenary. A whole series of events were designed to mark that anniversary, but the centerpiece of celebrations was to be a special Burns Night festivity.

In fact, for the three years leading up to the year of centenary (in 2011), the group had held a series of preparatory little Burns Night suppers. Hosted and conceived by Maggie, the policy director, and attended by staff and their families, these suppers test ran planned proceedings. On the face of it, the format was typical for such occasions. The evening began, for instance, with a formal welcome and a foreshortened parade of the haggis; in this case, an appropriately vegan haggis rather than the usual mix of spiced and minced heart, liver, and lungs with oatmeal in stomach casings. This was followed by the address, given in Burns's words, and then by a mock stabbing and toasting of the haggis. Again, as Burns Night convention dictated, the meal was interrupted by a series of entertainments. This included the recital or singing of selected poems, a rendition of the Immortal Memory, and other popular interventions such as the tongue-in-cheek Toast to the Lassies. (Burns Nights are usually full of sharp quips and wry humor; as several colleagues pointed out to me, part of the reason for its continuing popularity as an annual event in the Scottish calendar lay in the pleasure participants derived from hearing the playful irony of Burns's voice and the diverse uses to which that irony could be put.) But while these small gatherings very much kept the spirit of Burns Night, part of the point was to experiment with the event, especially by adding an occasional animal protection spin on the format. Apart from the deliberate veganization of the haggis, Maggie took the opportunity to introduce a few words of more serious intent. Into her Immortal Memory or formal praise of the bard, for instance, she added some reflections on Burns's animal protection credentials. The poems, she observed, consistently demonstrated an awareness of the individual experience and perspective of a range of critters, each one a fellow sentient being or earthborn companion.

On the centenary night itself, everything was glammed up. Invited guests arrived at the central Edinburgh hotel in formal dress. Men wore kilts or suits, and women wore evening gowns with the occasional tartan sash, hair band, or pin on display. Once everyone was seated, the animal group's chief patron, who came from a Scottish aristocratic family with long links to the organization, gave the formal welcome. Highlighting the same two lines from "To a Mouse" (now reprinted along with the original front-cover design for the annual reports on the menu cards in front of us), the dowager duchess provided a brief account of the recent history of the animal group and its current projects or good works. After everyone stood to clap in the vegan haggis, a retired history professor from Skye and old university friend of Maggie gave the address, this time partly delivered in Scottish Gaelic. In addition to a musical performance of some of Burns's songs, the entertainments took in a contribution from a past chairman of the board of trustees. First reflecting on the fact that he had been involved in the organization's original name change back in 1990, the man chose to read aloud a new poem in Doric (the common term for a dialect of Scots, not used by Burns, spoken in parts of the northeast of Scotland) that he told us was entitled "The Battery Hen." From what I could understand and later glean, the poem contained a critique of industrialized farming methods and a paean of sorts for the experience of free-range chickens. Once again, Maggie delivered the Immortal Memory. After a very animated and infectiously entertaining reading of "Tam o'Shanter" by another female guest, we headed into coffee and date balls. The latter were made by the IT officer, Iain, as an alternative to the customary but dairy-filled fudge treat known as Scottish tablet. The announcement of the results of a prize draw then drew the evening to a close. Each guest left with a goody bag containing further information sheets about latest campaigns, the most recent issue of the group's newsletter, and details about upcoming fundraising drives.

In some fashion, this celebration clearly aimed to represent a coherent version of the range and scope of Scottishness, at least in its most traditional linguistic registers. As a key annual event, Burns Night already signaled that national imagination; all schoolchildren in Scotland, for instance, learn to recite Burns poems and how to commemorate Burns Night, which itself is celebrated in a great diversity of forms and organizational settings. But, of course, the animal group never purely operated at the geographical or imaginative scale of a Scottish nation. As the history outlined in the previous section details, throughout the twentieth century and into the early twenty-first century the context for that national identity underwent many dramatic shifts. For example, the group began in the late period of empire when the nineteenth-century concept of Scotland as "North Britain" had not entirely faded from view. And in its most recent incarnation, in the age of the internet, the group flirted with the idea that it might become entirely deterritorialized, at least in terms of its supporter base. Although that notion drew heavily on quite specific Anglospheric idioms of philanthropy and networked activism, the example points to the willingness to periodically reconceive the relationship between organizational culture and a national outlook.

Aside from the shifts already articulated, overly stressing the significance of the group's Scottishness or the importance of its geographical base in Edinburgh ran risks. In particular, it could appear counter to the spirit of animal protection. For, as colleagues liked to regularly remind me, other species have no respect for national borders, "so why should we?" A genuine moral concern for animal suffering couldn't stop at the River Tweed or Scotland's coastline. It was true that specifically Scottish animal protection issues, such as the cull of mountain hare or the management of grouse shooting estates, did form an important part of campaign work. However, they had never exclusively done so, nor in any consistent sense across the organization's history. Recall that the group first split from NAVS not so much because of a perceived lack of attention to particularly Scottish welfare issues but rather because of concerns about the control and distribution of funds raised. Indeed, as a cause antivivisection itself required a wider outlook, since from a regulatory and hence lobbying perspective animal testing had always remained a Westminster or UK Parliament reserved issue.

Perhaps even more pressingly, then, the organizational scope of moral concern for animal suffering and the geographical scale of the group's work continued to be defined by the horizons of lawmaking and any historical changes therein that occurred. Central to the group's postwar generalist turn, for instance, was a notable shift in the target of their lobbying work and parliamentary petitioning. When Britain joined the European Community in 1973, campaigners of all causes in the UK suddenly had a new legislative body to lobby, especially after 1979 when Members of the European Parliament, or MEPs, became directly elected by universal suffrage. In the world of animal protection, this meant that Scottish and wider British-based animal groups had to immediately consider the legislative and judicial source of animal welfare regulation in two places: at West-

FIGURE 3. "Welcome to Scotland—a snare-free country." Image from 2010 magazine of animal group. Used by permission.

minster and at the new directive-issuing centers of Strasbourg or Brussels. It also increasingly meant that such groups had to become aware of and consider tactical alliances with animal protection organizations in other European countries. This new reality was consolidated by the foundation of the Animal Welfare Inter-Group in the European Parliament, which threw a fresh focus on Europe-wide concerns such as bullfighting and on genuinely pan-European protection issues such as live animal transportation.

At much the same time, the lawmaking landscape was promising dramatic change within the UK. In 1979 a devolution referendum was held in Scotland, which narrowly failed to approve the constitution of a Scottish assembly. However, by the mid-1990s it was clear that support was growing, and a second referendum in 1997 proved successful. Several years later, in 2004, the Scottish Parliament finally opened. With significant devolved powers, including over animal welfare legislation (excepting that pertaining to animal experiments), this was now a third and increasingly prominent legislative body on the animal group's doorstep. This body also soon hosted a Cross-Party Group on Animal Welfare, run by interested Members of the Scottish Parliament or MSPs but with participation from most animal groups in Scotland, including the group I worked with. Just as joining the European Union had created a new transnational legislative and campaigning space for considering the protection of animals, so the emergence of the Scottish Parliament decentered the United Kingdom as the legislative and campaigning space for animal protection in a new and different way. It made Scotland a meaningfully separate regulative space on a wide range of welfare issues, from seal culls to pet vending to snaring. That had the further effect of making the status of the animal group as a *Scottish* organization in some ways far more concrete than at its inception in 1911. This was the case despite the fact that the group's original name change in 1990 involved not just the loss of any explicit reference to the cause of antivivisection but also the removal of any direct reference to the group's geographical base in Scotland, a move left unaltered by the subsequent name change and the rebranding of the organization some twenty years later.

The Dog's Bazaar

Up until the Second World War, the annual reports of the animal group regularly featured an item titled "The Dog's Bazaar." Coming after accounts of the year's activities and notices of developments at Westminster, the item describes what clearly appears a major cause promotion and fundraising exercise but also social occasion in the group's calendar. Indeed, the bazaar always took place before Christmas in a large hall on George Street, one of Edinburgh's landmark thoroughfares. Decked out in banners and flags, the hired room seemed organized around a range of stalls manned either by branches of the newly independent Scottish organization (the Argyllshire Stall, the Renfrewshire Stall, the Midlothian Stall, etc.) or by sister animal groups such as the National Canine Defence League. In addition, there was a Refreshment Stall, a Parcel Stall, and an allocated space for a dog's photographer, where visitors could get their own dog's photo taken and then displayed on a "dog-screen" with others. Among the goods donated for the bazaar, we are informed that gentleman's socks and gloves, linen afternoon tea cloths, servants' aprons, pincushions, children's petticoats, silk opera bags, blotters, and white doilies all sold especially well.

The account obviously confirms the class origins of the organization; at the beginning of each report for the Dog's Bazaar, there is a long list of aristocratic patrons and of evidently bourgeois women listed either as key officeholders or as responsible for running stalls. However, the item really stands out because of the peculiar manner by which proceedings are narrated. Time and again, reference

CONVENERS OF THE DOG'S BAZAAR OF 1912

FIGURE 4. "Convenors of the Dog's Bazaar of 1912." Image from 1912 annual report of animal group. Used by permission.

is made to the fact that organizers and visitors are "dog-lovers" or "dog's friends," both in the sense of having pet dogs themselves and in the sense of being concerned about the plight of vivisected animals, again especially dogs. The item is at pains to stress that dogs as much as their human friends make the occasion. As well as typically narrating a tour of the stalls and telling us which "Miss" or "Mrs." or occasionally "Lady" runs them, we are regularly informed of the names of any dogs in attendance. From the very first set of annual reports: "You must speak to 'Sandy,' the Scotch terrier, for a minute at Miss Spencer's Stall—dear little Sandy, you know how he has worked for us." Indeed, after the report on proceedings, there typically follows a separate page laying out in long columns the names of both human and dog "helpers." This suggestion is taken a step further by the choice to annually identify certain dogs as the "convenors" of the bazaar and by the rhetorical move to present the item in the form of a letter addressed to the group's membership ("Dear Friends,") but signed off from "Ever your faithful friend, THE DOG."

This conceit of a doggy narrator is embraced with considerable relish across the annual reports. As well as lots of puns ("Loudest Barks of thanks," etc.), over the years it allows the authors to reiterate that sense of dogs and humans being in it together: "We dogs say how lucky they are to get Miss Balfour! [in charge of the Lanarkshire Stall in 1912]. She has dog friends by the score, who, by the way, seem *all* to have joined." The narrator then names a few of these dogs: "There is Billy and Toby Balfour, Pouf Guthrie, Punch Wilson, Paddy Gardner, Benny

Muir, Captain Dickie, Macduff, Bashful Bristles Peile—a dog of magnificent capabilities—etc., etc., etc." On one level, we are told that the simple presence of dogs at the bazaar serves the cause of antivivisection because it enables visitors to imaginatively extend their sympathies to less fortunate dogs (i.e., those subject to experimentation). But at the same time, dog helpers are seen to assist in a more literal fashion. The membership is encouraged, for example, to make donations or collect monies on behalf of their pet dogs, and photographic postcards with images of doggy convenors are widely circulated, often depicted as beseeching members to "please help our bazaar." In fact, the lettered report on proceedings is regularly presented as directly narrated by that year's convenors (rather than simply THE DOG), again accompanied by a photograph. These are usually the dogs of prominent officeholders or stallholders or occasionally of the aristocratic patron chosen to conduct the opening ceremony (for a few years, the convenor was "Vivisected Jack," an Irish terrier who in 1914 apparently escaped a University of Edinburgh lab after abduction from his home and whose case became a cause célèbre in local society). All dogs in attendance, it is assumed, have a natural interest in the plight of Jack and other vivisected dogs. Similarly, doggy convenors of the bazaar are held to be perfectly placed to give or receive thanks. "We dogs are *deeply* grateful to all our kind and generous friends," one early annual report advises before offering the following, as it turned out, overly optimistic prediction: "If we go on improving at this rate, we shall soon abolish vivisection with all its cruelty."

While colleagues chose to reproduce the front cover of these annual reports for the centenary Burns Night festivity, in truth they were somewhat embarrassed by the contents. Usually stored away in a disregarded filing cabinet, the back catalog was rarely taken out and read. When it was, individuals expressed unease about the class hierarchy and privileged assumptions displayed in the language and interests of the group's founders. They regretted what felt like the parochial nature of much activity, the sense of reading a version of society pages but also the evident amateurism or lack of expertise. Although there were expressions of admiration for aspects of its structure, for instance the apparent busyness of a volunteer membership operating at the local or regional branch level (the animal group today could only dream of such reach and spread across Scotland), in many ways the organization depicted in those early annual reports felt entirely foreign.

Colleagues struggled particularly with items such as the Dog's Bazaar. They were troubled by its portrayal of animals and by the frequent but unscientific reference to notions such as character. This included the pretense that dogs in attendance at the bazaar knowingly participated in such activities and that these dogs shared an interest with vivisected dogs or even somehow acted out of concern for the issue of animal experimentation. The claim appeared to be a straightforward breach of the dominant philosophical principle of modern ani-mal protection: that by definition other animals were moral patients, and only humans were capable of acting from a sense of moral obligation or from a per-ception of what the true interests of individual animals or species might be. Of course, colleagues also objected to the way that dogs appeared to be used at the bazaar. While certainly not averse to invoking the idiom of friendship to describe companion animal relations (see chapter 4), they were nervous about the pre-sumption that pet ownership gave these aristocratic and bourgeois women the right to put dogs on public display or to make them into workers for the cause. Likewise, colleagues were uncomfortable about the consequences of voicing dogs in this playful, all-too-human manner. When the animal group changed its name in 1990 to recognize a broader advocacy role and to remove specific reference to antivivisection, its new slogan read "giving voice, taking action." However, the point was precisely to highlight that animals had or could have no direct voice in moral and legal debate and that it was therefore the role of the organization to represent their welfare interests or act on their behalf. As with many animal activ-ists of their generation, that duty went hand in hand with the strong perception that anthropomorphic depictions of animals, including those that had the status of pets, were wrong or that they risked perpetuating regimes of mistreatment.

The concern returns us to the example of Paddy, Esme, and Amber. As already mentioned, it felt to me that these dogs had a place in the office that was partly symbolic of the organization's ethos and partly an index of the historical com-panion animal relationship that each dog had with a specific human other: that is, Paddy lived with Craig, Esme lived with Sarah, and Amber lived with Maggie. While each was only there because of those relationships and because the animal group allowed or even encouraged their presence as an illustration of organiza-tional values, colleagues were constantly anxious to stress that the arrangement matched these dogs' welfare interests. For instance, it suited Paddy to be able to stay with Craig or to follow him into the office. Indeed, if the workplace did not allow this to happen, that would be a breach of welfare. At the same time, colleagues recognized that these companion animal relationships were never entirely equal, that the realities of ownership or adoption introduced structural inequalities that did not go away simply because one stopped calling them pets. Hence the need to work on companionship, to attend to the presence of these dogs and ensure that what could be done to address that imbalance was done, the free passage of Paddy around the office being an obvious example of the willingness of colleagues to let themselves be interrupted. Hence also the gen-eral nervousness about how one talked about these companion animals. Unlike the confident voicing of doggy friends by the organization's founders, statements about Paddy, Esme, and Amber were full of conditionals. Colleagues expressed their feelings and thoughts toward them or colloquially described their recent behavior and then immediately worried that they might be introducing anthropomorphic language. Similarly, they invoked idioms to capture the quality of companion animal relationship and then warned of the dangers of imposing human-centric relational categories.

But to me the importance of dogs in the office had an additional value, this time linked to another kind of organizational anxiety. Although colleagues devoted themselves to promoting animal protection, they were very much aware that in a day-to-day working sense the connection to those animals that they helped was comparatively fuzzy and indirect. Unlike some animal groups, the organization was not an outgrowth of a rescue center or closely involved in sanctuary work, and as we have seen, it lacked the tangible connection to specific kinds of animals or specific kinds of abuse exhibited by groups with a specialist welfare concern. Instead, as a generalist organization, the difference it made had to be largely measured in legislative or regulatory change, public attitude monitoring, and single-issue petitions in the media. In fact, part of the challenge of organizational culture was precisely to convince various publics, including its own supporters, that actual animal lives were being positively impacted by its work. This was sometimes a question that colleagues also asked of themselves. If they worked for an animal group, where were the animals? Or, alternatively put, did they want to continue working *for* animals if they couldn't also work *with* them?

Seen in this context, Paddy, Esme, and the less frequent visitations of Amber became significant in their straightforward fleshy, intimate, and communicative reality. For here were individual animals not just talked about or advocated for in countless documents, meetings, and stakeholder consultations but uniquely able to be encountered in the workplace. Whether dozing under the boardroom table as a new campaign strategy got plotted out or forcing a staff member to interrupt a workstation conversation to rise and open a door or just snorting intermittently in the background, these dogs were vitally present. It was true that they were not there to be helped or to have their welfare circuitously improved by the organization and so were only ever partially satisfactory replacements. However, that too could be a bonus. For many colleagues, there was something equally comforting about the fact that these were not dogs in need. Instead, they were there to be dogs in an office and simultaneously to just be Amber, Paddy, or Esme.

Ethics of Relating

The perceived dilemma of working for animals in general while not working with any animals in particular had important repercussions. Most notably, it shaped

the diverse kinds of expertise that I go on to describe and some of the distinctive ways in which both before-the-fact and after-the-fact explanations of moral
agent and moral patient positions played out. But, for now, I want to take that
observation and apply it to a very selective reading of the work of Donna Haraway. This includes a brief look at the way in which Haraway's work continues to
influence anthropological accounts of animal-related forms of expertise as well
as the broader legacy for the anthropological study of human-animal relations
and what has come to be termed "multi-species ethnography" (see Kirksey and
Helmreich 2010). More specifically, I am concerned to examine how Haraway's
work affects our understanding of that literature as itself informed by a certain
kind of ethical project. The focus is intended to help further clarify the terms of
my own ethnographic task but at the same time to draw out some of the limits
that colleagues at the Edinburgh group did want to place on animal protection's
moral horizons.

Donna Haraway (2003, 2008) also has dogs. Indeed, that cross-species relationship, which includes specific relationships to two doggy companions, Cayenne and Roland, is meant to tell a story. Like the story about dogs in the office,
this one is ethically charged, self-consciously full of politics; in fact, it sometimes
takes the form of a manifesto. Haraway asks, "How might an ethics and politics committed to the flourishing of significant otherness be learned from taking dog-human relationships seriously?" (2003, 3). For Haraway, a central part
of that task requires an acknowledgment of their already entangled history as
companion species; it is "a story of co-habitation, co-evolution, and embodied
cross-species sociality" (2003, 4). Becoming aware of that history as it operates
at different scales, including the face-to-face interactions between Haraway and
Cayenne or Haraway and Roland, should at the same time produce the ethical
and political reorientation that Haraway advocates. "Dogs are not surrogates for
theory," Haraway warns, "they are not here just to think with. They are here to
live with" (2003, 5); and it is that "living well with" (2008, 11) that Haraway wants
to constitute as a good.

What Haraway presents, then, is an ethics of relating. Dogs, we are told, "are
not a projection, nor the realization of an intention, nor the telos of anything";
instead, "they are dogs, i.e., a species in obligatory, constitutive, historical, protean
relationship with human beings" (2003, 11–12). In fact, Haraway states that it is
precisely through these interactions that we (dogs and humans) make each other
and make ourselves (2003, 6). "Beings do not pre-exist their relatings," Haraway
declares at one point: "There are no pre-constituted subjects and objects, and no
single sources, unitary actors, or final ends" (2003, 6). Drawing on the work of
Marilyn Strathern and a wider legacy in both anthropology and feminist theory
(Haraway's own original training was in biology), Haraway states, "The relation

is the smallest unit of analysis, and the relation is about significant otherness at every scale. That is the ethic, or perhaps better, mode of attention, with which we must approach the long cohabitings of people and dogs" (2003, 24). In describing companion species such as dogs, one is therefore locating the proper basis for the kind of being one needs to ethically attend to; indeed, "who or what are in the world is precisely what is at stake" (2003, 8). Such a lexicon for ethics is entirely familiar to an anthropological audience, well used to the idea that the relation is a primary unit of analysis and that subjects become constituted through their interactions rather than before them. Indeed, within the multispecies literature, this is a common starting point for description and for the implicit and explicit invocations of an ethical register.

Although colleagues at the animal group did not know the work of Haraway (none of that author's books were on the shelves of the office library), they would, I think, have been sympathetic to much of what Haraway had to say. It would be possible, for instance, to locate analogies between the animal group's call for us to "make contact" (see the prologue) and the call of Haraway to live well with other animals, including companion animals. This was partly because such a call obviously connected to wider ideas not just circulating within the animal movement but in the literature that colleagues did consume. Indeed, while individuals did not read Haraway's work, they certainly did know the work of other authors in that general orbit. A quick internet survey reveals, for instance, that Haraway and the popular animal science author Jonathan Balcombe—a favorite of Craig, who ensured Balcombe's books did feature in the office library (see chapter 3)— were once both speakers at the same Minding Animals conference. So, one could argue, the degrees of separation were not always that great.

Likewise, I believe that colleagues would have responded favorably to those other aspects of Haraway's legacy that connected to broader themes—for instance, the invitation to more closely examine the ways in which the outlook and agency of animals impact the manner in which humans comprehend, value, and interact with them (see Noske 1997) and the concomitant observation that such an attention ought to result in "reconceptualizing what it means to be human" (Ogden, Hall, and Tanita 2013, 7). As both the biographical accounts of animal people and their accounts of specific companion animal relationships testified, colleagues were more than willing to concede a decentered role in their own lives or to stress the various nonhuman influences on them. In fact, one could clearly argue that the biographical emphasis on a before-the-fact moral patient experience necessitated such a reconceptualization. It could also be pointed out that the sci-fi film hoax produced by the animal group and previously discussed in the prologue rested on the very conceit that diverse sentient creatures *already* surround us and merely require our recognition.

However, colleagues would, I think, seriously struggle with the presumption that beings do not preexist their relatings or with the idea that this was a liberating principle. As we have seen, the theme of making contact often implied a transcendence of a prior state of disconnection or solitary being, as opposed to its simple collapse. Among colleagues, there was continuing mystery in the question of how that sentient being, human or nonhuman animal, got indexed or revealed itself in the world. But perhaps it is more accurate to say that what colleagues would have really struggled with were some of the postulated consequences of admitting the preexistence of relations or of becoming aware of subjects and beings as constituted by those relations.

By way of illustration, let us return to the ethics of relating offered by Haraway. I do so not just because Haraway's argument exists intriguingly just out of sight of the animal group's horizons but also because that argument sometimes seeks to challenge or speak directly to core concerns within the animal movement. Indeed, Haraway explicitly addresses the *Companion Species Manifesto* to what the author identifies as the dominant moral paradigm of animal rights and to the dominant interest in animal protection more widely, a focus on the relief of suffering. Partly inspired by the observations of Vicki Hearne (1986), Haraway proposes we turn (or return) to an idea of rights centered in "committed relationship" and its specific shared activity rather than in "separate and pre-existing category identities" (2003, 53). So, Cayenne and Roland obtain rights in Haraway and Haraway in them precisely through the companiable activities they do together. "In relationship," Haraway continues, "dogs and humans construct 'rights' in each other, such as the right to demand respect, attention and response" (2003, 53). According to Haraway, that notion of rights as emergent in the context of committed relationship leaves a genuine space for particularity, both in terms of species kind and in terms of animal individuality. It furthermore draws focus to the ongoing and mutually negotiated dimension of these rights (2003, 52).

From this perspective, Haraway argues, "the question turns out not to be what are animal rights, as if they existed preformed to be uncovered [in the manner advocated by moral philosophers like Tom Regan], but how may a human enter into a rights relationship with an animal?" (2003, 53). This move aims to decenter not just after-the-fact explanations of rights (and by implication of moral patient and agent positions too) but also any notion of an essential and generic human obligation toward animals such as the amelioration of suffering. It places emphasis instead on the obligations that do and can develop in specific activity, including activity that we might define as forms of animal-related expertise (note that Haraway's relationship to the two dogs, Cayenne and Roland, is largely described through their shared participation in the team sport of agility training). Indeed, Haraway offers the provocation that even the most problematic kinds of expertise (from the perspective of animal protection) such as animal testing should be assessed in precisely this fashion—that is, for the potential rights and obligations that can be generated from within its relational context. Haraway does not deploy the language of moral patiency, but the stress on the attentive and corresponsive dimension of that cross-species context suggests something close to it. In Haraway's alternative conceptualization of responsibility, for instance, classic ethical inquiries about the nature of human motivation and intention would become replaced by a new set of questions informed by the dynamics of interaction.[4] For Haraway, this includes the dynamics between scientists/technicians and "animal workers" in laboratories (2008, 70–71), as while those "co-respondents" might not be equal partners in that relationship of expertise, they remain nevertheless vitally conjoined.

Seen this way, Paddy, Esme, and Amber might each be said to obtain rights and obligations in their respective companion human (i.e., Craig, Sarah, and Maggie), and vice versa. Each might also obtain rights of respect, attention, and response in the organization, and the animal group might obtain rights and obligations in them, precisely through the relation that conjoins them as dogs in the office. Indeed, such a new appreciation of rights, in many ways exciting and radically divergent from the normative account of such concerns within the animal movement, might have its attractions for a group that takes a pragmatic view of animal protection. As we have learned, colleagues typically have some sympathy for an animal-rights-based approach, but at the same time, individuals often find its propositions somewhat inflexible and unhelpful when pursuing a gradualist agenda of reform. The latter precisely requires a willingness to consider the unequal relationship in front of them (between scientist/technician and lab animal, gamekeeper and fox, fish farmer and seal, etc.) and, by treating it on its own terms, to seek means of improvement. So, I think they would be intrigued by any redefinition of that language and particularly by the promise that such a redefinition might open a space to consider moral understanding as a capacity developed through committed cross-species relationship.

Colleagues would struggle, however, with Haraway's reworking of the moral category of responsibility. Specifically, they would strongly resist the implication that ethical obligation itself might no longer be perceived as the sole preserve of the human partner (2003, 53), and especially so in problematic animal-related forms of expertise such as animal testing. Haraway's suggestion that difficult human-animal relations might be countenanced or assessed in terms of those relations' autochthonous rights and responsibilities would therefore be a step too far. Indeed, as discussed in the previous chapter, despite foregrounding cross-species encounter and before-the-fact explanations of moral experience, colleagues ultimately insisted that sentient others could not or should not be considered as moral agents. In other words, they oscillated between a moral understanding that did appear emergent at least in part through committed relationship and an understanding based on after-the-fact and preformed category identities.

The distinction draws my attention back to the task I have set myself: to provide analysis and description of the moral dynamism of animal protection, as exampled in the case of this "Scottish" animal group. For there are other cautions to take into account when considering the legacy of Haraway's work and especially the influence of its overt ethical register, which often drives the insights of the human-animal relations and multispecies literature. As Matei Candea (2010) points out, an ethics of relating may enable new and bold forms of description to develop, which it certainly has, but it also risks foreclosing description. This is especially the case when human partners in animal-related forms of expertise want to place value or locate agency (or patiency) elsewhere. Indeed, Candea highlights a wider assumption in much of this literature, the fact that it is typically "connections, relationships, and engagements" that are taken to be the "key matters of concern" (2010, 243). For Candea, this reflects a broader normative opposition between engagement and detachment, which itself mirrors the role of that distinction as a key moral dichotomy in much wider Euro-American discourse about animals and about nature. In the context of seeking to provide an account of a British scientific study of meerkats in the Kalahari—these scientists are "self-consciously striving for objectivity"—Candea finds that assumed opposition a hindrance to the task of taking the moral stance of these subjects seriously. In response, Candea proposes that anthropologists need to start (or begin again) to "consider cultivated detachment as an ethical orientation" (2010, 244), not just the negative opposite of connection and engagement but its productive counterpart and complement.

It might be argued that Haraway and others leave room for detachment to be considered as an action within specific cross-species relations. This certainly seems to be the case in the way Haraway wants to re-ascribe the concept of responsibility—for instance, in that difficult relational context of animal testing. But it remains, I think, reasonable to observe that much of the implicit and explicit ethical thrust of this literature remains focused on practices of engagement. The call to live well with animals, to treat them "as parts of human society rather than just symbols of it" (Knight 2005, 1), continues to inform the task of much description and analysis.

As the previous chapter illustrated, it is also clearly possible to regard a version of that vision of the good animating the outlook not just of the animal group I worked with but of animal protection more broadly. Moral passions such as animal love, for instance, can clearly be read as the product of and prompt for further acts of engagement. But even there it is possible to identify moments when animal activists consider or make space for detachment as a cultivated ethical action. Song (2010, 39), for example, recognizes a form of detached and medicalized care, what he terms "gloved love," at the heart of the protests observed during an annual Labor Day pigeon shoot in the United States. In fact, for Song these acts of timid and apprehensive "rescue," driven by concerns about hygiene, contrasted oddly with the far more straightforward and unmediated forms of bird handling that Song witnessed among working-class shooters and their supporters. That observation leads Song to an increasingly unsympathetic account of the animal rights movement in North America. Song does not really take seriously this gloved love as cultivated virtue, and the examination of the ethical orientation of animal rights activists is heavily based on an analysis of the movement's published discourse rather than on activists' ordinary articulations of ethical work. Nevertheless, I believe that the example is indicative of one kind of alternation between engagement and detachment, or what one might term "detached engagement," that can also be taken to inform other areas of animal protection. Although I would want to undercut some of the other assumptions about that distinction—for instance, by exploring the invoked role of the moral patient position in the experience and articulation of both engagement and detachment (we note that in Candea's account each is still made to stand as a form of moral *action*)—these observations are a useful reminder of what an emphasis on an ethics of relating can sometimes obscure.

Looking more specifically toward studies of animal-related expertise and their associated moral projects in a UK context, I would argue that we do find other accounts concerned to describe processes of cultivated detachment. Beyond and before the example of Candea, there is the work of Sarah Franklin (2007), for instance. Franklin's study, which dialogues closely with Haraway and operates largely at a discursive level, invites attention on moments of both engagement and detachment in the story of the shifting intersections between historical traditions of sheep-breeding in Britain and contemporary scientific ideas and practice around cloning. (Note that Dolly the sheep, the first cloned mammal from an adult somatic cell, was conceived in a lab just outside of Edinburgh and that the preserved body of Dolly is now a permanent exhibit at that city's National Museum of Scotland, a fifteen-minute walk from the animal group's offices.) Likewise, in a more traditional ethnographic vein, one finds examples of detachment being positively registered as part of an ethical orientation toward the training of animals, especially of horses. Beginning with the work of Rebecca Cassidy (2002) on thoroughbred breeding and racing at Newmarket, this has most recently resulted in a study of "alternative horsemanship" and competing methods of training for home riding by Rosie Jones McVey (2023), which explicitly draws upon some of the points made by Candea.

What intrigues most about this study is the moral struggle and drama that Jones McVey describes around the very issue of partnership between horse and rider. Practices of engagement and detachment get played out in the context of a desire to ultimately demonstrate a cross-species conjoining of purpose and will. Indeed, key terms in the lexicon of Haraway and of much multispecies ethnography such as mutuality, responsiveness, and attention were invoked but also closely "entangled with conflicting sensibilities in the British horse world" (Jones McVey 2023, 9). In the case of alternative horsemanship, where concerns about human control or dominance were often foregrounded, this resulted in a training method with a very specific ideology. It first and foremost claimed that submission to the regulation of behavior, for human ends, could be achieved by teaching horses to freely choose that outcome (5). Here Jones McVey identifies ideals of both "autonomy and connection" (21): "For most riders, the crux of the task was about getting horses to want to connect as much as their humans did" (6), but this ambition required moments of detachment to enable the horse's capacity for choice or resistance to be regularly registered. In fact, across all training methods there was a "fierce scrutiny over what counted as a real connection" (3), with claims for expertise resting on competing interpretations, including sometimes of the very same observed horse-rider relationship. Seen in this light, Jones McVey argues, "relating well, for British horse-people, was largely a matter of knowing well" (5). The book's title confirms the emphasis by referencing not an ethics of relating but instead what Jones McVey terms an "ethics of knowing."

In terms of forms of expertise operating within animal activism that center on working with animals, it is possible to read equivalents to that epistemological concern with real connection. I am particularly thinking of claims to expertise reported in the animal sanctuary movement, where it appears that competing ideologies of rescue and care can open up similar kinds of scrutiny about knowing well (see Alger and Alger 1999; Abrell 2016a, 2016b; Hurn 2017; Parreñas 2018; Chua 2018; Sandoval-Cervantes 2023). And once the comparison between sanctuaries and animal shelters is included in the mix (Guenther 2020), those debates can become just as fierce. As Abrell (2016a, 59) reports in a US context, the sanctuary movement "is somewhat unique in its attempt to create environments that actively and regularly foster embodied interactions and socialization between humans and animals" and do so on the deliberate premise of granting those sanctuary animals apparently unprecedented levels of autonomy and respect. Indeed, those interactions were often explicitly depicted as grounded in an affirmation of animals as subjects rather than property (2016a, 59). However, ideals of autonomy and connection were constantly in tension with the reality of what Abrell terms a "captive freedom" (2016a, 88). Marked not just by restrictions on movement and behavior but also by regulatory practices of care that could include veterinary interventions such as spaying and neutering, this reality constrained sought-after expressions of animal subjecthood and the very ambition of sanctuary workers "to live differently with animals" (2016a, 88). The sometimes deeply felt contradictions resulted in arguments about proper forms of engagement and detachment that often came down to dispute about quite detailed practices and claims of animal-related expertise.

As already mentioned, numbers of colleagues at the animal group invested their free time in volunteer sanctuary work. For instance, Esme's human companion, Sarah, devoted hours each weekend to helping out at a local Staffie or Staffordshire bull terrier rescue center. Likewise, it was important for colleagues that Esme, Paddy, and Amber were rescue dogs; in many ways the ethos of having dogs in the office was presented as an extension of the ethos of the sanctuary movement. One could read similar concerns with ideals of autonomy and connection, and similar dilemmas. However, to repeat another earlier observation, there was also a crucial difference. These dogs were ultimately marginal to the core work of the organization. In fact, several years after I joined the group, there was a period when there were no dogs in the office: Paddy had moved on with Craig and Esme with Sarah, while Maggie had stopped bringing Amber into the office altogether. At the time, their absence caused a good deal of regret; however, the work of colleagues continued.

The point also highlights something important about this book. It is not, in any but the loosest sense, a multispecies ethnography. Nor is it centrally concerned to provide a study of the kinds of expertise and accompanying moral labor described above. It is not focused on training or sanctuary work (although we will return to the practice of rescue) or other ways of working closely with animals (with perhaps the exception of the peculiar disciplines of investigation). Instead, it describes forms of expertise and moral labor indigenous to a generalist and professionalized welfare orientation—that is, one committed to working for or on behalf of animals but in arenas where individual nonhuman animals are typically marked as not present.

A series of consequences followed. In their working lives, colleagues at the animal group were not centrally focused on the dilemmas or internal contradictions of seeking real connections with animals. In fact, in many ways connection was a given. As their moral biographies recount, colleagues generally assumed that they were already connected and that that sense of real connection in large part defined them and their moral purpose as both animal people and experts. Likewise, colleagues did not overly dwell on the moral drama of partnership or mutuality or any accompanying tensions between its respective patient and agent positions (such as the need to demonstrate that each cross-species partner was responsive to the other or alternatively that each acted freely). If colleagues did have epistemological concerns, these were not usually about knowing connection but rather again about knowing animals in general—for instance, through the issue of sentience. Just as importantly, when I joined the animal group, concerns were also informed by a newly clarified organizational aim to know human publics better. Why did other people lack real or "natural" connection with animals? And how might the animal group work well with the mainstream? These were some of the burning questions.

ENGAGING THE MAINSTREAM

Euan told me to take a break. So, I went in search of one of the coffee stations that lay between the long lines of stallholders in the large exhibition hall. The domed space lacked natural light and was therefore illuminated by a series of large hanging arc lamps, whose glow warmed the event's promotional banners. These hung down from steel rafters across the hall and read in a descending order and font size:

> Girls day out
> For every woman
> For every occasion

Everywhere I looked—on the banners, the top of the exhibitor stands, the carpeted walkways, and the complimentary gift bags that all visitors carried—were blocks of the same themed color, magenta pink. At the sides of the hall, in areas not adequately covered by the arc lamps nor supported by the additional white strip lighting incorporated into each stand, there were pools of relative darkness. Here visitors could sit down to chat with friends or rest their feet as they kept half an eye on the stallholder displays and activities spotlighted in front of them. When I took a seat to sip my coffee, I found myself watching numbers of visitors reclining to receive Botox injections. To my left I could also see a row of bodies seated and leaning over low tables to get their nails manicured. To my right was a brow bar, and a little further I could glimpse a raised stand where a couple of visitors sat upright on padded highchairs being given cosmetic makeovers.

This annual show, held during the month of November in one of the larger halls of Glasgow's main exhibition center, was described by its organizers as "Scotland's biggest ever health, lifestyle, fashion and beauty event." Over three days, a mix of several generations (generally groups of female friends or sometimes pairs of mothers and daughters) descended on the hall to visit exhibitor stands, watch trends and styling demonstrations, socialize over drinks at cocktail and champagne bars, and receive free or bargain-priced treatments and products. In the year I attended, the alphabetized list of exhibitors included names such as Art of Dentistry, Bling and Beautiful, Bustle Lingerie, Dream Cakes, Fakebake, Flash Diamonds, Glambags, Happy Feet—After Party Shoes, Heaven Hair & Beauty, Mint Cosmetics, Nailzone, Ocean Room Spa, Pout Photography, and Rapidlash. Interspersed among these stallholders were a small number of charities such as Arthritis Care in Scotland, Quarriers (specializing in social care for vulnerable children, adults, and families), and our own animal protection organization. Indeed, the decision to book a stand at Girls' Day Out was a deliberate manifestation of the animal group's aspiration to "engage with the mainstream." More than that, it was a relatively rare example of that engagement enacted in person rather than solely online or through social media, perhaps the most obvious instance that I witnessed of the aspiration being put into practice as a direct encounter between colleagues and an identified mainstream public.

For immediate release

SCOTTISH ANIMAL CHARITY TO MAKE "GIRLS' DAY OUT" SIZZLE WITH HOT SANTA'S GROTTO

—SCOTTISH WOMEN ASKED TO HELP MAKE AN ANIMAL "KIND WISH" COME TRUE—

Edinburgh-based animal welfare charity plans to warm up the chilly festive season for those attending this year's Girls' Day Out event in Glasgow by hosting its own Christmas grotto, complete with "hunky Santa."

Swapping his white beard and big belly for bulging biceps and a six-pack, Our Santa is bound to be a hit with the 15,000 women expected to attend Girls' Day Out 2010.

However, there is a serious message behind the fun. Santa will be asking those attending to help make his "Kind Wish" come true—by supporting our campaign to ensure that the proposed ban on cosmetics testing on animals comes into force across Europe by 2013.

The EU introduced a partial ban on cosmetics testing in 2004, but some animal tests still take place. The remaining tests were scheduled to be banned from 2013, but there is concern that this deadline will slip or be delayed. [The charity's] intention is to keep pressure on the European Commission to implement a full ban.

The grotto will be open from 10.30am till 4.30pm each day of the event and those attending will be given the chance to discuss our campaign and make a small donation. A range of animal-friendly beauty products will also be on sale at the grotto, to show that you can stay beautiful without harming animals.

Eilidh, Chief Executive of [the animal group] said: "We wanted to introduce people to our work in a fun way, in keeping with the spirit of Girls' Day Out. There is a serious message behind it all, which is that women wield tremendous power for positive change just through the contents of their shopping baskets. By helping to end cosmetics testing on animals across Europe, Scottish women will save countless animals from suffering."

[The charity] has also launched a new website—[website address]— which features a range of small simple suggestions on how to live a more animal-friendly life provided by members of the public and celebrities including [a list of celebrity names].

Visitors to the site can make their own suggestions as well as download an introductory guide to animal-friendly living.

Santa's grotto is open from 10.30am–4.30pm at stand A26 at "Girls' Day Out" in Hall 3 of the Scottish Exhibition and Conference Centre (SECC), from Friday November 5th to Sunday November 7th.

The press release, put together by the animal group's recently hired communications officer, was the usual mix of digestible information and quotable lines. Indeed, Euan had devised the idea of taking a stall at the 2010 Girls' Day Out, including the deployment of a "hunky Santa," as first and foremost a communications event. At a team meeting, he had tabled a proposal that included a list of "objectives and an outline of methodology." His core strategy was to try to involve a leading Scottish fashion and lifestyle magazine in a media partnership, comprising a discounted advertising slot, cobranding, the promise of an editorial piece on the group, and event support at the exhibition center. Euan hoped that the magazine might help identify a suitable celebrity to open the grotto, organize for its own Hunk of the Year to play the role of Santa, and provide one of its photographers to capture the spirit of the proceedings. But as the proposal laid out, the principal aim of the "action" was to exemplify and communicate the newly relaunched vision and values of the animal group. Taking a stall at Girls' Day Out, it read, would constitute "a fun creative to drive fundraising, data capture and membership" and at the same time "help to promote [animal-friendly] living as easy and fun." The emphasis on "fun"—stressed twice in these lines from the proposal and repeated in the words attributed to Eilidh in the press release— was no accident. Across the period of the relaunch, the group was concerned, in numerous ways, to demonstrate that it was approachable and particularly not po-faced (a British expression for an assumed seriousness or pious solemnity) about its advocacy. We may be serious about the cause of animal protection, but we also know how to enjoy ourselves or to understand what constitutes fun for the general public, was the intended message.

To colleagues, the fact that attendance at Girls' Day Out constituted an engagement with the mainstream—in this case defined as "Scottish women" and more particularly as Scottish female consumers but also more broadly defined as any publics not normally associated with moral priorities of animal protection—was obvious. In fact, it was immediately apparent in everything they saw around them. As colleagues pointed out to me, visitors at Girls' Day Out were able to try a range of cosmetic products or accept beauty treatments with dubious welfare credentials. Indeed, some of the promoted brands were known to have financed animal testing. Likewise, the fashion tips and styling advice available in the demonstrations and live shows clearly failed what they regarded as animal-friendly standards. In one "theater," for instance, visitors could sit in rows of chairs to watch a session on dress accessorizing, which when I looked in was running a segment on "the leather look." Numbers of models appeared onstage as someone talked the audience through the garments and styles on display. At one point the microphone voice beseeched us to remember that "leather is a statement in itself," before inviting consideration of a "statement necklace" to complement the look or the option of a "nod to the riding boot trend we have at the moment." When I left the theater, the session was switching to its next segment on spray tans. Colleagues reminded me that while most sprays were not directly animal tested, the regular use of amino acids in the ingredients meant few were vegan or welfare secure. In general, the exhibition, like the fashion and beauty magazine whose partnership Euan hoped to gain, was perceived as an alien or even hostile environment for animal people.

Although this participation was devised as a face-to-face engagement with the mainstream, the ambition remained to fully integrate this action within the animal group's overall social media strategy and in particular its desire for "movement building." Indeed, participation at the exhibition was also intended to demonstrate the successful integration of new organizational tasks and professional roles. This included showing how the expanded team at the relaunched organization could maximize that engagement by working in tandem or closely together.

Before the start of the event, Euan had allotted specific jobs to different members of staff. It was decided, for instance, that he and Laura, a temporary assistant on the fundraising team, would run the stall. Their task would include handing out leaflets about the rebranded organization and encouraging passing visitors to buy the donated animal-friendly beauty products displayed for sale. The two of them would also strive to collect donations for the group's campaign to ban animal testing and to facilitate the process by which visitors lined up to be photographed sitting on Hunky Santa's knee. (On the day, the male model sat perched on a padded red velvet chair between two white Christmas trees, his red robes left open to reveal a naked and well-oiled chest.) Meanwhile, David, the group's social media strategist, would patrol the exhibition hall's walkways, asking visitors to offer an animal-friendly wish. These would be periodically posted to a Twitter feed, reproduced alongside wishes from supporting celebrities on a live display panel at the exhibitor stand. David would also tweet prescripted information messages—for instance, telling online followers how to donate via text messaging or how to participate in new ways. One tweet read, "Let's not forget the boys," with a link to a cruelty-free male cosmetics company endorsed by the animal group. In addition, David would post original messages in response to what he was seeing and hearing at Girls' Day Out. Finally, it was planned that Sarah, the assistant fundraiser, would have a roaming brief to visit other stallholders. Her assigned task would be to assess the animal-friendly credentials of exhibitors' products or services with an eye to identifying viable potential corporate partnerships. Sarah would also introduce the animal group's new vision and seek opportunities to seed the message more widely across the exhibition.

Mainstream Movement

A month or so earlier, the animal group had organized a launch event at Edinburgh Castle. Devised as a platform for the formal announcement of the group's name change (see chapter 2), the commencement date for its accompanying new website, and a vehicle to begin the work of "sharing" its vision, this luncheon event welcomed a number of carefully selected guests. As well as notable figures from the animal movement, a diverse array of people from the broader not-for-profit or third sector and from the world of philanthropy were in attendance. The audience also included current and past members of the organization's board of trustees and a few local figures from Edinburgh's professional classes that Eilidh and her team wished to co-opt onto the board. Her short speech introduced the ambition behind the vision: "Over the last year we have been developing a new structure and new ethos—a paradigm shift in the way that animal welfare and animal protection are perceived and delivered. For many years we have been the voice for animals, inevitably focusing on the many harms that are done to them, and to some extent that made us a minority interest group. But now, we are setting out to engage with the mainstream—to put people at the heart of change by reconnecting humans and animals." Eilidh meant these lines to emphasize that the relaunch of the animal group marked a "radical reinvention," a new departure not just for this organization but also for animal protection in general. Her words were also intended to communicate that this was to be a deliberately inclusive vision. Instead of a minority interest group solely serving the concerns of animal people and the wider animal movement, it would be an organization that sought to bring welfare concerns into dialogue with the ordinary interests and concerns of members of the public.

Toward the end of the speech, Eilidh clarified the truly innovative nature of the reinvention. The group would not just be a representative of the animal movement strategically engaging the mainstream; rather, it would itself seek to become a "mainstream movement": "It's my hope that in years to come we'll look back on this day as the early beginnings of something quite extraordinary; the beginnings of a mainstream movement that achieves great positive change and benefits for animals, people, and the planet . . . I can sum up our message with one final thought. I said earlier we wanted to put people at the heart of change. Because it's people who make change—not only Ghandi, Mandela, Marie Curie and Bob Geldof—but committed individuals like you and our growing movement of supporters." These words make clear precisely how the mainstream movement was to be envisaged: that is, at the abstract level of the "committed individual." Indeed, at the heart of the new vision was a belief that all people had the potential to become committed individuals and that it was the relaunched organization's role to facilitate each person's individual ethical journey toward an animal-friendly stance. Perhaps most crucially, the group would be willing to do so at whatever point an individual happened to be. While some supporters might be at various advanced stages of commitment, others might have just begun their journey or even not yet be fully cognizant of any worked-out commitment to the cause of animal protection. A mainstream movement therefore would include a wide spectrum of committed individuals, not only animal people but also a range of other supporters, such as shoppers at a beauty and lifestyle exhibition, corporate and media partners, and local lawyers or accountants who might be persuaded to join its board of trustees. Many may never regard themselves as animal people or move beyond a minimal level of engagement with animal protection issues, but, Eilidh stressed, they would all nevertheless still be part of the same animal-group-inspired movement. Indeed, that flattening of conventional differences was identified as a core part of the inclusivity of the vision.

As we will see, part of what made this proposition coherent was its focus on incremental change and in particular on the measurement and communication of ever smaller and smaller units of intentional action—acts that could be adjudged to be animal-friendly and meet the minimal threshold of commitment when defined as a form of *doing* something. This was in large part possible because of the emphasis on "ethical choice" within existing frameworks of consumer activism, an important aspect of ethical lives within the animal movement that I will further explore in the next chapter. But it was also made possible by the opportunities detected online and especially by the advent of new social media. Indeed, Eilidh and her team admitted that the new vision of a mainstream movement of committed individuals was heavily dependent on the models of social networking already enacted on platforms such as Facebook and Twitter (by 2010, the microblogging service was beginning to have a significant impact in the UK's third sector). Drawing on the example of recent movement-building exercises, such as the impression left by the speed of support garnered for the Make Poverty History Movement a few years earlier, Eilidh and her staff reflected that it felt like old-style advocacy was on its way out. "The public," she and Craig advised me in our very first meeting, "now has the means to be their own advocates. They don't need an organization to advocate on their behalf." This assumption, reinforced by the advice given by a philanthropic consultancy firm that the group hired to help develop its vision, portrayed a radically new environment. The possibilities of self-advocacy online and through social media appeared to enable committed individuals not just to speak for themselves but to speak directly to each other, at least anywhere in the English-speaking world. Those same networks seemed to also facilitate the kinds of divergence in levels of commitment and degrees of participation that the group identified as characteristic of their mainstream movement.

As well as building a movement between those who now have the capacity to be their own advocates, Eilidh's speech stressed another basis for engaging the mainstream. In the new vision, the animal group would also be putting "people at the heart of change by reconnecting humans and animals." In an obvious reference to the making-contact narrative that I previously discussed (see the prologue), the mainstream movement would be grounded not just in self-advocacy and incremental change but in a process of opening up the human subject to the presence of nonhuman sentient others. In addition to encouraging individuals to cultivate specific relationships with animals in their vicinity, the group would facilitate a general awareness of diverse animality. This would include the development of an early-years education program. But it would also involve the regular publication of "amazing animal facts," the expectation being that if done right, animals will inspire people to make the change. In fact, as Eilidh's word choice highlighted, such a move was regularly envisaged as a form of "reconnecting."

She located the moral source for the relationship in the depths of nature, in feelings that humanity once had and that Eilidh and the rest of her team believed that people continued to be born with. They sometimes figured these feelings as akin to empathy; in Eilidh's case, this understanding was in in part drawn from her previous professional experience working in child protection charities, where empathy-building strategies were a core part of interventions centered on ideas about emotional literary. But colleagues more broadly presented those feelings as expressions of animal love. Known to be the distinguishing emotion of animal people, such feelings were also adjudged to be what all too often got lost growing up in mainstream society.

Although the presence of cruelty and cruel subjects remained vital to the imagination of colleagues and animal people more widely (see chapter 7), in this idiom of reconnection, the mainstream was more typically depicted as a state of indifference. Indeed, the action that Euan devised at Girls' Day Out was premised on the assumption that most of the "Scottish women" attending the beauty and lifestyle exhibition were apathetic rather than bad. They suffered from a form of moral passivity grounded in their disconnection from animals. It was the belief that the seed of animal love nevertheless remained present, if dormant, in a wide range of members of the public that made the action and the wider project of going mainstream feasible. This point returns us to my original discussions about the status of moral subjects as patients or agents (see chapter 1). For all the diversity of doings inscribed into a vision of networked and committed individuals, it was the issue of what befell or happened to that subject, or had not yet happened, that continued to matter most. Colleagues might be instruments of that reconnection, but the moral patiency assumed to be inscribed into the experience of making contact or of being inspired by animals to make a change essentially belonged to each member of the mainstream movement. As already reported (see chapter 1), it was important that those feelings of animal love felt uncontrolled or unbidden, that a certain mystery remained at the heart of the process of reconnection.

As well as describing this organizational project to engage the mainstream, it is important to give a sense of what that ambition meant to colleagues. The conversion of members of the public into supporters of a mainstream movement was conceivable only if those working at the animal group also accepted their place or role within the process. In other words, colleagues needed to be prepared to acquiesce to the idea of becoming mainstream. As animal people who typically imagined themselves operating outside the mainstream, often advocating against the impact of mainstream behaviors, this was far from a straightforward request.

Indeed, joining the animal group at a historic moment when it imagined itself undergoing such radical transformation—that is, plotting an entirely new course for the cause of animal protection—meant that I encountered colleagues in a heighted state of reflection. While they were not exactly undergoing a moral breakdown or rupture (see Zigon 2008), the changes implied in the new vision prompted many to ask a series of existential questions about the nature of their ethical lives and the basis of moral community. What would be the terms of the relationship, they pondered, between the animal movement and this new mainstream movement? How would they continue to distinguish themselves as animal people? On what basis would they relate to others within the mainstream movement? Could they tolerate the boundaries of their conventional ethical distinctions being so dramatically redrawn? Colleagues answered these questions differently and at different times felt very differently about them. But in essence they were conflicted. Tempted by the promise of massive gains for the welfare of animals if their movement and the wider cause of animal protection could successfully "go mainstream," staff remained deeply uneasy about the perceived compromises involved.

In large part, these anxieties were centered on the familiar issue of ethical alignment, the question of who colleagues would precisely be in a mainstream movement. For as Liu (2002, 183), drawing on observations by Macintyre, highlights, "There is always this question of the We that must be answered for us to know where we are in a moral or ethical sense." Indeed, the new vision implied a shifting relationship to the category of "ourselves" that to some members of staff appeared confusing and to others disquieting or even morally suspect. In the rest of the chapter, I want to look further at what the reconstituted We of the mainstream movement was striving to overcome. This includes a consideration of what that new relationship to Ourselves positively involved in terms of individual and organizational moral stance and commitment, especially when considered favorably but also when understood as a form of moral sacrifice or worse as a threat or betrayal of core values. Finally, I want to examine more closely the different versions or accounts of how this new We might appear and get constituted or recognized in the world. As well as the formal methodology laid out in the animal group's vision, I consider more implicit methodologies that in varying ways privilege alternative combinations of the experience of being done to and doing.

The Antis

It is hardly novel to highlight that movements of ethical reform must engage with moralities of the mainstream or that they contain theories about mainstream moralities explicitly and implicitly within them. Many accounts stress the tense relationship with what movements perceive to be their "outside," such as the market or the state, or specific "lay" or "religious" hegemonic moral mainstreams. Indeed, there appears to be a widespread assumption that ethical lives *should* unsettle the mainstream and equally that they should be unsettled by it (Fernando 2014; see also Benussi 2022). Accounts have also regularly described the ways in which, notwithstanding that unsettling, movements of ethical reform find themselves immersed in the mainstream. Antagonism toward the perceived mainstream is often combined with issues of entanglement. This is the case with animal protection too; for recent examples from wider animal activism in the UK, see Wrenn 2019 and Therese Kelly 2022. But what is perhaps novel about the situation I am describing is that this unsettling became experienced through a conscious organizational attempt to become the mainstream. To better understand some of the motivations for that move, we need to further explore the other side of the relationship—namely, ethical lives troubled by their oppositional stance.

Although a movement of ethical reform that sometimes places value on being moderate, animal protection has fairly well-established antagonisms. As previously noted in my brief account of the organization's history (see chapter 2), these are traditionally focused on the industrial agricultural complex held to be responsible for gross and systematic abuses of animal welfare, on shooting estates and institutions such as zoos, and on the breeding and supply chains linked to the retailing of certain species as pets or as laboratory subjects of scientific animal experimentation. Christianity was often held up as an ideological object of suspicion, and in a less problematic but still important way, so were the principles of conservation (see chapter 8). However, at least some of the colleagues and animal people I knew remained personally discomforted by the open expression of antagonism, while others struggled with the practical consequences of adopting oppositional stances. This was partly because of the distinguishing pragmatism of much animal protection and of this organization particularly, which meant that the arts of compromise and negotiation were prized and perceived to be central to the achievement of protection goals. But additionally, it was because colleagues often reported feeling compelled or trapped into taking up antagonistic positions by the expectations of the mainstream itself.

For many, the contours of this dilemma were exemplified by one popular adage. "You know animal organizations are often called 'the antis,'" Eilidh once explained to me, "because they are anti this or anti that, anti-bullfighting, anti-fur." Indeed, traditional opponents—for instance, spokespersons for farming bodies, gamekeeper associations, the pet industry, or supermarkets—regularly whispered the expression as a term of dismissal. More worryingly, from an organizational perspective, it was also known to be an expression historically uttered in the corridors of power, as part of a nod or knowing aside between politicians, government ministers, and their civil servants. One heard it in the lobbying process, for example, and in the consultative processes linked to lawmaking. "The antis" was also an expression regularly deployed by journalists, both in print and in conversation linked to getting or putting together a story. From the perspective of the animal group, the consequences could sometimes be benign. In selecting a team of favored stakeholders to consult at the preconsultation stage of a bill, for instance, or at one of its committee stages, bill teams and members of Parliament might think, "We need to ensure we hear from the antis." The same was true of journalists. Indeed, the animal group knew it was common working practice in British journalism for someone reporting on a news item to actively seek to demonstrate what was termed "balance" (see chapter 8), a requirement that was most efficiently done by trying to get quotes from two sides offering clearly opposed perspectives on the same issue. Less benignly, these practices tended to result in the public viewing animal groups as overly confrontational, aggressive, and judgmental. In fact, Eilidh believed that the overall result was the alienation of the mainstream from animal protection concerns.

This was not to say that the expression bore no relationship to the essential purpose of animal protection. Necessarily, she conceded, "our focus tends to be on the abuse or the cruelty or the exploitation that's happening, and that means you are always working in a very negative way." Pointing out or campaigning against what is wrong was part of the job or mission statement of the organization, and since systems of animal abuse were assumed to be built into mainstream life, the work of the animal group and groups like it was inevitably oppositional. However, for some, that reality continued to sit badly. These colleagues especially resented the implication that their ethical life was defined by what they were against and by the expectation that their professional role and expertise were to provide an oppositional perspective on what others said and did.

For Craig, Paddy's human and the science and research manager, such frustrations were central to the reasons he embraced the new vision. Indeed, I believe it is helpful to look a little further into the dilemma, as Craig saw it. This is partly because Craig was actively involved in the process of developing the vision of a mainstream movement, but also because as a long-standing colleague who had in many ways a very conventional advocacy background, he embodied a clear perspective from within the animal movement.

Craig remembered that as a young boy he often caught sticklebacks and tadpoles in local ponds and streams. From an early age, he also kept all sorts of pets. Growing up in a village in North Yorkshire during the 1970s meant that he had the space to own and breed budgerigars and guinea pigs and to keep rabbits and display them at local shows. Looking back on those days, Craig reflected, "I think I was looking for any interaction I could really have with animals." For Craig, these activities were an immature expression of animal love and of the feelings of fascination and wonder for animal life that continued to inspire him. But his first formalized awareness of the ethical issues around animal protection developed from the age of fourteen, initially prompted by someone handing him a leaflet about industrial farming as he left a McDonald's outlet in a nearby town. As a teenager, Craig went on to become interested in the antifur campaign and in conservation, joining both Greenpeace and Friends of the Earth. He also decided to "turn vegetarian," even persuading his mother, brother, and sister to join him, at least for a while. However, Craig was not heavily involved in campaigning or protests. Although he did join a hunt saboteur association when at university in Edinburgh (Craig read ecological science), he only went to a few of their meetings and never accompanied them on their early-morning actions. "At that time in my life I would rather go out and get drunk," he chuckled; "my priorities were enjoying myself I suppose, rather than being more proactive for animals." The real change occurred later, after he graduated, took a year out from studies, and then returned to study for a master's-level degree in applied animal behavior. Indeed, it was in that period that Craig began occasionally volunteering for the animal group. He formally joined the staff several years later and in 2000 became one of its then-few permanent employees.

At that time, Craig told me, the culture of the organization was at once both more "extreme" and more "conservative." Largely focused on campaigning and lobbying, its activities were indistinguishable from the advocacy work of many other small and large animal groups. At first Craig was assigned the role of press officer, responsible for getting the group's opinion cited in the media in connection with relevant news stories and for putting across an animal protection perspective. This usually took the form of a critique of the mainstream. In response to a press release from Edinburgh Zoo about a successful breeding program, for instance, Craig might be called on to question the claims of the zoo about its role in species conservation or to highlight the mental suffering that resulted from keeping wild animals captive and on display. Indeed, Craig explained that what journalists generally wanted was a "professionally angry" quote from an animal group. As well as being regularly asked to provide comment on news items about incidents of domestic animal cruelty or controversial farming practices or hunts or grouse shooting, Craig received more idiosyncratic requests. He remembered being contacted, for instance, to provide a quote for a news story about a Scottish restaurant that had announced it would be putting gray squirrel on its menu. Between these experiences and what he acknowledged to be a historical tendency of animal groups to define or even name themselves in terms of what they were against or concerned to stop (the League against Cruel Sports, the National Antivivisection Society, etc.), Craig increasingly felt that his allotted task in life was to continually point out what was wrong with something or to articulate how animals suffered at the hands of humans in innumerable contexts. This remained the case, Craig stated, when he later became the group's campaign officer and after that its acting director, before Eilidh arrived and took up the new post of CEO.

By then Craig was fed up with the pressure to be professionally angry and a little exhausted by the unrelenting nature of his role. "I have never really wanted to preach to people," Craig told me. "In fact, I think that's one of the worst things you can do." That reluctance was thrown into sharpest relief during his time as acting director when the work became "all absorbing." Not only in the office but also outside of it, Craig found himself continually invited to defend an animal protection position and justify an oppositional stance. When he was out with friends or social acquaintances, for instance, conversations seemed invariably to turn toward issues such as foxhunting or veganism; it seemed that they constantly wanted to have an argument with him. That experience led Craig to eventually reassess whether he wanted to work for an animal group: "I suppose I feel incredibly lucky to actually have a job where I am paid to make a difference in a way that is fundamentally, absolutely important to me. But sometimes I just wish I could be a postman or something, do a job with no serious consequences, no moral dilemmas, just a job to pay the money. Then I could do my bit throughout the rest of my life, speaking out for animals but just in a personal way." The development of the new vision was therefore timely. Craig was immediately drawn to the ambition of inclusivity and the nonoppositional stance implied in the creation of a mainstream movement. It seemed to promise the opportunity of finally escaping the label of "the antis" and the cycle of antagonisms that for him blighted an ethical life.

Positivity

But how was the animal group to avoid oppositional stances and enact this mainstream movement? Well, according to Eilidh, a key part of that methodology would be the implementation of a new ethos of "positivity." Indeed, the term was explicitly inscribed into the documentation outlining the group's vision, identified as a core aspect of both mainstream engagement and reformed internal organizational culture. Creating an ethos of positivity was to be an essential part of the route by which the new We of the mainstream movement would get constituted or recognized, both by those members of the public currently outside the animal movement and its orbit of welfare concerns and by animal people themselves. Central to that ethos was not just animal love and the principle of making contact but also the well-trodden category of "sentience." Indeed, it was a shifting relationship to that category, in both its scientific and more commonsense evaluations, that in many ways marked out what positivity meant and how the "radical reinvention" of animal protection was to proceed.

This was most powerfully brought home to me during an organizational review meeting I attended at the animal group's offices a year or so after the new vision was launched. As was quite common in the third sector, an external review facilitator had been hired to walk team members through a reflection on campaign goals and performance indicators. The terms of conversation were therefore in part phrased to enable someone from outside of the organization to grasp its workings. This process was relatively straightforward until we began discussion of another previously identified goal of organizational activity: "to increase public awareness and understanding of animal sentience." The problem here, interjected Maggie (the group's policy director), was that the recent history of the organization was "all about animal sentience." For her, the issue therefore could not be collapsed into a campaign target; it was rather an aspiration that should be integrated across all their activities. Eilidh immediately concurred. "Yes, sentience is really our USP [unique selling point] or brand," she explained; "it's what we believe and value." Clearly confused, the facilitator asked again why the promotion of animal sentience could not just be listed as one of the organization's goals. "But everything we do is based on the knowledge that animals are sentient," another team member responded.

There was, on the face of it, a rather odd concatenation of claims being made here. Sentience was at once an object of knowledge, crucially open to being evidenced by science, but also a belief and a value. It was simultaneously what defined the spirit of the animal group (at another point in the same meeting, a colleague claimed sentience was also an "ethos") and its public-facing brand or unique selling point (USP)—that is, how they wanted the organization and its activities to be recognized. That included the basis for making a distinction from other rival animal groups. As colleagues were quick to acknowledge, there was nothing particularly original about stressing the importance of sentience; it was a widespread and essential category across animal protection. However, they believed that a crucial difference did arise in the manner of emphasis. For unlike other animal groups, which highlighted sentience in its negative definition as a measure of the capacity to experience pain and suffering, the organization of the new vision intended to place equal weight on other measures of sentient capacities. This included the diverse abilities of nonhuman species to perceive and feel things, to be richly aware of their surroundings and relationships. Although the wealth of that emotional, cognitive, and communicative range clearly rendered the suffering that animals continued to experience at the hands of humans ever more tragic, it also spotlighted a more-than-victim "positive" existence that the group wished their mainstream movement to embrace.

As colleagues accepted, being all about animal sentience in this fashion tapped into a wider emergent mood not only among animal people but just as importantly among animal scientists increasingly concerned to research sentience in the round. Appeals to this kind of positivity were also heard in more popular science and the publications of scientist-activists. Indeed, not long before this team meeting, held in August 2011, the group had posted an online interview with Jonathan Balcombe, a US-based ethologist, who specializes in writing about the positive dimensions of animal sentience. That interview was prompted by the release of the author's then-latest book, *The Exultant Ark: A Pictorial Tour of Animal Pleasure* (2011), which acted as an accompaniment to Balcombe's earlier and widely read trade book *Pleasurable Kingdom: Animals and the Nature of Feeling Good* (2006). What attracted staff members, and in particular Craig and Eilidh, to these volumes was the equal value Balcombe placed on scientific and common-sense or everyday appreciations of animal sentience and on the inspiration those encounters provided. In the promotional blurb for *The Exultant Ark*, a book that principally curates a set of photographs of positive animal experience or feeling, readers are promised that by the final page, "old attitudes [will] fall away as we gain a heightened sense of animal individuality and of the pleasures that make life worth living for all sentient beings." Crucially, as Balcombe references and repeats in the interview, it is claimed that acknowledging or recognizing those pleasures carries moral implications for the human observer or participant.

This idea, that an enthusiastic understanding or experience of the positive expressions of animal sentience could itself hold moral implications and by extension lead to moral renewal (and perhaps reconnection), was in large part the source of the ethos of positivity. What could an animal group achieve if it engaged the mainstream on this basis rather than by lecturing people on what they were doing wrong or on how they were contributing to the systemic abusive treatment of animals? How might an invitation to consider the capacity for pleasure instead of the capacity for pain and suffering broaden the appeal of the animal protection cause? And just as importantly, at least as far as Craig was concerned, how might such an emphasis enable animal protection to escape its oppositional reputation as the antis? Could the ethos of positivity built into the new vision reinvigorate the work of this reluctant preacher?

As the organization's science and research manager, Craig was at the center of the attempt to engage positively or inspire the mainstream through the communication of the latest findings on the positive capacities of animal sentience. Indeed, since his anticipated role involved liaising with scientists—in the documentation for the new vision, the position was also titled "knowledge-builder"—and his own educational background included training in applied animal behavior with a strong focus on the science of sentience, Craig was regarded as the group's natural expert. Leading up to the relaunch, much of his time was taken up trying to locate and provide inspiring content for a planned large and important flagship section of the new website. Known as "Animals A-Z," it promised the public "fascinating [and constantly updated] facts about some of the world's most amazing animals," gleaned from Craig's reading and popularized summarization of the current scientific work on animal sentience. Each couple of weeks, throughout this period, Craig uploaded a new entry, which was heavily trailed and promoted through the organization's social media accounts.

I recall a posting, for instance, on dog vocabulary, that carried the headline "A border collie recently set a new world record for nonhuman animals by learning the names of more than 1000 items." Other notable entries explored facial recognition in pigeons ("New research has shown that feral, untrained pigeons can recognise individual people and are not fooled by a change of clothes"), the remarkable powers of seals to identify prey ("Harbour seals can detect the fattest fish using just their whiskers"), the numeracy of mosquitofish ("According to a new study, fish can differentiate between quantities as large as 100 and 200"), the complex language of whales ("Biologists interpreting the language of sperm whales have found that these animals have both accents and regional dialects"), cooperative skills among elephants ("Elephants work together as a team and understand when they need help from a partner"), the communicative range of chimps ("Scientists have discovered that wild chimpanzees use at least 66 different gestures to communicate with each other"), and the capacity of chickens to feel the emotions of another ("Researchers have gained new insights into the minds of domesticated hens, confirming, for the first time, that they show empathy for their baby chicks"). Each time, Craig reminded his audience that scientific findings verified the display of positive sentience and that these emerging facts should be a source of celebration: "I have been captivated and enthralled by animals for as long as I can remember. In fact, most of us can probably remember the feelings of awe we had about animals when we were children. Perhaps many of us also felt compassion. I believe that by re-connecting with animals we can create a better world for all. To quote the Scottish-born American naturalist and writer John Muir, 'Any glimpse into the life of an animal quickens our own and makes it so much the larger and better in every way.'" These words, scripted by Craig for the opening entry of his Animals A-Z, were pure new vision. Indeed, they directly echoed some of the phrases uttered by Eilidh in her launch speech. This included an emphasis on the positivity generated by allowing oneself to be captivated by the talent and emotional intelligence of nonhuman species, a theme that also ran through the sci-fi teaser film the group produced. As Craig's quotation from John Muir highlighted, becoming enthralled by the sentient powers of animals should be a self-enlarging experience. Of course, it should also produce compas-

Be inspired about animals!

The diversity of the animal world is astounding. From the tiny ant to the mighty blue whale, the rich variety of different animals is truly mind-blowing.

ANIMALS have evolved to live and flourish in every conceivable environment. To do this they have adapted by becoming smaller, or bigger; by growing more legs or losing their legs completely; by growing wings or even fins. The physical differences between animals are endless.

Sentient beings

But there are also many similarities. Increasingly science is showing what many people instinctively know: that, like humans, other animals also have feelings. They can feel happy and excited, or sad and depressed. We're not that different.

Learning more every day

There is so much about the natural world that we have yet to discover. Every single day we are finding out new things about animals. The more we study and observe them the more we learn and understand. The more we understand the more we admire. The more we admire the more we want to protect.

Did you know?

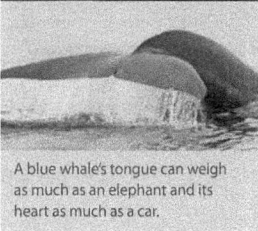

A blue whale's tongue can weigh as much as an elephant and its heart as much as a car.

Did you know?

A dung beetle can pull 1,141 times its own body weight.

Did you know?

Rats are extremely social animals and tend to become lonely and depressed without companionship

FIGURE 5. "Be inspired about animals." Image from 2010 magazine of animal group. Used by permission.

sion. Like Eilidh, Craig invoked the notion of a journey, in which the committed individual's ethical concern grew in response to the developing knowledge and admiration for the extraordinary powers of animals. His entry continued, "There is so much about the natural world that we have yet to discover. Every single day we are finding out more and more new things about animals. The more we study and observe them the more we learn and understand. The more we understand the more we admire. The more we admire the more we want to protect." Again, the idea that the impulse to protect could arise from the sensations of admiration, this time initiated by study and understanding, rather than those of distress or anger or guilt was crucial.

Alongside their reading of Animals A-Z, colleagues and supporters were encouraged to post content on the website and on the group's Twitter and Facebook pages about those moments of admiration or captivation. Indeed, Craig was careful to emphasize to his readers the validity of their commonsense understandings of sentience. Much like Balcombe, Craig consistently stressed the lag between what we must already know about the capacity of other animals to feel pleasure and the ability of animal science to evidence it. That understanding was only ever a few steps away from animal love, in particular from the appreciations that arose from the cross-species relationships colleagues and, it was assumed, members of the mainstream knew best: those with their household animal companions.

"I absolutely love Paddy with all my heart," Craig told me by way of elaboration. "You know, he's my buddy; we have spent every day, all day, together solidly for ten years; it's a very close relationship." Indeed, Craig did not need animal science to tell him that Paddy was sentient, that Paddy experienced pleasure as well as pain and had a rich engagement with his environment:

> He communicates with me; I communicate with him. We can look at each other, he can smile at me, you know, and I can tell if he is in a good mood. I can tell if something is wrong with him. If he is not feeling well, he will come over to me for a cuddle, you know, regularly throughout the day. . . . I mean, in some ways I have more in common with Paddy than I have with someone in China because I can understand him, do you know what I mean, and he can understand what I am saying.

The final words were uttered with considerable nervous laughter. Craig was keen to stress that the barrier he was suggesting might exist with people in China was linguistic rather than a matter of essential difference indexed in a concept such as race. Likewise, Craig wanted to acknowledge that as a species he knew that humans shared some things in common that dogs could never access, and vice versa. "Ultimately, I must remember to treat Paddy as a dog because that's what is best for him," Craig added in conclusion, "and dogs are basically wolves, so I kind of think of him as a wolf still because that's the way his mind works." Previously frustrated by the lack of space within animal protection for the formal expression of these relationships and the positive sentient existences that they could be taken to index, Craig and many other colleagues at the animal group welcomed the opportunity to showcase what they knew.

But if the ethos of positivity allowed colleagues to express their commonsense understandings and experiences of animal pleasures and make that the basis for engaging the mainstream, it also imposed new strict constraints on animal protection's organizational culture. In fact, as Eilidh saw it, her biggest struggle would be the reeducation of colleagues. For the ethos of positivity inscribed in the new vision was intended as a direct challenge to long-standing conventions of animal protection advocacy. It required a new form of engagement from animal people working at the animal group.

When I started coming into the office regularly, in the months just before the new vision's launch, the effort involved in creating an ethos of positivity was front stage and obvious. At that time, team meetings were dominated by discussions about organizational language and attitude, the new tone and emphasis that the website, fundraising appeals, press releases, and briefing statements should strike. Eilidh, often with the support of Craig, would keep reminding colleagues of the need for the elimination of negative frames of reference in the animal group's future communications. The consistent message was to downplay abuse and cruelty, any specific and detailed allusions to animal suffering, and instead to stress positive engagements with animals and the difference to welfare that could make.

By way of reminder, large poster-size strips of paper had been taped to the walls above some colleague's desks. On one poster I read, someone had written the title "Negative Words and Phrases," in what was clearly intended to act as a prompt for the drafting of future publications and communications. Below that title, the poster listed examples of language to avoid in the future. This included verbs such as "slaughter, exploit, imprison, campaign, boycott, ban, fight, attack, battle, oppose"; adjectives such as "cruel, barbaric, painful, bad"; nouns such as "suffering, violence, vegetarian, vegan, carnivore, abuse"; and, finally, the phrase "cruelty-free." Although the clear ambition was to identify and excise vocabulary that might alienate the mainstream (and remind them of the label and stance of the antis), the effect was to remove nearly all the key registers of animal advocacy and self-expression. It is true that this instruction might not have been followed to the letter; the team did experiment for a brief while with trying to enact a program of action without reference to the term "campaign," but the suggestion of no longer referencing terms such as "vegetarian" or "vegan" was never seriously implemented. However, the poster demonstrated intent and the kind of radical change of mindset that Eilidh felt was necessary to push the new vision through.

Indeed, by way of contrast, on an adjoining strip of paper another title read, "Good—Positive Words and Phrases." In this case, the list began with some general words of advice: "Make people feel good—create positive feelings and emotions; Create a point of contact—make it personal so that people can relate; Make people feel they are part of the majority; Show the benefits." Examples of positive language to be considered or used were cited beneath. This included a diverse choice of recommended verbs ("help, achieve, join, influence, advise, drive, push, support, work, respect, lobby, provide, keep, cooperate, defend, inform, care, share, save, unite, educate, deliver, protect, change, understand, investigate"), recommended adjectives ("Amazing: dynamic, responsible, quick; Wonderful: creative, humane; Beautiful: effective; Positive: real, good; kind: holistic, easy"), and recommended nouns ("Compassion: empathy, diversity, energy, opportunity, movement, goal; Positivity: truth, biodiversity, confidence, majority, society, alternatives; Harmony: balance, welfare, awareness, together, future, replace; Reason: justice, protection, reality, action, home, initiative; Peace: nature, behaviour, vision, success, freedom, family"). At the bottom of the poster, there was also a reference to recommended pronouns ("you" and "we") and to a series of notes on illustrative phrases. These included, "You can make a difference: meat-free, animal-friendly, real change; We can make a difference: no one can do everything, but everyone can do something, brings you together; Be part of: every little helps, issues that matter to you."

The same reeducation was deemed necessary when considering the organization's use of images. Indeed, in some ways this was an even more dramatic change. Eilidh instructed that the animal group should stop using photographs and videotape that showed animal suffering or illustrated acts of cruelty. Once again, this was a deliberate attempt to switch from negativity to positivity and to avoid alienating the mainstream. It was also another highly visible instantiation of the clear difference between the animal group and other animal protection or animal rights groups. Perhaps the most obvious contrast here was with the use of images by PETA, who pioneered the photographic or videotaped exposé of abuse (e.g., footage of hens trapped in battery cages and unable to lift a wing, crying lambs having their tails docked, panic-stricken mice submitted to the "forced swim test" in labs, and distressed calves being removed from their distressed mother cows in the dairy industry) as a central part of its self-presentation and campaigning strategies (see Song 2010). However, images of animal suffering as a result of human cruelty had been the normative tool of public messaging across animal protection for many years, including at the Edinburgh animal protection organization. Such images featured strongly in all its past communications—leaflets, press releases, campaign materials, lobbying presentations, and so on—and was assumed to be a vital instrument in both mobilizing the support of animal people and raising funds.

As well as removing such images from its literature, the animal group took them down from its office walls. Up in their place went posters and other images that celebrated positive expressions of animal feats and sentient capacities (similar kinds of images featured across the new website). Above the photocopier, for instance, there appeared a large photograph of an elephant's toenailed foot, raised just above the ground as if in midstride, communicating both extraordinary power and flexibility. On another wall someone had placed a photograph of a swinging orangutan, its arms lifted and extended above it to athletically grip a forest branch. Beside the desks of David and Iain, there was a poster of a peregrine falcon at rest after flight, with a strip of paper taped to the wall above it and the printed words "We build too many walls and not enough bridges—Sir Isaac Newton." On the walls of the fundraising office, several other images hung. This included photographs of a leaping hare, puffins in courtship, and a close-up shot of a green snake's head, its tongue flicked out to pick up chemical particles in the air that might signify the presence of prey. The selection of images clearly referenced the influence of Jonathan Balcombe (2011); in *The Exuberant Ark*, series of photographs are organized according to the positive animal capacities or pleasures that they are taken to depict (section headings in the book include "Love," "Play," "Food," "Comfort," "Touch," and "Companionship"). But in this case, the images were also meant to remind colleagues of the ethos of positivity that would define their labor going forward.

Kind Acts

The idea that a mainstream movement could be built out of or from an ethos of positivity—for instance, that admiration for animal sentient powers rather than distress and outrage at human-inflicted animal suffering could be its mobilizing stance—returns us to the issue of moral patiency. It also reminds us that in a wider context of ordinary English usage, where there is still no common word for the opposite of actions or for what Collingwood terms "instances of being acted upon" (1944, 86; see also chapter 1), that experience of the patient position struggles to get fully expressed. In fact, its articulation often had to work through dominant languages of moral agency or doing—for instance, the "making a difference" register in which both old-style advocacy and the new-style self-advocacy largely operated. "The more we study and observe them [animals] the more we learn and understand," Craig told his website readers. "The more we understand the more we admire. The more we admire the more we want to protect." Such statements represent a series of enchained actions—things that "I" or "We" can do—that culminates in the ultimate act of animal protection. Yet, as we have seen, admiration itself was perceived to be far less the result of something one determined to do, and far more the result of something that impacted or worked through the moral subject. Elsewhere in his posts, Craig connected admiration with states of enthrallment, captivation, and awe.

Indeed, learning about diverse animal capacities for pleasure and sentient feats, whether through animal science or via commonsense observation, was understood to be part of the same process of reconnection. Remember that in the end it was animals that will inspire people to make the change—but not animals in their classic after-the-fact definition as suffering moral patients or objects of human concern, instead animals as subjects that amaze or demand our attention simply by being positively themselves.

As we have seen, in the moral imagination of the organization's new vision, doing or moral agency was itself greatly complicated by that ethos of positivity. A whole raft of actions—including ordinary speech acts—suddenly became problematic or reexperienced as things that colleagues needed to stop doing. They should not criticize or judge, for instance, or offer an oppositional argument or professionally angry quote. Instead, they should help, respect, educate, share, understand, support, advise, and so on. "Make people feel good," the poster on the office wall urged, "create positive feelings and emotions." And just as importantly, "Make people feel they are part of the majority." These newly sanctioned doings or modes of action also found expression in another essential strand of the new vision, its revalorization of the virtue of kindness[1] and in particular "the kind act." More so than any other action, this one defined what the moral subject of the mainstream movement did and how that doing connected to welfarist concerns. Additionally, the kind act clarified the question of the We, or what the highly diverse acts of the members of this new mainstream movement might have in common. Finally, it represented perhaps the purest instantiation of the kind of positive moral action that Eilidh believed encapsulated the ethos. Although kind acts are further explored in the next chapter through the prism of ethical choice, I briefly introduce its significance here to give some sense of what sorts of doings flourished in the context of the condition of moral patiency discussed above.

As well as citing kindness as a key virtue, internal documentation about the new vision referenced it to help characterize other important organizational values. Compassion, for instance, was said to mean that "we behave in a way that is kind." The virtue also featured strongly across the animal group's front-of-house publications. In fact, for a period before the vision's launch, it was thought that the group's new name would make explicit mention of it. The compound title "AnimalKind," for instance, was until the last moment a strong contender. It immediately appealed to Eilidh and the other members of the working group because it seemed to aptly communicate their ethos and the tone of the new vision. In particular, she liked the emphasis on the double meaning in "Kind": that is, the inference not just of compassion but also of a group that shares characteristics. In the end, the name was discarded for two reasons. First, the working group felt that it ultimately reinforced the very separation that the principle of reconnection was working against. "You know," Eilidh explained to me, the name implied that "AnimalKind are over there, and We are over here, and that They need our help." But more decisively, the group discovered that an animal protection organization specializing in rescuing stray cats in the state of New York was already using the name.

However, both before and immediately after the new vision's launch, references to kindness were everywhere. The virtue featured strongly in the form of celebrity endorsements that the animal group solicited at that time and then published on the new website, across social media, and in press releases. It also featured in the early strategies deployed to try to encourage prospective supporters to narrate and share their animal-friendly commitments. Recall, for example, that at the Girls' Day Out event the group invited those visitors paying a pound to sit on Hunky Santa's knee to also "make his 'Kind Wish' come true" (i.e., by supporting the organization's campaign to ensure that the proposed EU ban on cosmetic testing came into force). Likewise, the team invited visitors to the exhibitor stand to write down their own kind wishes for animals so that they could be posted online. Such requests replicated wider practices on social media platforms. "What is your kind act," one oft-repeated prompt on Facebook and Twitter asked, "that little or big thing that you have recently done to help animals?" Or more prospectively, prompts asked supporters to consider the kind acts that they could next perform. "Hope everybody is enjoying their Friday," one invitation on Facebook began, "what could you do this weekend to make a difference to our animal friends?" Responses appeared in the chain below the post but also sometimes got collected and reproduced in illustrative fashion in promotional materials:

> We feed a hedgehog every day, who comes in our garden. His name is Prickly. We got advice on what to feed him from a local hedgehog rescue centre. I have three cats, but that does not seem to bother him!

> I write to my MP, whenever an important issue regarding animals arises. It is important to keep them informed, after all they have the power to make changes, and you voted for them! Children also have to be made aware, that animals have feelings and can feel pain just like humans and have to be treated with respect.

> My employer is always trying to have a staff night out "at the dogs." As a senior manager in the company, I vetoed the suggestion. They think I refuse on the grounds of gambling but when I tell them I object to animals used in sport, they do not know what to say. I have a voice and I use it as much as I can. They had no idea of the horrors of dog racing. I made sure they now do.

> When I see a spider in the bath or shower, I pick them up and put them somewhere dry, and somewhere where my cats will hopefully not see them.

> At 75, I'll continue to eat a wholefood vegetarian diet, which I started at age 17. Not only have I prevented a lot of cruelty and unnecessary killing, but I have enjoyed exceptional health compared to my contemporaries.

One recent event came to mind when one of the feral cats I adopted brought in a bird. I did manage to get the bird off her. She was still alive but very stressed. I left her in a cage all night and the next morning she was fine, so I opened the door and away she flew. That was such a good feeling. The cat now has a collar with a loud bell on it.

My act is being vegan—and talking about it as much as I can! I love telling people about all the wonderful things that are vegan and exactly why it is that I have chosen this lifestyle.

When I find an exhausted bee in my house (that can no longer fly), I give it a drop of jam or syrup to eat. It can suck some of it up, becomes re-energised and after a few minutes can fly back home to its hive.

I vow to always stop—as I do—to check any injured animal I see on the roadside—if they are not dead, I'll take them to the nearest vet for treatment, there is sometimes a happy ending—a rabbit I suspected had a broken back (as usual, after being hit by a car) spent the night at the vet and was surprisingly well enough to be able to be released by me the next day!

I help an elderly neighbour by offering to look after or walk his dog occasionally.

My act is rehoming rescue hens. If you have a garden with a bit of grass (a huge garden isn't required), please consider rescuing battery or barn hens. These wee chooks have had a hellish life stuck in a tiny cage smaller than an A4 sheet of paper or crammed in a barn with no room to roam and getting the living daylights pecked out of them. . . . And if you aren't able to keep rescue hens, please don't eat battery eggs or food containing them—go organic free range.

This selection highlighted the extent to which supporters responded to requests by offering acts of animal rescue. It also demonstrated how much work there was to be done to turn traditional supporters—from their tone most of these posts appeared to be authored by animal people—into members of a mainstream movement eschewing reference to animal suffering and oppositional stances for an affirmative emphasis on positive sentience and reconnection. Nevertheless, the invitation for supporters to consider these actions as expressions of kindness stood. Indeed, Eilidh believed that reframing them as kind acts could itself lead these supporters toward the desired ethos of positivity. For the stress on kindness was a deliberate attempt to define the committed individual by positive action and a noncondemnatory outlook. From this perspective, for example, the commitment to become vegetarian or vegan should be interpreted not so much as a statement or protest *against* factory farming but instead as a kind act toward farmed animals. Bearing in mind one of the above posts' references to "exceptional health," such decisions could also be interpreted as a kind act toward oneself.

The shape of the posts also emphasized how kindness was to be chiefly interpreted: as small, singular, ordinary, but largely disconnected virtuous acts. Advice on the animal group's website spoke of the "simple things people can do every day to protect animals." Indeed, the diverse published responses successfully reiterated that formula. Typically told in the first person, though sometimes containing a second-person address that implied the author and reader of the post were part of a We, entries began with "I help . . .," "My act is . . .," "When I see . . .," and so on. As Eilidh and Craig readily admitted, the stress placed on kindness was already familiar to mainstream audiences through fundraising appeals. In fact, it had become a common way for all sorts of charitable causes to invite public support. Most obviously, the working group's idea of renaming the organization AnimalKind was a conscious redeployment of Oxfam's high-profile strapline "be humankind," also begun as a campaign in 2008. But the idea of highlighting the importance of "everyday" acts of kindness, as a way of effecting societal-level social change, was also a specific feature of philanthropy and social activism online.

At the time of the relaunch, the popularity of websites that promoted, reported, and celebrated quotidian expressions of kindness, usually toward other people, was evident. Colleagues recognized, for instance, the existence of a North American weblog OneKindAct.com (now defunct), which described itself as a "social movement designed to change the world by motivating others to leap in and really live, just One Kind Act at a time." By publishing constantly updated reports of citizens' ordinary kind behavior, the weblog hoped to demonstrate that "small acts of kindness are extremely contagious, and it is those small acts that make the biggest difference." Members of the new vision working group of course also knew of the existence of another US-based website, RandomActsofKindness.org. Still existent and apparently thriving, it strives "to change schools, the workplace, families and society through kindness" and by providing its visitors encouragement, challenges, and tips on being kind to others. This includes the regular publication of "kindness ideas": currently listed suggestions include "be polite on the road," "organize a clean-up party," "find out something new about a co-worker," "wheel out your neighbour's trash bin," "help out in the kitchen," "plant a tree," and "have a judgement free day." The last suggestion speaks back to the animal group's own ethos of positivity, its insistence that kind acts were ideally nonjudgmental models of doing.

Instead of preaching, these sites claimed to be reproducing the public's own self-authored testimony of exemplary behaviors. In the working group, the thinking was that if the organization could effectively appropriate the moral sentiment of these human-oriented kind acts and make them animal-oriented and likewise "contagious," then it could radically transform the welfare of animals. In this comparison, a further shift was implied. For online sites such as Random Acts of Kindness do not just wish to draw public attention to acts of kindness; they also want to "make kindness the norm." In similar fashion, the animal group hoped that spotlighting animal kind acts might itself result in animal kindness becoming an assumed mainstream virtue.

Fellow Feeling

Although the virtue of kindness and the practice of kind acts or doings were central to the way the animal group envisaged a mainstream movement emerging and expanding, there was a great deal of imprecision in the new vision's formulation. It was not altogether clear to colleagues, for example, how the model of contagion would operate. What made these kind acts spread? How did they inspire the doings of others and build a sense of moral community? Was it simply through reportage, the publication and viewing of each other's posts online? Or was there a wider principle of emulation at work? I think Eilidh believed that people wanted to be good and that if one gave space for the public expression of goodness, in this case directed toward animals, it would in turn attract goodness. Certainly, as we will see, the group's new concern to promote animal-friendly lifestyles assumed that this was the case (see chapter 4). However, kindness always appeared to most colleagues as a rather vague mechanism of mobilization, especially for an organization without a hands-on relationship to helping animals (or humans).[2] It was notable that a majority of kind acts posted by supporters centered on examples of rescue, not an area the animal group specialized in. In fact, after the initial enthusiasm for the relaunch, the stress on kindness somewhat dissipated, at least as an explanatory apparatus for movement building.

But what other perhaps more compelling theories of moral community could be read into the new vision and the animal group's plans for its implementation? To answer this, I want to return to the principle of making contact and to the solidarities implied by such encounters. Instead of working from explicit theories contained in the new vision, though, I here start with an explanation sourced elsewhere, in the work of Mick Smith (2011), an environmental philosopher concerned to propose alternative terms for understanding ecology's sense of moral community. I believe that Smith's explanations also provide a prompt for reassessing the nature and strength of moral connection within animal protection. Before I begin, it is important to stress that to my knowledge colleagues were entirely unaware of Smith's project. Indeed, as previously mentioned (see chapter 1), although some colleagues did read academic works (for instance, on the philosophy of animal rights), they rarely cited them in conversation or made them the explicit basis of organizational action.

With that in mind, what is perhaps immediately helpful about Smith's observations is the fact that they emerge via a general skepticism and suspicion of the efficacy of moral formalism or rationalist conceptions of moral knowledge. Indeed, for Smith, current explanations for ecology's moral community fail precisely because they tend to reduce the "unique being" of what appears or is encountered in environment "to a formal and formulaic abstraction" (2011, 33). In a corresponding fashion, they also tend to understand ecology's community as a "social arrangement constitutionally dependent upon these same formal abstractions" (33). While these explanations operate with an assumption, which Smith shares, that "ethical relations are irreducible to instrumental approaches," they differ radically from the explanation that Smith prefers "in terms of how entities come to be regarded as ethically significant" (33). First and foremost, Smith distinguishes a stance by making a case for the primacy of what Smith terms "value-feelings." Helpfully, Smith defines these as "the actual experience of concern, fellow feeling or love for the Other" (33) that arises as an outcome of encounter. And most relevantly, for our purposes, Smith proceeds to argue that those value-feelings may also be read as responsible for the genesis of ethical society.

In this regard, Smith draws inspiration from the work of the late nineteenth-century philosopher Max Scheler. In particular, Smith picks up on Scheler's insistence that the "fellow feeling" born from encounters (unlike Smith, Scheler talks about only interhuman interactions) can expand into "a wider community of [ethical] significance" without the need to abstract axioms or to deny the continuing importance of the original phenomenal experience of the other (2011, 28). In fact, Smith quotes Scheler as saying, "Once benevolence has gained its impetus from pure fellow feeling, it can, by its own activity, enlarge the scope of the latter to an unlimited extent" (28). Smith is interested in how this human benevolence grows or escalates to establish senses of moral community and in how one can describe that process without providing an explanation that completely "overshadows the phenomenal role of fellow feeling" (28). In Smith's terms, Scheler succeeds because the philosophy of ethics proposed contains a much better or more convincing sociological account of how moral concern becomes shared.

Of course, the idea that it might be value-feelings rather than moral axioms or first principles that form the basis of moral community would make complete sense to the animal people I knew. The centrality of phenomenal experiences of encounter and their link to that essential value-feeling—animal love—has been well documented. As well as expressing a fellow feeling that emerges across species divides, most notably but not exclusively with companion animals (Craig insisted that Paddy was his "buddy" etc.), it was clear to them that the mutual recognition of animal love was in large part the basis of fellow feeling within the animal movement. Although the animal group's ambition to go mainstream challenged aspects of that animal love—in particular the concern expressed as a result of the phenomenal experience of encountering human-inflicted animal suffering—its movement building still rested on the presumption that the escalation of fellow feeling would depend on the quality of encounter, albeit now defined as a purely positive experience. It is worth repeating that in the mainstream movement imagined by the animal group, animals would (continue to) inspire people to make the change. The fact that this could occur just as much by engaging the public with "amazing animal facts" as by reconnecting with the sentient animal life around individuals points to another important emphasis in Smith's work, the insistence that in ethical society fellow feeling was the result of both direct and mediated encounters with nonhuman being.

Smith's evidence for this observation partly falls on an analysis of the mobilizing effects of a classic image, indigenous to the traditional campaigning spirit of animal protection as much as or even more so than to that of ecology. If one imagines a photograph of a seal pup about to be clubbed to death on an ice floe, Smith at one point invites, it is clear that such an image can generate value-feelings in the human viewer and that this emotional albeit mediated contact with the seal pup can also initiate further fellow feelings (2011, 26). As Smith goes on to point out, "the single seal which appears in what is, after all, only a photograph, also elicits concern for a much wider constituency of beings many of whom we will never encounter on film, still less 'in the flesh'" (27). There is, then, an escalatory logic not just to the growth of fellow feeling within a community but also with regard to its objects of moral concern. "Being unable to save the actual seal pictured," Smith elaborates, "we might still find ourselves campaigning against industrialized seal slaughter, to conserve marine mammals like whales, or concern ourselves with the destructive exploitation of marine environments in general" (27). While the escalatory logic of animal protection might scale up and across in different ways—it is unlikely, for instance, to conclude at the level of cross-species habitat protection—the parallel is evident.

Indeed, I believe that a version of that kind of escalatory logic can be read into the ways that the animal group assumed and envisaged the development of their moral community. This was the case in terms of both the ethical society of animal people that it was imagined colleagues already belonged to and the mainstream animal-friendly community that they hoped to create through an ethos of positivity. In either case, it was taken that a singular encounter had the potential to generate moral concern and fellow feeling not just for the nonhuman being that they made direct or mediated contact with but for an expanding range of sentient beings and kinds of animal life. Just as in the past it was hoped that the publication of an image of a fox caught in a snare might generate fellow feeling for foxes or trapped animals in general and hence a commitment to campaign against snaring, which might in turn lead to a commitment to campaign for a ban on field sports or for an end to the culling of red deer or mountain hares, so it was hoped that the publication of an image of puffins in courtship or of an orangutan swinging through a forest canopy might inspire an admiration for and attachment to all sorts of animals and in multiple directions. This was assumed to be equally true for the people brought into the same moral orbit or wider community of ethical significance by those mediated or direct encounters. In all cases, something close to Smith's value-feelings were identified at work; in fact, one could claim that it was often assumed that those feelings had an in-built capacity for escalation.

This interpretation goes some way to explaining the peculiar dynamics of moral patiency-cum-agency within animal protection. Certainly, the stress of Smith (and Scheler) on the nonintentional nature of ethical solidarity—the fact that fellow feeling can appear to escalate without coordination or even direct communication between human members of its community—helps illuminate aspects of colleagues' relationship to and sense of participation in the animal movement. It also assists our understanding of how Eilidh and her team imagined a mainstream movement developing. Perhaps as importantly, the capacity of value-feelings to contain their own escalatory logic or sense of moral community helps further pin down what might have been at stake in such a transition, and especially for animal people. As we will see, colleagues sometimes viewed the animal movement and the envisaged mainstream movement as two competing regimes of fellow feeling. Indeed, their attempts to engage the mainstream led some on occasion to invoke a crisis of fellow feeling. This could sometimes play out through internal debates about which form of encounter—that is, the phenomenal experience of animal pleasure or of animal suffering—better mobilized support or generated the purer protection ethics. But as the last section of this chapter briefly illustrates, it could also reveal itself through clarificatory moments of re-antagonism or actively *not* sensing a fellow feeling with others.

"These Women Were Not Really the Right Crowd for Us"

When I arrived in the office one morning soon after the Girls' Day Out event, I noticed that Sarah had stuck an "inspiration poster" on the wall behind her desk. Sarah explained that she had read about these posters on Twitter earlier that week and had immediately determined to make one herself (at that time inspiration posters were commonly used as motivational tools in both corporations and the third sector). Organized on a large A2-sized piece of paper, the poster was essentially a collage made up of various pasted words, messages, drawings, and images that Sarah had carefully selected. Placed in the very center of the poster, in the largest font size, was the word "Sincerity." Arranged around it there were a cluster of other words, in smaller font sizes: "pure," "open," "clean," "genuine," "honest," and "believable." Sarah had drawn the head of Esme, her Staffie rescue dog, in one corner. She had also cut and glued a photocopied still from the group's sci-fi teaser film and pasted key phrases across the poster. "Always remember why we are here," I read, "Free from pretence or deceit," and "Never assume." Another phrase was placed in quote marks and highlighted in blue: "Sure, no one can write like you." Sarah informed me that this was something her granny back in Northern Ireland used to say regularly by way of encouragement. In addition, the poster displayed the animal group's new name, printed in bold black lettering. Set below it there was a photograph of a walking herd of elephants, cows, and calves in close train. Finally, Sarah had drawn a lit bulb with the word "donor" written inside.

Sarah disclosed that the inspiration poster was up there to give her strength, to help her recall why she did her job, and to remind her what was important. As an assistant fundraiser, Sarah felt the pressures of the role, especially to meet the ambitiously set monthly income or donation targets. However, the prompt to draw up the poster was due to a broader uneasy sense of the direction the relaunched group was taking and of what place there would be within it for animal people like her. Although attracted to aspects of the ethos of positivity—for example, the emphasis on engaging the public through positive expressions of animal sentience—Sarah worried, like other colleagues, about the compromises involved in attaching herself to a mainstream movement. For Sarah, not just her moral integrity but also the integrity of the community mobilized by the experience of animal love was potentially at stake. Could one accept the new expanded definition of "ourselves" that a mainstream movement implied and remain a sincere animal person? Sarah wasn't sure. Certainly, the recent trip to Girls' Day Out had done little to reassure colleagues.

Toward the end of the first day at the Glasgow exhibition hall, I had found Sarah and another member of the fundraising team, Laura, standing at the back of a packed "theater" watching a psychic in action. As we witnessed queues of people patiently lining up to approach the stage, typically to ask questions about whether a husband or grown-up son or daughter would find or keep a job or whether someone would be able to sell their house or have a baby, Sarah told us that she hated psychics because they just manipulated people's misery. While from Sarah's perspective the day had not been entirely unproductive—the tour of other exhibitors had resulted in positive conversation about copartnering and an exchange of cards with a company specializing in aloe vera products—this show had made them both a little depressed. Earlier in the day, I had taken my lunch with David, the group's movement builder or social media strategist, and he too had seemed a bit dispirited. Partly this was because he had to spend less time than he would have liked posting tweets, since they had needed him to help attract visitors to the group's stand. However, David's mood was also due to the general experience of the exhibition. Like Laura and Sarah, David complained that their participation at Girls' Day Out felt like an affront to his values, not just those connected to animal protection but also his self-identity as a feminist. Even Euan, who organized the action, had been less than cheerful. His temper had not improved since the last-minute news that the planned media partnership with *No. 1* magazine had fallen through, and so the animal group was likely to lose money on the event.

But most of all, the growing sense of gloom was due to the shared realization that very few of the visitors at the exhibition had any genuine interest in animal protection. Laura told me, for instance, that while plenty of them seemed keen and willing to pay to sit on Hunky Santa's lap, few showed any real concern for the group's advertised campaign goal to end animal testing on cosmetics. In fact, Laura was unsure whether any of them whispered an animal-friendly message into Santa's ear. To Laura, they seemed far more focused on getting their friends to take their photographs. Likewise, there was very little evidence of visitors engaging with David's attempts to have them tweet a kind wish. Indeed, when I talked with participating colleagues a few days later, they painted a grim portrait of the proceedings. Apparently, as the event dragged on into the late afternoon, groups of visitors became more and more intoxicated (as well as the pop-up champagne bars, some exhibitors offered complimentary glasses of wine) and rowdy. There were even instances of open hostility. It was revealed, for example, that at the end of the second day one drunken visitor had leaned into Laura and hissed that she hated animals. Hearing this report, Iain reflected that it sounded like "these women were not really the right crowd for us." It was clear from what the others said, Iain continued, that many visitors to the exhibition hall were never going to respond to the animal group's attempt to engage them positively. Instead of reconnecting, they remained indifferent, seemingly unconcerned with issues of welfare and unwilling to alter their consumer practices in an animal-friendly direction.

In general, colleagues felt somewhat bruised and dejected by their encounter with this version of the mainstream. Coupled with the clearly limited success of the exhibitor stand, there was a sense of resentment toward the organization for putting them in that situation. To some it felt like their ethical identity as animal people (and feminists) was being actively placed in jeopardy or disrespected. More broadly, the experience affirmed a growing suspicion about the viability of the group's new vision and its strategies of mobilization. For there was no possibility, in their minds, of a fellow feeling emerging with these women. Indeed, some colleagues feared that any continuation of the organizational attempt to become a mainstream movement might end by alienating the animal group's traditional supporter base: animal people like them.

Even among staff members with a more sympathetic attitude toward the new vision, practical concerns soon arose. Craig, for instance, reported that he found it hard to engage other animal people in the ethos of positivity. This struck Craig most markedly when he tried to organize an art exhibition during the Edinburgh Fringe Festival to showcase the rebranded group's vision. All participating artists were invited to develop artworks that highlighted the theme of animal pleasures or positive animal sentience. But Craig was disappointed to discover that the vast bulk of submissions ignored this instruction. Instead, what the organization received was a range of artworks either vividly depicting human-inflicted animal suffering or offering symbolic critiques that Craig knew would be read by mainstream audiences as further indexes of the "anti" tendency within animal protection.

To understand further why animal people, inside and outside of the animal group, were so resistant to the new vision's call for inclusivity and to embrace an ethos of positivity, we need to further consider what was at stake for them. This means looking again at the historic nature of their personal and professional engagements with the mainstream. Indeed, it is important to stress that any rejection of the call to become a mainstream movement was not a rejection of engagement itself, which all colleagues saw as a necessary project. But it also means looking more closely at those doings marked as intentional moral actions, which they did identify as self-defining aspects of an animal protection stance. In the next chapter, we turn more squarely to examine the issue of ethical choice.

THE ETHICAL CHOICE

I now want to examine the experience of moral agency in more detail. This includes a closer consideration of those moral actions that colleagues at the animal group typically marked as intentional human behavior. More specifically, I investigate kinds of decision-making understood to be informed by freely made "ethical choice." As well as raising the question of the nature of that freedom, the focus takes us squarely into the realm of those after-the-fact explanations of moral agent positions normatively accepted by animal people. It is worth, I think, reminding us of the definition provided by Regan (see chapter 1), versions of which colleagues broadly reproduced and often deployed: "Moral agents are individuals who have a range of sophisticated abilities, including in particular the ability to bring impartial principles to bear on the determination of what, all considered, morally ought to be done and, having made that determination, to freely choose or fail to choose to act as morality, as they conceive it, requires" (2004, 151). As Regan's definition assumes, "normal adult human beings" are the paradigm example of moral agents, who are further distinguished by the fact that they are held accountable for their actions (152). "Since it is they who ultimately decide what they do," Regan adds, "it is also they who must bear the moral responsibility of doing (or not doing) it" (152).

From the perspective of animal people, this appreciation of moral action as a specific kind of deliberate doing or not doing, shaped by a prior decision-making exercise that introduces a responsibility to act, was far from sufficient. As we already know, such understandings existed in tension with before-the-fact explanations of moral behavior grounded in patiency. Rather than decision-making, the emphasis here lay on the mystery of moral commitment and purpose and on something far nearer to states of unfreedom. The latter term is helpful in elucidating the way that colleagues often felt that they had no choice but to love and help animals; it is less helpful, though, if it obscures the alternative senses of freedom that could ensue from submitting to such passions.

However, in this chapter I want to background the patiency attached to moral feelings such as animal love and instead consider the tensions inherent within and between experiences of exercising moral agency or occupying moral agent positions. This includes a consideration of the ways in which colleagues reported the centrality of making ethical choices as well as of the ways in which those choices periodically became reconfigured as only *apparently* freely made. Either through the speculations of individuals themselves or as a consequence of unexpected outlooks introduced by other colleagues, that loss of confidence could sometimes appear to render the decision-making process suspect or alternatively convert it into a form of patiency. All of this returns us to issues of expertise as well as to the organizational aspiration to become a mainstream movement; it also begins to highlight some of the ways in which observed differences between colleagues mattered. Yet, as I argue, the musings of colleagues on their status as moral agents did something further. It opened the possibility of a different interpretation of what it might mean to engage the mainstream. That alternative methodology drew on forms of expertise finessed at home rather than in the office, and just as importantly on experiences that did not depend on the implementation of the animal group's ambitious plans.

Throughout the chapter, I especially concentrate on one key decision, the choice to become vegetarian or vegan. Of course, the place of vegetarianism and veganism in the history of the animal movement is well documented (see Spencer 1993, 2016; Walters and Portmess 1999; King et al. 2019). It features prominently, for example, in key manifestos of the 1970s such as Peter Singer's *Animal Liberation*. Indeed, that choice consistently appears as an answer to the Aristotelian question of how one should live. It represents UK-based animal activism's perhaps most convincing example of an arts of existence or measure of self-cultivation (see Therese Kelly 2022; Krøijer 2015b) and in that regard features prominently across the rest of the global animal movement too (Weiss 2016; but see Davé 2014). In terms of the animal group that I worked with, the importance of that choice was illustrated through the arc of moral biography typically narrated by colleagues (see chapter 1). This presented that move, alongside the care work of rescue, as among the most quintessential forms of enacted responsibility available to animal people, in part because of the obvious impact on their ordinary lives. But what particularly interests me about that ethical choice is the quality and sheer quantity of acts it supported. For as all colleagues knew, a decision to turn vegetarian or vegan was not just a singular, definitive choice; it was a prompt to ongoing choices and moral action in general. A whole horizon of choices and comprehensive regimen of acts ensued. These enabled animal people to feel that they were free and to keep demonstrating to themselves (and to others) that they were doing something for the cause of animal protection. That spectrum of choices also provided the animal group with one kind of template for envisaging how a mainstream movement might work in practice and most significantly a basis for modeling the new vision's core principle of inclusivity (see chapter 3).

But the examples of vegetarianism and veganism were influential in other ways too. Their emphasis on abstention or renunciation, for instance, tied to a more traditional principle of do no harm still widely invoked by colleagues. As we will go on to see, that principle could easily translate into a broader ethos of noninterference where doing or moral agency itself needed to be first and foremost manifest as radical inaction (but see Reed 2024). Here, senses of moral responsibility often became entangled with imperatives to actively not do anything about the welfare of animals, especially when that welfare did not appear to be the outcome of human action. Likewise, those senses of responsibility became caught up in occasional dreams of a future without human moral subjects of any kind, an aspiration clearly at odds with or at least in ongoing tension with the equally compelling impulse to make contact.

Vegetarian Birth

In reaching for a metaphor that best conveyed the attitude of the rebranded animal group toward its new publics, Euan sometimes liked to equate the condition of the mainstream with the plight of the alcoholic. "You know you are not going to get an alcoholic to stop drinking unless the alcoholic wants to stop drinking," Euan observed. He continued, "We are not trying to be an organization that is telling people they are naughty, and they have to stop it; it's more about getting them to realize things themselves and to change their behavior." His chosen metaphor was an unusual way of accounting for the familiar claim that most members of the mainstream—including those that colleagues loved or knew best—failed to conduct themselves in an animal-friendly fashion. It certainly provided a striking rationale for understanding mainstream apathy; the metaphor of addiction had the added benefit of succinctly conveying why conventional critiques of mainstream behavior might prove counterproductive (i.e., you can't just tell an alcoholic to stop drinking). It also paralleled some of the assumptions behind the empathy-building project so close to Eilidh's heart. For instance, in both cases there was a shared diagnostic tendency, an inclination to reread unethical behavior as a psychological or even physiological issue or as a problem with affect. But what did clearly distinguish Euan's observations was the unusual angle of inquiry often adopted toward the cause of animal protection itself. For Euan frequently seemed able to take up the position of a neutral bystander, so that to me and to many of his colleagues, it was not always clear what his animal-friendly values or credentials were. Similarly, it was not always self-evident where exactly Euan stood in the existential debates about the organization's future that at this time animated staff discussions.

In fact, at times there appeared something slightly off-kilter about Euan's stance on the world, especially from the perspective of those colleagues who had been at the animal group a long time. As they conceded, this may in part have been due to the nature of his expertise as the organization's communications officer or to the very specific ways in which Euan chose to interpret that organizational role. Coming from a professional background in corporate public relations, which included working for a long period as the press officer in a financial services company, Euan repeatedly insisted that his core responsibility was to act as a "gatekeeper" between the animal group and the outside world. To effectively carry out the tasks allotted to him, Euan explained, he needed to develop an understanding of what different news agencies wanted from a story, "to work more closely with journalists, to try and get a better idea of what will work for them." That in turn necessitated presenting himself as a "channel of communication"; it was vital to the effectiveness of his role, Euan argued, that journalists never saw him as the author of the organizational messages, quotes, and opinions that he passed on to them. Likewise, Euan was resolute in his view that he should not be involved in the internal process of determining policy or campaign goals, even if he might well be responsible for drafting the press releases that flowed from those decisions. Indeed, in Euan's mind, any expression of a view was not just an irrelevance; it was a hindrance to doing his job well.

But in the eyes of many colleagues, there was something else amiss. Like some of the other new hires made during the period of organizational expansion, Euan lacked a professional history in animal protection. He also lacked many of the signs of a personal commitment to the cause, such as past membership in other animal groups or volunteer involvement in high-profile animal activist campaigns. This was not in itself necessarily a problem. Other colleagues too lacked aspects of a typical moral biography, yet their animal-friendly credentials were rarely questioned (see Reed 2017a). No, what was particularly disarming, even confusing, about the moral status of Euan's role and outlook was that in one crucial degree he absolutely did conform to the normative biography of animal people. For Euan was a long-term vegetarian. In fact, as Euan liked to regularly stress, he had been vegetarian since birth:

So, you were raised vegetarian?

Well, there was no pressure. Dad would have been disappointed, but certainly when I was young, I didn't think any more about it. When I got to school there was more of, "Why don't you try McDonalds?" But to be honest, at that point, even at that age, why would I? I didn't feel I was missing out on anything. I wasn't looking at meat and thinking, "Oh, succulent, tasty." If you've never had it, you don't see what the attraction is, you know, you might as well have said to me, "Why don't you eat a bag of nuts and screws!" It just didn't appeal to me, to be perfectly honest, and it just never has. Obviously, as I got older, I actually started to feel a certain amount of superiority about it [Euan chuckles], you know, to think, well, I do quite like animals, and I don't necessarily want to eat them. You know, some people are just born with a way of looking at the world and just going, I am not going to eat this because it just doesn't make any sense.

And, in terms of your dad, then, were you ever presented with an ethical argument . . . ?

I was never presented with one. To be honest, I have always wondered about why. Dad was in the Second World War; he enlisted when he was fifteen because at that point in the war, they didn't really care how old you were. I mean, I have never really known the point at which he became vegetarian, and because he is not alive anymore, I never really got the chance to ask. My mother, she didn't really know either. She sort of attributed it to what happened in the war. People didn't talk about post-traumatic stress and that stuff, but I've been told he saw some really horrific things. You know, people dying next to him, in quite horrible ways. Perhaps he just associated [meat eating] with all that slaughter and killing. It's a compelling argument, but, knowing my dad, I don't necessarily know that would have been it.

So, he didn't sit you down and say . . . ?

No, never. He never did that kind of thing anyway. I mean, to be honest, he was the kind of person that was interested in alternatives. After the war, he traveled a lot, he didn't settle, he had a kind of wanderlust. But he wasn't, like, a hippy or a spiritual person, anything like that. I always associated his vegetarianism with his massive streak of individualism. You know, he was perfectly amiable with people, well integrated, you know, people really liked him, but he had that thing about him, he was definitely determined to be different from everybody else. He wasn't going to be conventional. You know, one of my aunts used to say, "Oh, he's contrary." That's what it was in those days, it was contrariness. That's what they called it. No, I never really saw him pass judgment or anything. I mean, my mother is a terrible cook as well, absolutely abysmal [Euan chuckles].

Is she vegetarian?

No, she wasn't. And when my sister was born, my dad said, "Well let's raise her vegetarian too." But my mum said, "I've had enough of this, it's bad enough with you two. I'm just going to give her meat." So, my sister went for years and years as a meat eater, and we never thought any more about it. There's two in the family that weren't, and two in the family that were. And then I remember, The Smiths have got a lot to answer for [Euan chuckles]; she got really into *Meat Is Murder* [a popular studio album by the band] and turned into a vegetarian pretty solidly for a number of years. Then she married a carnivore and gradually got back into that. She eats meat now. But we never set it up as, like, a betrayal of family values. I never got the sense that my dad was, "Well, I like him more than you because he's vegetarian like me." I mean, I think he did have a slight smugness about it as time went on. Because I think he did begin to realize he was kind of ahead of his time, and I think he was quite pleased that I had stayed vegetarian.

Right.

So, I never got an ethical thing. However, the pressure of the time was very much this was a weird thing to do. But then that just added to my general strangeness at school. So again, I sort of embraced it, probably in the same way my dad did. But we never had a real big discussion about it. I mean, I don't eat fish either, so I wasn't like a pescatarian.

But you are not vegan?

I am not a vegan, no. That was partly again the upbringing thing. I mean, if I had been raised vegan, I probably would have remained vegan and not thought any more about it.

So, it doesn't sound like you have the same context as some [colleagues]?

No, they had to make a choice. But what's odd to me is the amount of people who defend themselves to me. I have found this all my life. People go, "Oh, I don't eat a lot of meat myself, actually, only a little bit now and again." Almost as if by saying I was vegetarian it was some kind of accusation. You know, "You are a murderer!" This was especially the case after the 1980s, thanks again to The Smiths, when people started realizing what vegetarianism was about. Before that, people just thought you were a bit odd. You know, "Do you live in a commune?" "No, I don't; I live in a council house in the West of Scotland!" But The Smiths' message is a bit extreme. I mean, I have cooked meat for people, and I've eaten meals with people that are having meat.

You married a meat eater, right?

Yes. I mean, to this day there are certain things I won't have anything to do with, like liver. And the other thing is fish. I am really sickened by the smell of it cooking. Sometimes, especially with school dinners, little bits of meat would creep into my meals. Then, I wouldn't have been aware of it, but now if that happens, I become really violently ill. My dad had really bad reactions too. He always said he had been vegetarian for so long that his body found it alien if something like that went into his system. But, you know, nothing has ever shown up [medically speaking]. I literally just think that if I was to go and eat a bag of nuts and screws, I would have exactly the same reaction. My body just doesn't know what it is. I could be glugging poison, for all it knows! So maybe I haven't had that kind of [moral] leap, but then I haven't had to make a conscious choice.

From the perspective of many colleagues, Euan's vegetarianism was a troubling anomaly. Although, as we will see, they too might sometimes reflect on the extent to which moral choices were self-willed and on occasion draw on third-order explanations for their own decisions, that skepticism was always underpinned by a strong faith that one choice was freely made. Euan understood this very well. Indeed, in the account above, Euan regularly marked that contrast. Unlike him, "*they* had to make a [conscious] choice." Likewise, vegetarianism never signified a crucial "leap" in moral development, as it did for them. But Euan's awareness of that difference was precisely what unsettled other colleagues. It pointed to the fact that Euan's vegetarianism was not about being good, or at least not a straightforward example of action informed by moral decision-making and reflection. There was no deliberative renunciation, no sense of self-sacrifice or struggle, and hence no need for self-monitoring and discipline.

   In fact, as the dialogue above illustrates, Euan seemed far more exercised by the question of his father's decision to become vegetarian than by his own "choice." While colleagues typically described the turn to becoming vegetarian or vegan as a marker of freedom, including freedom from parental influence, Euan stressed the opposite but in other circumstances very familiar form of kin reckoning—that is, the assumption of the individual's derivative status as a composite of parental lines of influence (Strathern 1992, 14–15; see also Carsten 2007; Cannell 2011). In this case, an action confirmed an overdetermining relationship between father and son and in the process distinguished that connection from other sets of relationships—between mother and son, husband and wife, father and daughter, and siblings. But what was equally significant, as far as the oddness of Euan's stance went, is the fact that he also insisted that his father's decision was ultimately a mystery, made without the giving of reasons or the articulation of any kind of moral case for vegetarianism. Perhaps most disturbingly, as far as his colleagues were concerned, Euan seemed to have the same casual attitude toward veganism; he told me and others that had he been raised a vegan, then in all probability he would not have "thought any more about it" too. It was true, as Euan liked to reassure us, that over time both father and son came to an awareness of some of the emergent, soon-to-be-conventional ways in which vegetarian living might be attached to virtue or moral decision-making and that they both came to enjoy that association. Similarly, Euan was not shy of telling stories of occasional family conflict, including one famous battle over a beef burger his mother had once cooked for him, which sounded a lot like the typical household tales colleagues narrated to index evidence of free will. But the suspicion of a sense of disparity between his vegetarianism and the vegetarianism (and veganism) of others remained jarring.

It was almost as if Euan's *ordinary* vegetarianism challenged the emphasis given to that moral choice in the lives of animal people and in both the traditional and new-vision versions of the organization. (I use the term "ordinary" or "everyday" here in the manner meant by Lambek [2010a; 2010b] and Mattingly [2014], who want to speak of an ethics grounded in unspoken agreements and in practice that takes place "without calling undue attention to itself" [Lambek 2010a, 2], though with the strong caveat that in this case the "ethical" dimension is itself a moot point.) Others might well look forward to a distant future in which animal-friendly behavior was unthinkingly adopted by the mainstream, but it seemed that colleagues were less sure how to deal with someone for whom vegetarianism (or veganism) was not an obvious moral commitment at all, and most particularly did not appear underpinned or motivated by a special consideration for the welfare of animals.

Catering for Our Values

Euan's vegetarianism flummoxed colleagues in other ways. Most notably, this was because Euan also resisted the most recognized alternative explanation for turning vegetarian or vegan: that it was a choice made out of the desire to improve one's health.[1] Although they never presented this as a primary reason, some colleagues did passionately believe in that argument, and everyone identified the claim as a useful strategic resource. But as the communications officer regularly pointed out, he was considerably overweight—hardly a walking advertisement, therefore, Euan quipped in innumerable office conversations, for the physical benefits of a meat-free diet. Indeed, it was that consistent failure to have his vegetarianism index positive change of any kind that jarred the most.

However, the anomalous status of Euan's vegetarianism did have some advantages, especially for an animal group undergoing such radical redefinition. For alongside the peculiar neutrality Euan claimed as communications officer, that outlook enabled a more dispassionate assessment of challenges. Euan was very good, for example, at diagnosing the nature of the dilemma facing colleagues and the transforming organization, particularly the fact that the new vision essentially problematized the question of values. Becoming a mainstream movement, Euan advised, necessitated encouraging individual and collective reflection on those values, a process that Euan anticipated would be "difficult both externally and internally." In fact, Euan often highlighted what he regarded as the most pressing set of questions for the animal group in these circumstances: "Should we be more vegetarian or vegan? Should we just allow for the fact that some people eat meat? Where do we stand?" The urgency behind such questions was a consequence of the attempt to implement the new vision, but it also laid out the terms of a debate internal to the organization to which colleagues kept returning.

Before the launch event for the animal group's new vision (see chapter 3), Eilidh and her fundraising director, Fraser, were locked in a prolonged dispute about what kind of food should be served at the planned luncheon. Fraser strongly argued that the menu should include high-welfare-sourced meat dishes as well as vegetarian and vegan options precisely to demonstrate to invited guests that this was a mainstream movement, "open and inclusive to all." Indeed, Fraser was concerned that to do otherwise would communicate an image of the organization as extreme or judgmental, thereby reinforcing the stereotype of animal groups that they were trying to combat. As the CEO and hence one of the chief proponents of that new vision, Eilidh took his point. However, she ultimately demurred. "It is vital," Eilidh told colleagues, that on this occasion "we are seen to be catering for our values." Her statement highlighted an inherent struggle. If Fraser was concerned that the absence of meat from the menu would alienate potential mainstream supporters and hence contradict the new inclusive message, Eilidh was worried that animal people on the board of trustees and in the audience would simply not accept the group serving meat at one of its own functions. Her decision indicated one kind of limit to the new vision's ambition, the fact that in the end the organization needed to continue to set and exemplify core values of animal protection. But as Eilidh explained to Fraser, this did not mean that they were prescribing choices outside of the luncheon or that the organization was intolerant of meat eating as the basis for making ethical choices within a mainstream movement.

From one perspective, the incident appeared as a test of the animal group's absolute values. Certainly, this was how many colleagues figured it. Yet with Eilidh that determination was far less clear. Although a vegan herself, the CEO seemed as much concerned with impression management. As her qualifying statement indicated, catering for the values of the organization was not in her mind necessarily at odds with supporting choices beyond the traditional moral framework of vegetarianism or veganism. Indeed, in her view, this was precisely the split perspective that the new vision enabled. What the incident really highlighted, one might read Eilidh as saying, was the fact that the choice to become vegetarian or vegan now operated in two very different relative scales of value. On the one hand, there was the scale already familiar to them; on the other hand, there was a new scale to be defined in partnership with mainstream supporters. Colleagues would have to learn to switch between them.

As we have seen, for most colleagues the normative scale of value was typically contained within the parameters of a vegetarian or vegan arts of existence. The range of possible acts or choices within vegetarianism and veganism were what largely occupied the attention of animal people, with an in-built assumption that the moral subject aspired to eventually become vegan. Within veganism itself, that relative scale allowed for different levels of commitment to be constantly displayed. This included the choice between simply following a vegan diet and embracing a whole vegan lifestyle. Commitment could be demonstrated by diligently scanning product labels in shops for clearly marked animal ingredients but on another level by also investing time and energy into researching hidden chains of animal derivatives. Similarly, some might just commit to investigating the other products that a company manufactured and sold, while others might commit to additionally checking the range of products manufactured and sold by that company's umbrella corporation.

If each act indexed the freely made choice of the moral agent, then it appeared possible to keep ratcheting up the scale and sense of that freedom too. As well as the progression normatively marked by the shift from vegetarianism to veganism, there was a whole host of micro-shifts to consider and an almost dizzying number of accompanying decisions to make. Obvious examples included the choice of leather and wool in the wearing of clothes or in the use of furniture or bedding. But there were also possible choices attached to watching television or going to the cinema, especially where animals might have been used for entertainment purposes. Likewise, the choices attached to personal grooming were not necessarily restricted to the issue of whether cosmetics had been animal tested; colleagues knew that there were further possible choices to be made around care products that contained elements of beeswax, lanolin, keratin, or musk. In terms of household cleaning products, the same kind of breakdown could be made. Choices existed around whether to use detergents, bleaches, and washing-up liquids that contained animal products but then again around whether to use cleaning agents that included harder-to-trace elements such as caprylic acid (sourced from milk) or oleyl alcohols (sourced from fish), as well as ingredients without explicit animal references such as tallow (i.e., rendered beef fat). The list of choices sometimes seemed inexhaustible, and, as colleagues often pointed out, there were always new kinds of choices emerging.

Nevertheless, there were clearly defined limits to that scale. By contrast, the relative scale of value operating within the planned mainstream movement seemed both massively enlarged and at the same time far more indistinct. Although evidently drawing on models of consumer activism and decision-making taken from their vegetarian and vegan practice, colleagues recognized that in this case the choices or acts concerned worked across those boundaries. There was a marked redefinition of the kind of intentional action assumed to index moral commitment and freedom. Most dramatically, as the terms of debate between Fraser and Eilidh over the animal group's relaunch luncheon illustrated, that now included choices made within the parameters of meat eating. For instance, it was held that commitment in a mainstream movement could just as well be demonstrated by a decision to switch from consuming factory-farmed meat to consuming high-welfare-sourced meat products, or by a decision to shift from buying battery-hen eggs to buying free-range or organic eggs. The new vision also required that one admit into the mainstream movement someone who continued wearing leather but decided to stop buying cosmetics that were animal tested, or parents who still took their children to the zoo but committed to the family eating less meat in their weekly diet. These decisions' admittance into the moral orbit of animal protection disturbed many colleagues, especially where they appeared to decenter the primacy of the choice to become vegetarian or vegan or at least pushed that choice toward the end of a newly stretched spectrum of ethical choice. However, at the same time, most recognized the cumulative value for animal welfare of these shifts and the value for the individuals concerned in beginning to orient themselves in some fashion toward the protection of animals.

Instead of sitting in judgment, Eilidh reminded us in one team meeting, the task of the animal group was now to help members of the public identify their own animal-friendly "set of self-ethics." That fact was demonstrated not just by the ways in which the group supported and encouraged individuals to make such decisions but also through the organization's hiring practices. For among the new staff members appointed by Eilidh, there were at least a couple of unabashed carnivores. Most notable among them was Fraser. As well as eating high-welfare meat, though usually at a respectful distance from the office, the fundraising director sported leather shoes and tweed jackets. Almost as provocatively, he came from a background in conservation. Unlike Euan, whose nonchalant vegetarianism unsettled colleagues precisely because it sat within the conventional parameters of an animal person's ethical choice, Fraser was definitively locatable outside of the animal movement. In this regard, he was not just a spokesperson for the implementation of an inclusive vision but also its living example within the organization. Indeed, Eilidh left me to assume that this was an appointment she deliberately engineered. If the group was truly to become a mainstream movement, it needed to not only positively engage the mainstream and hence affirm the new vision but also contain the mainstream within its swelling ranks.

There was no doubt that for some colleagues Fraser presented an uncomfortable presence (see Reed 2017a). Soon after my first day in the office, I remember one of them whispering to me in an aghast tone that now some of the team "were not even animal people!" Eilidh, I think, would not have been entirely displeased by such a reaction. That tension certainly disguised some of the commonalities and continuities between the two regimes or scales of value. As already discussed, in both instances there seemed to be a tacit agreement that choices were intentional acts and that regardless of how those choices were defined, they could be taken to index the freedom and individuality of the moral subject. However, the differences also remained striking. For example, while the terms and rationale for choices made within a vegetarian or vegan existence were clearly understood by colleagues, laid down by long-established discussion within the animal movement, this was not at all the case for choices made within the mainstream movement. Furthermore, the emphasis on self-advocacy (see chapter 3) and the setting of "self-ethics" seemed to suggest that the determination of what constituted an animal-friendly action was entirely in the hands of the moral subject. There was no obvious external arbiter or reference point to monitor choices. It was true that the role of the animal group was to support the animal-friendly choices already made or about to be made by individuals and to encourage a consideration of more choices, but the organization was not there to prescribe what those choices should be or where that journey of decision-making should end.

At times this led some colleagues to accuse Eilidh and other keen supporters of the new vision of something akin to moral relativism. But, for the most part, colleagues were able to process the split perspective introduced by that organizational vision by appealing back to the tradition of pragmatism. As moderate animal activists, they were well used to weighing the pros and cons for animal welfare of any given strategic plan or policy, and they were familiar with the notion that the optimal benefit for animals might result from a compromise or sacrifice of absolute values (see chapter 9). In fact, it was that very principle that informed their reflections on what they regarded as the most obvious limit to moral action within a conventional vegetarian or vegan practice of ethical self-cultivation.

I am referring to the decision-making process around the consumption of medicines. Even though there was still no labeling information provided on medical products in the UK, all colleagues knew that the vast majority of medical treatments were at one point or another tested on animals. So, the choice to take medicines was a clear infraction of the do-no-harm ethos behind their vegetarianism and veganism. Colleagues recognized that fact—a widely reported dilemma within the wider animal movement—but nonetheless consistently made the decision to take medicines. "If I need to take that medicine to keep me healthy, even though it is tested on animals," Craig, for instance, observed, "then I will do." For, Craig continued, "my personal health would override [animal welfare concerns]; it would win the moral dilemma." While Craig acknowledged that some animal activists have called for people to stop taking medicines either identified as tested on animals or without attribution, he entirely dismissed the prospect. "It's not going to be very beneficial for animals," Craig responded with a wry smile, "if all the people who most passionately care about them are dead or too unhealthy to do anything!" In this case, then, the usual terms of ethical choice got suspended to ensure that the moral subject could keep doing the right thing, making other animal-friendly choices and generally enacting an ethic of care.

From Habit to Awareness

But there are, of course, other ways to interpret or read the freedom and moral agency of the animal activist. Davé (2017, 37), for instance, suggests that within the animal movement, subjects often experience that emphasis on choice, and especially the choices involved in an ongoing commitment to vegetarianism or veganism, as a form of domination. Davé particularly speaks of a "tyranny of consistency," which places constant pressures on the vegetarian or vegan not just to keep making the right choices but to iron out inconsistency or contradiction (2017, 37; see also Davé 2019, 2023). That demand is felt to be exercised by forces outside the animal movement—for instance, through the casual interrogations of friends and family or more broadly of mainstream society—yet also from inside it (2017, 38): "The ethically otherwise [Davé's generic term for the animal movement and other nonmainstream movements of ethical reform] . . . is a closet tyrant itself, always demanding more, asking, 'if this thing, then why not that thing . . .,' until eventually all the world can potentially make a moral claim on it" (2017, 38). For Davé, that dilemma is a straightforward result of the fact that despite its oppositional stance, the animal movement necessarily "still belongs to, and emerges from, the world as it is." Activists are therefore caught up in its moral logics, an observation that leads Davé to propose an alternative ethics that aims to escape what Davé regards as the trap of moral expectations such as consistency (see chapter 9). However, for the purposes of this chapter, I merely want to register the ethnographic observation that choice-making can also appear to "exhaust" animal activists (2017, 38) and further that it can be experienced as something rather closer to a state of patiency (i.e., something that happens to and in this case coerces the moral subject) than a state of agency.

Alternative readings could also be located within the practice and vision of the animal group that I worked with. These might not challenge the premise of freedom associated with choice-making directly. Nevertheless, those organizational interpretations did sometimes complicate an understanding of the relationship between moral agency and intentional action and between individual outlook and collective response. By way of illustration, let us return to the metaphor of addiction, earlier invoked by Euan to explain mainstream attitudes and to justify the assumption of "self-ethics" at the heart of the animal group's new mainstream movement. I particularly want to focus on Euan's claim that "getting them [i.e., mainstream publics] to realize things themselves" was crucial to the ambition of changing individual moral behavior.

There is a clear and obvious resonance here to ideas of consciousness-raising and to popular slogans such as "the personal is political," which have exercised a long influence in the history of social movements, especially in the United States but also in Britain. Indeed, at the time of the animal group's relaunch in 2010, all eyes were on the recent spectacular electoral success of Barack Obama, who ran a whole presidential campaign explicitly drawing on that tradition. The centrality of Obama's call in 2008 for American people to "be the change," alongside the innovative use of social media and the internet in rallying support for that message, had received a great deal of prominence. In the UK it was often cited as a model for new ideas of movement building not just in politics but in the not-for-profit or third sector as well. Eilidh, for instance, regularly stressed the example of Obama's campaign, which she saw as a direct inspiration for the new ethos of positivity that she wanted to instill in the organization and to make a defining feature of the mainstream movement. While other colleagues might not have entirely shared her level of enthusiasm for that project, they too embraced consciousness-raising as a model for their animal activism. This was most notably the case through the choice to become vegetarian or vegan, which was frequently invoked as the most obvious expression of that commitment to be the change.

As Keane (2016, 198) elaborates, a vital component in ideas of consciousness-raising is the assumption that societal ethical transformation is best achieved by "demanding a shift from habit to awareness." It is the daily life and unexamined ordinary acts, assumptions, and choices of people that are usually presented as the chief object of reflection. In the American feminist movement, for example, Keane tells us that consciousness-raising was squarely centered on that work of problematization. Keane argues that concern fell on making the personal or everyday explicit and knowable, able to reveal its link to abstract categories of ethical judgment within feminism (such as "patriarchy" or "sexism") and hence amenable to reshaping or alteration (2016, 188). "Consciousness-raising," Keane explains, "was meant to form a bridge between change in individuals, on the one hand, and large-scale social transformation, on the other" (190), but with the crucial distinction that "the process started in the reverse direction, from the personal to the social" (191). In fact, Keane presents this move as a form of "exculpation," essentially grounded in the adoption by feminist subjects of a "third person perspective on their own lives" (194). It was typically manifest, for example, in the ability to recognize habitual action or personal experience as oppression, or in the individually felt but shared communication of powerful feelings (such as a previously unacknowledged but justified anger), which were taken to result from that realization.

In the case of conventional animal activism, the processual movement from the personal to the social usually took a slightly different route. For unlike within feminist consciousness-raising, the oppression that was taken to underscore habitual action had its chief consequences beyond that habitual actor. It was suffering inflicted on other kinds of subjects altogether that mattered most, and while those nonhuman animal subjects might well be sentient, they were not evaluated as in need of having their consciousness raised. Indeed, in the terms of normative after-the-fact explanations of animals as moral patients, that kind of consciousness or capacity for realizing the need for change was precisely what was absent (see chapter 1). Moral patients, to requote Regan, "lack the prerequisites that would enable them to control their own behaviour in ways that would make them morally accountable for what they do" (2004, 152). So, in contrast to the participants in the feminist consciousness-raising groups that Keane reflects on, who learned to identify themselves as oppressed and others as oppressors, participants in the animal movement generally aimed to exculpate by enabling animal people to purify themselves of their own previously unexamined agency to do harm. Strong feelings generated by the shift from habit to awareness principally operated, then, in the moral subject who self-identified as witting or unwitting oppressor rather than in the moral subject identified as oppressed. Adoption of a third-person perspective by the animal activist did occur, typically expressed

through abstract categories of ethical judgment peculiar to the animal movement such as speciesism or through common critiques and growing awareness
of the way assumptions of human exceptionalism or human dominion played out
in their lives. However, the resultant feelings from consciousness-raising, which
included anger or outrage at the welfare consequences for animals, were just as
likely to be guilt and a growing other-oriented empathy or moral concern.

A shift from habit to awareness also suggested an increasing sense of responsibility. As colleagues' knowledge of the links between their unexamined everyday lives or habitual actions and animal suffering expanded, so, it was assumed,
ever new and unexpected responsibilities would continue to emerge. Once again,
this was a moral journey in large part played out through the choice to become
vegetarian or vegan, which in turn significantly relied on forms of consumer
activism. While consciousness-raising might start with a decision to stop eating
meat or dairy, those choices were just the beginning of the realization process.
As well as the progression normatively marked by the shift from vegetarianism
to veganism, colleagues typically discovered or found themselves made aware
of the oppressive or harmful dimensions of an increasing range of consumer
products and linked habitual actions. Various vegan societies and animal protection organizations such as PETA might produce "cruelty-free" lifestyle guides,
which certainly facilitated the process of becoming aware of responsibilities, but
the expectation remained that there was no straightforward endpoint to animal
peoples' shift from habit to awareness.

Keane, as in the author's commentary on ethical life more broadly, wants to
stress that this work of consciousness-raising does not simply operate within the
person. For Keane, that process is not simply about the strivings of self-fashioning. It is not even just about the transformations in first-person perspectives
wrought by adopting a new third-person perspective on one's own life; a move
that certainly complicates the image of moral agent and an understanding of how
that agency unfolds. Rather, it is also crucially informed by the second-person
perspective that the moral subject achieves through interaction. The legitimacy
of a feeling like anger and the growing awareness within the feminist subject of
alternative ways of being a woman, for example, required recognition from others. Indeed, Keane (2016, 196) points out that within the traditions of feminist
consciousness-raising, it was expected that those feelings were vitally cultivated
through a shared articulation within the consciousness-raising group.

Colleagues at the animal group made very similar observations. As well as
adopting a third-person perspective that included a vital sense of the (nonhuman) lives impacted by their own lives and of course acknowledging the second-person perspective achieved through making contact, individuals regularly
referenced the work of mutual recognition between animal people that for them

was at the heart of the animal movement and their animal protection organization. In fact, for several colleagues, the vital transition from vegetarianism to veganism was absolutely achieved only upon joining the animal group. This was partly because the workplace was set up to support and affirm that shift, not just ideologically but in a range of practical ways too, from the organization of a staff kitchen that operated on the assumption that it was a meat-free and optionally non-dairy space to the organization's animal-friendly procurement policy. But it was equally acknowledged to be the result of simply being in the constant company of fellow vegans, whose support and knowledge provided the final impetus needed to make that commitment happen. Joining the organization, then, was presented as a further education in the raising of a colleague's awareness, drawing their attention to innumerable previously unexamined links between habitual behaviors and the exploitation of animals.

Despite the antagonism expressed by some colleagues toward aspects of the new vision, most notably its downplaying of the traditional emphasis on animal suffering (key to the way in which consciousness-raising and a third-person perspective on their own lives developed within animal activism), it was clear to everyone that a second-person perspective on the lives of mainstream subjects would also be crucial to the vision's success. Most obviously, Eilidh envisaged that after the creation of a mainstream animal protection movement, the group's role would shift toward a kind of recognition service. Colleagues would become responsible for recording and positively responding to each small animal-friendly change made by mainstream subjects and at the same time for mediating the process of mutual recognition between those subjects (most obviously via communications hosted on the animal group's various social media platforms). In fact, the new vision of the organization anticipated that these self-advocating subjects would not just belong to a mainstream movement but also eventually begin to naturally constitute their own mutually affirming and self-advocating animal-friendly "communities," or consciousness-raising groups. Over time, these communities would enter partnership with the organization, which would continue to provide guidance and to offer support and recognition but from a position of increasingly stepping back from the work of advocacy itself.

Being Good

We might recall that Euan's vegetarianism caused a certain amount of consternation among his colleagues in large part because Euan insisted that it was not about being good. In fact, if we take the consciousness-raising paradigm into account, Euan's stance clearly also resisted any attempt to link that vegetarianism to a transition from habit to awareness. Although we have examined how the consciousness-raising movement itself complicated the issue of the free or unfree status of human behavior—for instance, through the cultivation of a retrospective third-person interpretation of one's previous life and choices as a moral subject—I want now to consider another instance of certain conditionalities being attached to the freedom of the moral agent. For at various times, other colleagues were prepared to question whether their own choices were free willed. They sometimes expressed skepticism about the motivations behind those ethical choices, even to the point of distinguishing between choices on the very basis of their status as truly intentional acts. On occasion, these reflections threw up the possibility of internal differentiations within the agentive condition of the moral subject, often considered by invoking shifts in perspective on the attributed source of doing. At other times, the reflections depended on highlighting what were taken to be essential distinctions between the moral nature of humans and that of other animals. But in this section, I want to explore those musings by looking particularly closely at the self-narrated ethical journey of one colleague, the organization's IT officer, Iain. Indeed, my inquiry is partly prompted by a question Iain once raised in conversation: Do we choose to be good?

Of course, such questions have provoked long histories of debate in moral philosophy. They have also inspired twists and turns in forms of psychological explanation for moral action. One might acknowledge the attempt to demonstrate that grounding categories in moral philosophy such as free will, acts of reasoning, and self-awareness may in practice be irrelevant or illusionary when considering why subjects are good or act ethically. Take, for example, those third-order explanations that have emerged from experiments to test the micro consequences of situational differences for moral decision-making, a field of inquiry also examined by Keane (2016). But what interests me is the way in which colleagues identified with or switched between such orders of explanation themselves, including between the first-person perspective of the moral agent and explanations that superseded that perspective entirely. As well as looking at the dynamic oscillation between choices at one moment envisaged as self-willed and at another moment presented as predetermined, this section and the next look toward a returning concern with the assumption that choices were also in some fashion co-constructed—for instance, through the well-rehearsed principle of making contact.

My focus here obviously borrows something from the methodological approach that has come to dominate anthropology's "ethical turn." I am particularly thinking of the late Foucault-inspired concern to trace specific cultures or histories of ethical self-fashioning and the shape of freedom within those identified socio-ethical regimes by which the reflective subject measures and experiences choice. In fact, opening a space to ethnographically describe freedom is listed as one of the necessary ambitions behind that ethical turn (see Laidlaw 2002, 2014; Faubion 2011). It is also presented as a differentiating mark from the previously dominant Durkheimian-inflected anthropology, which is said to be principally concerned to explain how moral obligations get socially reproduced. Likewise, it is taken to be what distinguishes the turn from a Kantian moral anthropology grounded in a freedom exercised through rational argument. As Laidlaw highlights, in recognizing and describing freedom, the insistence remains that moral subjects do not naturally seek to act in accordance with reason or self-interest and that their choices are never helpfully conceived as absolutely voluntary acts—that is, as operating without constraint and beyond influence (Laidlaw 2002, 323). Yet within this Foucault-inspired methodology, that freedom is closely associated with deliberate thought and reflection (2002, 324). For it is during those moments when subjects consciously withdraw from action to meditate on it or to problematize behavior that it is assumed specific freedoms become practiced and articulated. As we will see, there were moments when colleagues at the animal group did appear concerned with the issue of how their moral stances and choices got socially reproduced and when they offered accounts of moral life somewhat akin to Durkheimian-inflected explanation. But I want to begin with an investigation into those occasions when colleagues practiced and articulated specific constraints on freedom or refigured moral action as *really* self-interested.

Iain grew up in Loanhead. As he explained, it was one of several small Midlothian towns just south of the Edinburgh bypass, originally founded on coal and shale mining. Nowadays, though, Loanhead was more familiarly known for being located near the Ikea superstore and accompanying retail parks. Iain joined the animal group in 2008, appointed to be the organization's first IT officer. His responsibilities included maintaining digital systems as well as the group's digital communications; this involved the design and production of the website, social media accounts, electronic petitions, appeals and newsletters, donation buttons, and campaign action applications. Initially, Iain had taken the position as a part-time post, intending to devote the rest of his working hours to contract jobs as a self-employed IT and design consultant. But in 2010, Eilidh had persuaded him to join the team full time. By then in his midthirties, Iain saw the invitation as a once-in-a-lifetime opportunity, the chance to marry his professional skills with a moral cause that had long been close to his heart.

Yet in contrast to many other long-standing colleagues, Iain had no previous background in animal protection or even any active experience in supporting campaigns run by any of the UK's diverse animal groups. Instead, his past commitment had been solely measured by private choices, most centrally by the decision at the age of nineteen to turn vegetarian:

Do you have a history of being involved in animal protection issues?

No, not really. Just my own lifestyle, but no structured, formal campaigning as such.

Why did you decide to become vegetarian, then?

Well, when I was a kid, I was quite uncomfortable eating meat. Probably from the age of about twelve or something, because I knew what I was eating. And I loved animals. So, if I was given lamb, I would just think that that's a lamb; it's been sitting in a field. . . . And I used to not eat my food. My mum and dad, and my sister, would leave the table, and I was meant to stay there until I finished, but what I used to do is I used to just get the meat and put it under the stand-up piano [we both laugh].

It was just the meat you were not finishing?

Just the meat, yeah. Then I would sometimes stuff, like, [beef] mince into my pockets and go outside. When I went out to play football, I would just chuck the mince out in the field.

This was, like, twelve years old?

Yeah. So, I was going through a bit of a funny phase there. And then I sort of started eating meat again, but not really enjoying it that much. I didn't really have a varied diet when I was a kid. It was quite a traditional Scottish diet, mince and tatties [potatoes], stew and steak. I think we first had a curry when I was about fourteen.

Right.

So, I started eating chicken and fish. I always liked fish. But when I went to university, I ate less and less meat because I had to sort my own food out. Then there was one night. Me and my girlfriend at the time, we were watching this film about animal cruelty, a documentary about some farms. There was footage of, like, a cow being lifted by a crane and the crane breaking and this cow being dropped onto her legs and the legs kind of snapping underneath her. We both decided not to eat meat again, and I've not eaten meat since.

That was at [the University of] St. Andrews?

Yeah. So, it was just an awareness of the cruelty. It was just due to watching that one program. I was like, "All right, that's it! I've had enough."

So it wasn't that you started to dislike the taste? It was more that you could see the animal behind the slab of meat, so to speak?

Yeah, yeah. I mean, I did really like fish and steak and things like that. I enjoyed it, and I think before I put the cruelty thing to the back of my mind. I certainly missed the fish. I gave the fish up about four months after everything else, actually. I found it really hard to let go of fish [Iain chuckles].

In terms of the ethical issue for you, is it that you just don't want to eat an animal, or is it that you don't like industrial farming methods?

For me, I wouldn't want to eat an animal again. But I personally don't think there is anything morally wrong with a human being eating an animal because, to me, other animals eat other animals. We have evolved to eat other animals. It's just the nature of the production process and commercialization.

So, it is industrial farming.

I think it's completely abhorrent. But I wouldn't say to someone who has a little bit of land and has got some chickens which have a good life and who kills them in a humane way and eats them, I can't say that's morally wrong, even if I wouldn't do it myself.

And you wouldn't do it because . . . ?

Because it's like a personal pact you make with yourself, isn't it. You make a decision on what's right for you. I can survive without killing an animal because a vegetarian diet is perfectly fine for me. So, if I can go through my life without actually killing an animal, then I will do that.

In this dialogue, Iain does relate a transition from habit to awareness, partly mediated by a documentary film that he watched with his girlfriend. But the emphasis is consistently on a decision he made for himself; indeed, Iain speaks of his animal activism as a "personal pact." It was especially the initial choice to turn vegetarian that shaped a sense of moral commitment, which, Iain stressed, continued to unfold. In 2008, for instance, Iain decided to become vegan. As for several other colleagues, this decision was partly motivated by joining the animal group and being reassured by vegan colleagues that veganism was a viable option, both practically and nutritionally. Those colleagues helped Iain feel that the decision was a logical next step, a realization closely tied to Iain's growing familiarization through work of the realities of dairy and egg farming. That decision had in turn led him to think about further choices that might confront him in the future—for example, the question of how he would raise children; Iain and his wife were then expecting their first child. Iain speculated that they would raise any children as vegetarian rather than vegan, just to ensure that they "didn't muck around with the development of their bodies." Likewise, Iain had become more committed to actively avoiding animal by-products in his everyday consumer lifestyle.

Iain remained wary, though, of turning his choices into criticisms of other people's behavior. This was partly because his choices were first and foremost *personal* decisions. But it was also because Iain was not entirely convinced that ethical choices were always freely made.

The IT officer's musings were prompted by discussion of a contrast very commonly drawn in the office, between the free will and hence unique moral obligations of humans and the absence of free will and corresponding lack of capacity for moral reflection among other species. "I mean, most animals don't make conscious choices," Iain told us; instead, they make "choices based upon what evolution has taught them. . . . I see animals as innocent." The latter claim was clearly connected to familiar after-the-fact explanations of moral agent and moral patient. It also invoked what sounded like a traditional Christian doctrine of innocence. Iain was very much aware of the irony of that undertone, especially given his own recanting of a Scottish Presbyterian upbringing and the suspicions that he shared with colleagues about the role of Christianity in animal exploitation. While describing himself as "probably agnostic," adding with a sly smile that "I won't say that there is definitely no God, because there could be a wee one," Iain persisted with this line of thought. "The majority of animals don't make a conscious choice to do wrong," Iain insisted; "they don't have free will, so they can't be criminals, they can't sin, I suppose." The use of words such as "most" and "majority" left open the possibility that some sentient critters would in the future be recognized as moral agents; Iain suggested, for instance, that it might be possible that chimpanzees could understand a difference between right and wrong. However, in conformity with the strong consensus in the office, Iain disavowed the general idea of ever holding other animals morally to account for their actions. In fact, justifications for *the* ethical choice (i.e., to become vegetarian and then vegan) were often couched in these terms. As Iain explained to us, "A fox needs to eat a chicken in order to survive, whereas we have many choices of ways to survive that don't involve keeping other animals captive or giving them a poor quality of life." In this respect, only humans have freedom.

But having made that distinction—between free-willed humans and other animals who cannot be held to account because they lack conscious choice—Iain then sought to explore the distinction's collapse. In doing so, Iain's concern was not to liberate animals from arguments of human exceptionalism (for instance, by allowing them entry into a moral realm of responsibility) but rather to cast doubt on the conditions of that human freedom. Iain did so by returning to the issue of survival and by reinvoking a third-order folk explanation, this time to challenge the motivations behind choice or moral decision-making. "Well, we [humans] tend to think around things which are possibly just evolutionary things," Iain posited, "you know, like why are we good to other people?" The question led him back to Christianity and to a historical understanding of religious and ethical forms. "I suppose religion came along," Iain continued, "and said, you know, do unto others as you would have done unto you, but that's just a simple evolutionary thing, isn't it." Puzzled, I asked him to explain. "Well, you have to sort of coexist with your species, cooperate, otherwise you could be left out by yourself and die," Iain responded, "so I suppose things like religion and ethics elaborate on what are primarily an evolutionary necessity."

Iain then turned the argument to more immediate self-reflection. "I think of my way of living, where I try to be good to people," Iain offered. "Now, you think you are doing that because you are a good person, but, you know, you need to be good to get good back." Iain turned the argument again to question his motivations from a different angle of self-interest. "If you look deep down," Iain told me, "you can probably see that you are benefiting because you feel good inside, from having done it." This further suspicion led him to suggest a new measure for selflessness, to be defined as an act "where you can do good but not have time to go over about how good an act it was." While Iain hoped he wasn't being good just to feel good about himself, he admitted that when he studied his actions and the actions of other people around him, it often seemed so. What interests me is the dynamism of that switch back and forth—not just the constant recalibration of the same action as either self-interested or other-interested, which carries the corollary concern that only other-interested action is ethical (as Laidlaw points out, itself a peculiarly Christian view), but also the possibility that the moral agent was defined by that internal tension and movement, or even that moral action was simply both self-interested and other-interested. For even if it was not the case that people did good to feel good about themselves, the same hidden, unconscious motivations behind choices remained. "I would say that a lot of the time it does come down to an evolutionary principle," Iain reiterated. But this was not just because we have to be good to get good back. With a slightly altered emphasis, Iain threw in a further basis for the human distinction from other animals. "In another species," Iain concluded, that survival "might depend more on physical strength, ability to hunt, etc., but for us, it's a lot about being likeable . . . because to survive, people need to like you."

The latter claim introduced a wider unanswered question. For although Iain generally talked of ethics and of what prompted individuals to be good or to do good, he always seemed to specifically refer only to moral actions toward other people. The examples listed above notably avoided instances of individuals being good to nonhuman animals. Indeed, in our conversations I was never entirely sure whether those musings also referred to that decision. Neither was I clear by what terms being good to other animals could be understood as self-interested (other than for that benefit of feeling "good inside") or as conforming to an evolutionary principle of likability and hence in some manner predetermined.

In the office and among colleagues, Iain's own likability rested in part on his much-commented-on sense of humor. As well as the delicious homemade date balls that he made and regularly distributed at work, Iain was renowned for his witticisms, certain dry turns of phrase, and for his love of enjoyably bad puns. It was no coincidence that Iain's first job after leaving university was as a graphic designer for a company specializing in humorous postcards and T-shirts. That same humor, which Iain liked to describe as an ironic stance on the two-line joke style of the British lollipop stick or Christmas cracker, served him well on social media. In fact, Iain enjoyed the discipline of scripting the animal group's Facebook posts and Twitter lines: "Feline bored? Stuck in a rut? Why not paws a while and visit [hyperlink to the group's website] and read our amazing animals A-Z." Indeed, although never made explicit, Iain's thesis that being likeable was an evolutionary necessity threw a different kind of perspective on the role of social media strategist, which Iain shared with his workstation partner, David. Their shared mission could in essence be read as making the relaunched animal group attractive online by precisely targeting an identified compulsion for friending, liking, following, and so on among the various platforms' mainstream users.

But in these musings, Iain's chief concern was not to explain motivations for human behavior online, something he and David were much animated by at other times. Rather, Iain aimed to outline the scope of his own freedom. In a sense, that freedom was demonstrated or revealed by these musings and the conscious choices that both inspired and flowed from them. Yet, Iain also considered the possibility that moral reflection and any accompanying commitment to being good was itself an elaboration of evolutionary necessity, less an index of freedom or of the subject's problematizing relationship to self and more an outcome of the survival instinct in humanity. Indeed, for Iain, the notion that different registers of obscured self-interest might principally drive moral decision-making threw the whole notion of conscious and hence freely made choice into question. At his most skeptical, Iain proposed that goodness could be truly appreciated or identified only in the absence of moral reflection or meditation on choices, a claim that might inadvertently and positively recalibrate the moral status of Euan's stance. Though in this case, Iain's emphasis fell more on choices made without time to think about their moral basis instead of on choices simply made without moral reflection at all.

The idea that ethical choice operated most convincingly when akin to something more like an unconscious or instinctive impulse might lead us back to the attributed source of value feelings in the specific human-animal encounters that often enlivened the biographies of Iain and his colleagues. However, it is also a useful entry point back into one of the early animating debates within anthropology's ethical turn. As previously mentioned, the influence of Foucault's late work on freedom within ethical subjectivation has been challenged by those wishing to draw attention to the tacit, less explicit forms of moral action, taken to be more typical of everyday life (see Lambek 2010a; Das 2010; Mattingly 2014). Here, ethics is said to be grounded as much in unspoken agreements as in freedoms defined by thought and reflection, in practice as much as formulated positions, choices, and beliefs. From Iain's perspective, it was as though that analytical stance was itself a measure of goodness. But there is a further sideways connection that I want to explore.

Much of the ethnographic work informed by this approach tends to describe the ethical dimension of ordinary working and household lives, of familial-level struggles and dilemmas, where musings and the giving of reasons are often eclipsed by quieter, less elaborated but more persistent decision-making and pressing concerns (see especially Mattingly 2014). While Iain's suspicions of moral reflection and of the extent to which freedom was ever truly expressed through conscious choice did not exactly lead him to privilege such ordinary ethics, they did bring him to a renewed consideration of the household and of family as the sites in which those choices often developed and played out. And they did so in at least two directions—on the one hand, from a desire to better understand where his choices truly came from and hence the degree to which they were freely chosen, and on the other hand, from a concern to successfully negotiate the consequences for his social relationships of the ethical choice that Iain knew he had made. For regardless of his ongoing skepticism, Iain ultimately wanted to insist that his vegetarianism and subsequent veganism were a conscious decision or intentional moral action. In fact, Iain claimed that it was the one choice that he could with some confidence assert "I created for myself."

Iain's faith in this assertion was in large part because he grew up in a meat-eating family and within a wider nonvegetarian and certainly nonvegan community.

As Iain liked to point out, there were no obvious reference points in his upbringing. "Like, for me, I didn't know veganism was a possibility until I left home," Iain explained, "because I never came into contact with any vegans until I was about seventeen." This observation was made in the light of a broader conversation about the nature of other moral decisions. Iain referenced, for example, his attitude toward touchstone public issues in Scotland such as voluntary euthanasia and more long-standing debates around abortion and the death penalty, all of which he was prepared to accept might have been informed by inherited values. The contrast was also drawn in the context of attitudes to sex and gender identities that Iain suspected "can be quite cultural." Such reflections on the inherited or cultural status of moral choices further arose as part of arguments Iain regularly deployed when seeking to highlight sets of mitigating circumstances for what could be read as the non-animal-friendly behavior of others. For instance, they often surfaced as an explanation for the continuance of meat eating among family members or among friends and acquaintances. "I don't think people are being deliberately bad to animals when they do it," Iain once told me; "they are just trying to survive in the way they know best, persisting with habits they have grown up with." This kind of claim was often backed up by an articulated awareness of the challenges of following a vegan lifestyle, particularly in nonmetropolitan areas, or of the economic costs incurred when switching one's diet.

However, it was the emphasis on habituality and therefore on the lack of intentionality or deliberate cruelty behind much mainstream behavior and especially meat eating that always mattered most to him. "If I saw Dad punching a dog or something, then that would be a big problem," Iain expanded, "but him eating meat or my friends still occasionally going to KFC or McDonalds just isn't." This remained the case, Iain stressed, even if the cumulative negative effect on animal welfare of such unthinking mass consumption of factory-farmed meat was demonstrably far greater than the long-term consequences of individual acts of cruelty, such as a man punching a dog in the street (see Reed 2024).

The question of how much blame to attribute to the actions of other people one knew, and especially to friends and loved ones who ate meat, arose for nearly all colleagues. Everyone had relatives and many had partners or spouses who were neither vegan nor vegetarian. David, for instance, explained that his wife was a meat eater. "I don't think she has great issues with eating animal products and using them," David told me, "because she is a biologist by training, and so she kind of sees it as a circle of life." David continued to try to outline her perspective: "You know, we are the civilized race, and we have the upper hand. . . . That's probably a crude way of putting it, but she sees it in a more scientific way." However, David insisted that his wife's views and actions didn't cause him any problems. "I know people for whom that does cause friction, you know, results in the breakup of relationships or in lots of arguments," David admitted. But for them it didn't. The reason seemed to lie in David's attitude to confronting ethical difference and in his wife's willingness to precisely countenance small changes: "You know, in my mind everybody has to kind of reach these decisions by themselves. I can encourage people to be more ethical, but I don't think everybody *should* be vegan. Saying that, I do encourage my wife to buy higher-welfare [meat] products, and she does that. You know, she also regularly eats vegan meals. So, I think that if every partner of a vegan who isn't vegan does that kind of thing regularly, then it is making a contribution. You know, I think that's probably the best you can hope for, unless the person is entirely like-minded." As both David and Iain readily acknowledged, there was here a model of sorts for mainstream movement building and for the wider tactical engagement with the mainstream embodied in the animal group's new vision.

Indeed, this was perhaps where colleagues came closest to fully embracing the ethos of inclusivity—that is, at the point where it intersected with the ordinary strategies of moral struggle, with what they had already long been doing at home or in their everyday dealings with friends and family. Iain, never prone to overt expressions of enthusiasm when it came to the new vision, was in this respect quietly hopeful of progress. "Dad," Iain once pointed out to me, "is now eating less meat than he used to, and he's eating higher-welfare meat too, because my mum will buy him that now. So, I am not going to have a go at him." Iain continued by making direct reference to the new vision: "You know, we [the organization] talk a lot about how people making small changes can collectively make a big difference. . . . Well, to me, a big change has happened already with him [his dad]." Iain reiterated his point: "I feel like my parents have already made quite a leap."

Like David and many other colleagues, Iain was very reluctant to appear too preachy at home. "I don't tell others what to do," Iain insisted. This was partly because it was not part of his nature to do so but also because Iain thought such preaching was counterproductive. "You know, people fall out very quickly if you start telling each other what to do," Iain advised. However, Iain was willing to stand as an exemplar of sorts. "I do try and market the lifestyle, the choice I have made," Iain explained. To friends and family, for instance, Iain would often highlight the health benefits of veganism. "I have not had a cold in three years," he would regularly mention in passing, or "I am a lot fitter than I was when I ate meat." His hope was that the appeal of veganism might therefore "rub off on them," a strategy for which he claimed some past success. "You know, an old friend of mine who used to eat meat all the time," Iain offered by way of illustration, "he went vegan for about three weeks." Although the friend insisted that the change wasn't for ethical reasons but rather because he was "training for some martial arts thing," Iain claimed to notice subtle changes in outlook. "Actually, listening to him talking," Iain stated, "I knew it was more than that. . . . I could tell he started thinking about stuff, about animals." Indeed, while the friend did return to eating meat, Iain noticed he now consumed a lot less—that he had, for instance, a lot more tofu in his diet. "So that's a massive difference," Iain asserted, especially "if you take it over somebody's lifetime." Iain claimed to have observed similar changes in other friends as well and to be partly responsible for subtle shifts in their attitude, which taken collectively might constitute a mobilization of sorts.

So, while shot through with doubts about the extent to which his doings indexed freedom and with doubts about the motivations behind his own strivings to be good, Iain exhibited a quiet confidence about his ability to influence the choices of those nearest and dearest (at least in terms of dietary change). Indeed, his suspicion that ethical behavior might be determined by evolution or by culture or by the choices of his parents, and hence that most of his moral decisions were prefigured, did not seem to impact his faith that in terms of veganism Iain could be a prompt or example for small change in others. That effect, claimed by Iain, was in a sense the strongest proof that this choice (i.e., to be vegan) was authentically his own, perhaps the only truly self-willed moral action.

Me and Caley

Throughout our many conversations, Iain consistently argued that other animals should be treated as counterexamples in any discussion of human freedom. For him this was first and foremost because they were survival driven and hence incapable of making conscious choices in a moral sphere; even if sometimes predetermined, human choices were still identifiably moral in dimension. However, Iain often ended his musings on ethical choice by suggesting that encounters with sentient others inspired his decision to become vegetarian and then vegan. Specifically, Iain regularly invoked the experience of his relationship with Caley, a twelve-year-old ginger cat Iain had adopted many years back from a local rescue center. In fact, that relationship was the subject of several posted reflections. On one occasion, for instance, Iain shared with me the draft of a guest post that he intended to submit to the weblog of an online business specializing in sourcing and selling "ethical pet products." In part conceived to promote awareness of the animal group's new vision and its plans for a mainstream movement, the script was composed in Iain's inimitable social media style:

> Caley is a rescue cat who came to live with me when I was a single carefree young man. I suppose she was my de-facto girlfriend—well she certainly had misgivings about any new women in my life and would make her feelings clear by leaving a "welcome gift" at the bedroom door.
>
> Was she jealous? Or was she just protecting me? I thought I was looking after her, but she probably thought it was the other way around. Indeed, there are many occasions where she has been a nurse to me, albeit a nurse who licks my arm with her raspy tongue and meets my eyes with a knowing look.
>
> I believe cats can read minds. Really, I do. I also believe that when a cat forms a bond with you, they feel what you feel. This is also true in human-human relationships, but our many layered conscious thought streams get in the way. Babble. Babble.

Although his tone might be humorous, Iain intended to make a serious point. Instead of regarding our companion animals as vulnerable, just in need of human protection and care, Iain wanted to suggest that we should consider the mutuality at the heart of positive cross-species relationships. It might be hard to read signs of moral reflection in Caley's behavior, but, Iain countered, we might certainly recognize an independence of will or being and reciprocity of sentient feeling. Iain meant to imply that at some level this could be understood as a form of co-constructed ethical regard. Certainly, in his draft post, human-to-human relationships suffer negatively by comparison, precisely because of "our" capacity for endless babble or the interference caused by layers of conscious thought.

"There is communication," Iain once told me. "There is recognition between species of another living being, which is quite a powerful feeling." Iain continued, "That connection is one of the things that made me become vegetarian." The observation returned him to the question of him and Caley. "With my cat," Iain mused, "I feel there is definitely a strong bond between us. I wouldn't want to call it friendship, but, you know, you look at each other's eyes, and there is contact." That invocation of a second-person point of view, the sense of in some way being addressed by Caley and of Caley being addressed by Iain, redirected his explanatory perspective on moral action (see chapter 7). Instead of the first-person outlook of a moral agent who enacted a choice that "I created for myself," or the retrospective outlook of that agent who considered that they might in fact be acting out of predetermined principles, Iain oriented us back to the essential interaction of making contact. "I know that she is pleased to see me when I come in the door," Iain noted, a response that Iain felt confident was "not just because she wants food." As Iain reiterated, "I feel that there is recognition of a kindred spirit, maybe not another person but another being on the journey of life."

Do No Harm

While other colleagues would not necessarily feel the need to add some of Iain's caveats—for instance, his note that Caley was not quite a person, more kindred spirit than friend—they would, I think, respect the rationale behind the caution. Indeed, colleagues remained wary of some of the consequences of highlighting cross-species mutuality, especially where it led to the inference that responsibility or blame might be attributed beyond the human. For them, despite the experience of making contact, there was ultimately no ambiguity about the fact that the moral load or burden must fall exclusively on "Us." To suggest otherwise, colleagues regularly advised me, risked diverting attention from their core obligation: to improve the lives of animals, as they were impacted by human action.

That emphasis connected to what I have already highlighted as a key principle of animal protection, the commitment to do no harm or to what we might term "positive inaction." It also informed the sometimes-utopian ambitions of noninterference and the various scenarios that colleagues sometimes associated with it. For instance, after a holiday visit to Majorca, an island famous for its large colonies of feral cats, Iain returned to work energized with the virtues of letting domestic animals go wild. Partly prompted by a story that local authorities on the island and some of its residents wanted to introduce managed culls, Iain mused on the current situation as a future model for what leaving nature well alone might look like. Other news stories could spark similar thoughts. When an outcry on social media broke around a report that a large red stag had been shot on Exmoor, some colleagues reacted with fury. "Why do we feel we have to manage these animals!" Iain retorted. "Other animals are fortunate enough that we don't feel we have to manage them; the choice is so arbitrary!"

Here, as so often with animal people that I knew, the principle of do no harm was crystallized through a negative comparison with the actions of conservation (see Reed 2017b); in this case, the stag was shot as part of a local wild deer management plan. "We spend too much time meddling with other species," Iain elaborated, "thinking that we are somehow helping the environment." Addressing his concerns to me and David, Iain launched into a longer rant against conservation, by then quite familiar in form:

> I think they [conservationists] are just like landscape gardeners, because they are trying to mold the countryside how they want it to be. I just don't agree with that. Imagine some other species coming in and meddling with the way we are. You know, for example, going to India and saying, "We are going to have to kill four million Indians because Calcutta is a hell hole!" Well, I don't think we should do that either in a

forest just because all of a sudden there are held to be too many badgers. You know, nature has to be allowed to take its course. A badger population will only grow so long as there is sufficient food in that woodland to feed them. Eventually, nature sort of balances everything out. But what we are trying to do, what *they* are trying to do, is a very direct way of manipulating the environment. At least when people eat meat it's more indirect, more natural in a way, because they are not doing it just for the sake of making the landscape or the ecosystem look the way we want it. With meat eating, it just so happens that the land being overfarmed is a side effect.

Such comments placed conservation at the center of human wrong thinking, in some ways more immediately culpable for animal suffering than the habitual behavior of the meat-eating mainstream public. Colleagues argued that this was because conservation killed with such managed purpose. Indeed, the chief objection to conservation was often the arbitrary nature of its interference. This included the impression that ecological concerns, while admirable in themselves, resulted in doing intentional harm to individual sentient beings and to certain species—and most importantly, from an animal protection point of view, the fact that harm was enacted in the name of preserving other species and particular habitats designated as worthy of conservation protection. Colleagues consistently rejected or cast suspicions on those designations, especially where they legitimated culls and other forms of landscape management. The labeling of certain species as "non-native," for instance, was regularly highlighted as an example of conservation mystification. But as Iain's words demonstrate, the underlying problem was consistently perceived to be human interference itself and the animal suffering caused by that intervention. If only humans would let nature "take its course," Iain pleaded, then natural processes would naturally resolve any habitat imbalance or interspecies competition. That resolution might result in species loss and other forms of animal suffering, but at least the loss and suffering would not be human-inflicted, and there would therefore not be a moral problem to speak of.

But, of course, conservation was never regarded as the real enemy. In fact, any animus toward the "meddling" carried out in the name of environmental protection was always partly informed by a frustration that animal groups and conservation groups could not work together more successfully (see chapter 8). The principle of do no harm was instead developed in the historical context of practices such as blood sports and scientific vivisection, though first and foremost in the context of industrial farming. Although both Iain and David stated that they did not want to kill or eat animals in any circumstances, the choice to become vegetarian or vegan was nearly always articulated as a response to the suffering and unnatural deaths of farmed animals.

"The ethical issue for me," David once explained to us as we sat around their workstation, "is that I don't feel animals are there for humans to abuse in that way." A cataloging of specific examples usually followed such statements. "For instance, look at the treatment of male calves; they just slaughter them pretty much right away," David elaborated, "or, you know, the treatment of cows, subjected to being put into small compartments for hours on end whilst they are milked." These observations would then often lead David and Iain back to reflections on the negative impact on human physical and spiritual health of this interference, most specifically as it related to the consumption of meat or dairy products. "My issue with the dairy industry," David told us, "is that biologically while cows do need to release milk, it is meant to be for their young; it's not meant to be for us. It causes a lot of health problems because it's not for us." Like Iain, David then deployed a classic situation-reversal scenario to push his point home: "You know, if an adult female human was giving their milk to other animals, it would probably cause issues there too, because it's not designed for that species either. So that's my issue. We have these endless lines of factories and farms pumping out this stuff and animals being abused in the process, for a product that we don't need, and which isn't intended for us, you know, by nature." The same kind of appeal drew David to reference models of alternative community, where he imagined that the do-no-harm principle and the general ethos of noninterference were already integrated into social life.

More specifically, David invoked the commonly used and largely rhetorical example of Buddhism:

> Most Buddhists follow a vegetarian, largely vegan diet. It's a very civil, compassionate society where people respect animals and aren't doing damage to each other. I think we could take a leaf out of their book. I don't wholly agree with their philosophy of karma, but I do think there is something to be said for eating a vegetarian diet in order to avoid eating something that was traumatized when it was killed. I agree that eating meat from an animal that has suffered trauma can't be good for your system. You shouldn't be ingesting something like that. While I am not totally au fait with the scientific evidence here, I do think that there is something to the spiritual idea that if you eat something that has been traumatized, then it will cause some anxiety in your self.

Iain nodded. As ever, Buddhism stood as a graspable foil to what they took to be the ideological complicity of most Christian traditions (see chapter 1). While neither of them expressed any serious intent in practicing Buddhism or visiting

Buddhist communities, as a tangible embodiment of a viable vegetarian or vegan society, it romantically appealed.

Other colleagues invoked different models of alternative community grounded in different examples of the social integration of the do-no-harm principle. As already noted, it was common for them to suggest that indigenous peoples practiced noninterference (see chapter 1). Indeed, the small, self-sustaining indigenous community, typically in a deep forest environment, featured strongly in the distant-future envisioning of animal people. But here the romantic appeal was often accompanied by more rupture-filled, less human-tolerant utopian visions. Usually expressed in moments of frustration, these visions were prompted by far bleaker assessments of the human moral condition and capacity for change. I am particularly reminded of a conversation with Craig, one of the more enthusiastic endorsers of the organizational relaunch and of the ethos of inclusivity and positivity that defined it:

Is there an animal-friendly future that is important to you?

Yes, but if I am honest, it might be one without human beings.

So, a future in which you are not involved?

I think that could perhaps be the only kind. I mean, ultimately, all of our work is to try and change things that human beings do. We don't campaign to change anything that animals do; it's always humans. So, we are the problem and must be the solution. So, far from an idyllic world for us, my ideal future as far as animals are concerned would be a world where animals are never caused unnecessary harm or suffering. . . . I think for me it's fundamentally to do with respect. I think my fundamental problem with the earth at the moment is that there are far too many human beings.

Overpopulation?

Yes, I think that's clear. So, I think to create a sustainable better world we are going to have to remove, say, nine-tenths of the human beings on the planet, however that is done [nervous chuckles]. It sounds a bit regressive, a backwards step, but maybe with more tribal cultures in it where animals are treated with respect. Animals may be eaten, caught from the wild; I don't have a problem with that. You know, like a South American Indian catching a monkey or someone catching a wildebeest and eating it, because that's part of their culture, that's what they do to survive. Generally, in these cultures I think they kill the animal with respect. But we don't in our society. We treat them as units of production. We hide it all in factory farm sheds, and we process it, by a lump of red meat wrapped in clingfilm and intentionally create this dissociation. We don't want to know. Ignorance is bliss! That's the one thing I have found most frustrating in our society. We are happy for this to happen and to legalize it. We buy the products that keep it all happening, but we don't want to know the reality, and we turn away from the reality, and we hide the reality, and we think it's acceptable and that it is ethically OK. And I think it's absolutely, morally bankrupt!

However, as proper to moderate animal activism, these rejectionist outpourings and posthuman future imaginings nearly always coexisted alongside a more pragmatic constitution. Craig's instinct was to ultimately find a workable compromise—to open up dialogues with animal science, for instance.

Likewise, after sharing his musings on the physical and psychic damage done by eating meat or consuming dairy products, David felt the need to reassure me. "You know, I am not a radical in my views," David stated. "I accept that it is wholly unlikely that one day everybody is going to be vegan." In fact, later in the same conversation, David offered a far more sober assessment of what might happen if everybody stopped eating meat:

I think this is a really difficult question to answer because we haven't had that state yet at any point in history. I mean, other than small pockets of societies, sects, religions. On a global scale, I really have no idea what it would look like. Would people have no pets? Would there be any difference in how we interacted with animals that are wild? I guess there would still be ethical questions because obviously there would be a lot more wild animals if nobody was killing them or everyone was vegan or vegetarian. I assume that farmed or captive animals would have to be released. It would be very interesting to see how frustrated we very quickly got with sheep just roaming across the roads or cows wandering about because they weren't anybody's responsibility anymore. You would be releasing quite a few animals that haven't been in their natural environments for a long time. What impact would all this have on ecosystems? They [the released animals] would probably die, or they would kill off a lot of smaller species, or they might find that they couldn't digest their food because they hadn't done so since their forefathers' forefathers, you know. So, I think there are a lot of ethical implications of having a society like that, and I think that it wouldn't necessarily result in a very good outcome.

Similarly, Iain's dreams of noninterference inevitably came up against his own down-to-earth outlook. Releasing Caley to run feral, for instance, never entered his mind. Indeed, in other moments Iain would also be quick to point out the potential welfare hazards of such a move: exposure to the winter cold, lack of a regular food supply, and of course the fact that Caley had no experience of how to survive in a wild or semiwild state. Apart from anything else, Iain would miss her too much; and, Iain trusted, Caley would miss him.

All these reservations highlighted that it was no easy task to draw a circle around interference or human-impacted animal suffering. When Iain first adopted Caley, for instance, he ensured that she got neutered, a measure advised by most animal protection groups precisely in the interests of feline welfare. Was this meddling? It certainly was not an example of letting nature take its course. Such actions—of which there were many—were normally justified based on other animals being in human company and hence incapable of a return to natural existence. Another much cited example was the case of zoo animals. While colleagues generally welcomed the idea of zoos ceasing to operate, they remained extremely skeptical of the idea that captive animals could be successfully returned to the wild and hence also of the typical conservation claims made by zoos. This was partly because in many circumstances colleagues doubted that a viable wild environment still existed, and even if it did, they doubted that captive-bred animals had either the biological or mental resources to survive. If zoos closed, therefore, colleagues imagined that most of their animals would need to see their lives out in specially created sanctuaries. Those dilemmas, the fact that human influence on nonhuman animal life was in a practical sense almost impossible to place a limit on, partly explained the attraction of posthuman utopian visions. For in the end, those visions offered the securest promise of a freedom gained through the pious labor of noninterference or doing no harm.

A BRIEF NOTE ON THE TOTEM OF PERSONALITY

How can one elicit moral regard for the animals or species one wants to protect? Carrithers, Emery, and Bracken (2011) proffer that question in response to the reported dilemma faced by officers of an environmental body operating in the catchment of a major English river who sought to preserve the freshwater pearl mussel. These officers needed to convince other agencies and local partners to provide match funding and at the same time coax the surrounding landholders to alter farming practices. Although acknowledged as endangered across its ecological range, the eyeless and furless pearl mussel was apparently not an easy candidate for public sympathy. Particularly, as Carrithers, Emery, and Bracken closely observe, it resisted straightforward integration into some of the more common forms of "moral suasion" by which care and a sense of shared responsibility for nonhuman life could be garnered in the UK (2011, 673). It is hard to directly or even imaginatively conceive of making contact with or encountering the pearl mussel, and environmental officers struggled to deploy the usually most effective equation in such circumstances: "individual animal = individual person" (2011, 665). It was possible to shift gears and relocate the value of the pearl mussel through the idiom of possession rather than personhood, a strategy often used by the wider governmental body Natural England, which liked to invoke the idea of species as a form of national wealth (2011, 673). However, this sort of move did not elicit the kind of "moral energy" (2011, 666) that made a conservation project work. So, instead, the officers proposed that personhood be reinscribed away from the individual pearl mussel and toward its status as species, an innovation that finally enabled the project to be taken up as a joint undertaking.

For the authors, this example is indicative of personhood's continuing flexibility and power as a "sacred" or "guiding value," or what they prefer to term as a "central argumentative resource" (2011, 665). At one level, the observation is hardly novel; the literature on Western traditions of anthropomorphism and the shifting moral and political figuration of animals as either certain kinds of persons or nonpersons is well known. This includes rich ethnographic explorations of personhood's deployment in specific British human-animal contexts (see Knight 2005; Marvin 2005; Jones McVey 2023). But on another level, the essay intrigues precisely because of its descriptive attention to the way in which invocations of personhood operate or fail to operate not just as forms of moral suasion but also as the generator of moral energy, including within an organizational project or campaign. While their essay is principally concerned with the entanglement of tropes that can make the equation between animal and person such a dynamic and compelling form of moral imperative, it further directs our attention to an aspect of personhood that perhaps receives less attention in these debates. I am referring to the diverse roles ascribed to personality.

At once an informal category within animal protection, permeating, for instance, the intimacies of relationships to companion animals, and a formal category that structures the science of animal sentience as well as common forms of moral suasion, personality can be slippery to grasp. As we will see, it is also a category that vitally informed how colleagues understood their interactions with each other, including the moral effects of persons not just on others in the office but additionally on broader, more amorphous entities such as "office mood" or "office culture." Looking toward the chapters that remain, the prompt provided by Carrithers, Emery, and Bracken also leads me to consider the complex intersections between expertise and personality. Are there personalities that colleagues regard as *naturally* disposed toward key expert positions? Or do they believe that certain expert positions cultivate or shape personalities? Such questions take us back to the issue of moral agent and moral patient, both to the assumption that personality in some fashion animates moral agency and to the equally compelling assumption that the actions of personality reinforce a sense of moral subjects as patients.

* * *

As one can infer from the wider connections drawn by Carrithers, Emery, and Bracken, at the level of those tropic entanglements the argumentative resources of conservation and animal protection strongly intersect. This was the case even though from an animal protection perspective, the failure to respect the universal equation between individual person and individual animal exactly distinguished the conservation outlook. For instance, it enabled agents of conservation to justify the killing or sacrifice of one type of animal to conserve another (see chapter 4).

Yet in both fields, the authors argue, one finds a compelling enchainment between person and notions of uniqueness and irreplaceable value (2011, 668), as well as links to notions of interests and rights. Carrithers, Emery, and Bracken cite the influence, for example, of the utilitarian propositional logic offered by Peter Singer and of notions of human and nonhuman animals sharing a status as "subjects of a life" (see Regan 2004). Their essay also draws attention to the very versatility contained within the popular notion of person. As the authors elucidate, the kind of equations made between individual animal (or species) and individual person can generate very different concepts and relationships of "moral amity" (2011, 674) depending, for example, on whether the sort of person invoked is depicted as like kin or more like a friend or like a stranger.

As should be clear by now, colleagues at the animal group made regular appeal to a principle of sacred personhood. Idioms of friendship, for instance, were rife in their accounts of the special quality of amity between themselves and their companion animals. Recall that Craig chose to describe Paddy as his "best buddy," an expression that for him captured the depth of affection he felt toward this dog as well as the matey spirit of the "very close relationship" between them. Slightly more tongue-in-cheek, Iain spoke of Caley as his "de facto girlfriend." The romantic idiom may have been a joking nod to the rescue cat's sometimes overprotective behavior; however, it also referenced what Iain read as Caley's nursing tendency and the nature of their general household partnership. Other colleagues liked to refer to companion animals, especially when remembering the cats, dogs, and other pets that they grew up with, as "part of the family." They particularly invoked idioms of siblingship to describe that childhood bond. Just as importantly, colleagues regularly figured other animals as strangers. Indeed, the broader organizational narrative of making contact rested on that premise; especially with wild animals, part of the power and impact of the encounter was perceived to derive from a stranger-like unfamiliarity. As the sci-fi teaser film scripted by Iain demonstrated, the animal group could also use that idiom to strategic effect. An aspect of the playful appeal of the film lay in the way it refigured the wild, domestic, and farmed animals around us as aliens. This included refiguring companion animals as strangers in the home. In such a reckoning, mainstream publics rendered cats, dogs, cows, and other animals as strangers through the disregard shown for their extraordinary sentient powers.

Unsurprisingly, then, individual allusions to sacred personhood were matched by their deployment at an organizational level. Group campaigns on specific animal protection issues invariably revolved around trying to communicate the consequences of human action for a sentient creature who was precisely envisaged as being the subject of his or her own life. As well as the naturalization of this language in office discussion and organizational communications and appeals, the strategic valence of the tropes around personhood was often an object of professional interest. Like the English environment officers described by Carrithers, Emery, and Bracken, colleagues strove to understand how the equivalence might work. Although they rarely faced a species as challenging as the freshwater pearl mussel, they did sometimes have to consciously reflect on how tropes of personhood might vary in effectiveness depending on the individual species at the center of a campaign—the difference it made, for instance, whether they were dealing with mountain hares or harbor seals or foxes or deer or mice and rats (in the lab or as targets of pest control) or farmed salmon. This could include reflections on why they failed to garner widespread support for a particular campaign.

But while I remain very much interested in how colleagues defined the pragmatics of moral suasion and in how they accounted for what instilled moral energy in human others (see chapter 8), I want here to bring our attention back to one of the key claims of the relaunched animal group—that the organization was now "all about animal sentience" (see chapter 3). For at its heart, that claim revolved around a specific definition of personality.

In the context of their exploration of entangled tropes, Carrithers, Emery, and Bracken also draw passing inspiration from the work of Claude Lévi-Strauss. This includes a reflection on one of Lévi-Strauss's dictums, the observation that "everything takes place as if in our civilization every individual's personality were his totem: it is the signifier of his signified being" (1966, 214). Although the authors do not really draw out the specific relevance of that assumption of personality—for instance, its exact contribution to the moral suasion generated by the working equivalence between individual animal and individual person—the quoted lines serve as a useful provocation. For in an organization that aspired to be all about animal sentience, personality was valued not just as an individualized expression of sacred personhood but as recognizable evidence or proof for that nonhuman sentient being. Indeed, in his capacity as the animal group's science and research manager, Craig once advised me that the category should be read as a crucial "sign of sentience." This was because, Craig explained, personality was what potentially allowed one to demonstrate the presence of a sentient being to others, as well as an aspect of what instinctively allowed oneself to know that an encounter with a sentient being had taken place. To paraphrase Lévi-Strauss, being all about animal sentience certainly involved learning to read personality as the totem for each individual nonhuman animal (and species), yet it also involved reading personality as the totem for sentience itself.

Colleagues understood that it was the compelling or self-obvious nature of human sentience and human personality, both to themselves and to the publics that they addressed, that made the achievement of a wider recognition of sentience and personality beyond the human possible and powerful. In fact, Craig made this point explicitly in the content pages of his Amazing Animal Facts section of the organization's website. As well as continuing to update readers on the latest scientific findings about the personality of over sixty species, including species of primates, of fish, birds, and mollusks, Craig regularly referred readers back to the ordinary features of human personality. On one page, for instance, he told them that "human personality is something with which we are all familiar, in that 'we know it when we see it' . . . it is what distinguishes one individual from another." The point was made not just to provide a familiar ground but also to introduce a series of caveats, including points about the difference between scientific and popular ways of understanding the category of personality and about the challenges of measuring sentience in nonhuman species.

Craig warned, for instance, that defining even human personality "scientifically" had always been difficult. It was therefore only relatively recently that animal scientists had started to investigate personality in other species, using models from psychology as a starting point for their investigations. Craig referenced and briefly outlined for his website readers the five-factor model before cautioning that such human-oriented measurements might inevitably be challenged by radical differences in species behavior and environment. By way of example, Craig explained that it made less sense for scientists to try to measure "inventiveness" in hermit crabs than in chimpanzees. He also pointed out that species behavior sometimes required scientists to stretch the utility of a measurement category. A favorite example among colleagues was the idea that octopuses might display "episodic personality," variant or only sometimes recognizable to human observers. Nevertheless, as Craig repeatedly informed readers, we can still be reassured that "like people, animals can be described as having personalities because individuals consistently differ from one another in behavior in such a way that these behaviors can be described as individual traits."

While most other colleagues got no further into the animal science of personality than an occasional perusal of Craig's posts, the effect of thinking in this fashion about what constituted a consistent difference could be defamiliarizing. It sometimes resulted, for instance, in greater reflection on the "we know it when we see it" status of human personality. This could lead colleagues to muse on their own encounters with psychological models as well as on their ordinary understandings of personality as sets of dominant traits or characteristics, including when assessing their own distinctiveness or the distinctiveness of other people that they knew well. In fact, I would want to argue that one of the effects of assigning personality to be a sign of animal sentience—an action that as we have seen was itself possible only because of the new organizational emphasis on animal pleasures or positive forms of sentience—was the reanimation or intensification of personality as a totem for individual human beings. Each appeared to play off the other.

* * *

In part, this intensification was also a consequence of making personality into a tool of both team and self-evaluation. Icebreaker exercises, for instance, were a common feature of office meetings, especially during the period of relaunch. These directly invited employees to consider how attributes of their own personalities could be adjudged to enhance or weaken the performance of professional roles. On occasion, this was done through the use of animal cards, knowingly engaged with as symbols for human personality traits. I can recall one exercise, for example, where colleagues were invited to select the animal card that best represented them. After some lighthearted suggestions and friendly jibes at each other's expense, a selection was made. Eilidh started things off by announcing that she had chosen the giraffe card. This was because in her position as CEO she must have "overall vision" for the organization, view it as it were from a height, and yet at the same time "keep my feet firmly on the ground." Next, Iain picked the cheetah card. He did so, the IT officer explained, because he liked to work in "fast bursts and then slow down." Maggie, the policy director, then announced that she had selected the lioness since she was "hardworking and maternal" but also, she added with a wry smile, at times a little bit "grumpy." Euan made his choice and then offered us a self-deprecating aside about the fact that he really should have been allowed to select a wild ass. Mairi, a part-time colleague who worked in supporter services, identified with the beaver card because she was happiest "working away" by herself, and finally Shelia informed us that she had opted for the elephant. This was because, Shelia explained, as the office manager and longest-serving member of staff in many ways, she continued to just "plod along" in the background.

    This kind of invocation of personality, like its more formal use in the annual staff performance appraisals that Eilidh also introduced, was of course already familiar. Many colleagues had conducted similar exercises before, including in previous places of work. However, bringing that wider office culture of self-evaluation into conjunction with discussions about the scientific signs of animal sentience seemed to enliven reflections. In fact, the latter also easily segued into more casual office conversation about personality. After discussing an Amazing Animal Facts post about the personality of red squirrels, for instance, a group of us sat around and mused further on how we read each other's personality traits. The discussion was specifically prompted by Craig's report that animal scientists in Canada had discovered a "range of [squirrel] personalities, from exploratory and aggressive to careful and passive." David began by offering the view that Iain, his workstation partner and office pal, was in his opinion loyal, compassionate, funny, and warm but also sometimes quick-tempered and irritable. Elaine, the personal assistant to the CEO, then provided an assessment of her own personality, informing us that if asked, she would probably describe herself as effervescent, open-minded, adventurous, and instinctive. Elaine admitted that this sometimes led to her getting frustrated with others, especially those colleagues and friends who were indecisive or who looked too closely and long at an issue. The observation made Iain laugh. For as Iain explained, other colleagues were always noting what a "calming effect" Elaine had on the office. Our conversation then smoothly switched to the personality traits of household companions. Once again, it was Elaine who took up and ran with the topic. "Now, Marley, my first cat," Elaine explained to us, "had a very quiet personality. He kept to himself but was very laid back." By contrast, "Theo was a bully, quite aggressive and very demanding, not at all like Hamish [her present rescue cat]," Elaine added. "He is opposite again, highly sensitive and retiring."

A few days after this conversation, Elaine came across to the desk I was currently occupying to tell me that she had once taken a comprehensive psychometric test as part of a job application process for an executive personal assistant position. Elaine revealed that upon receiving the test results, she was amazed to be informed that she was not really suited to such work! Specifically, the assessment concluded that while Elaine might be strong in sociability, communication, and conversation skills, she was weak in logic, analysis, and decision-making. What surprised Elaine most of all about this was that in all her past employments, senior colleagues had consistently praised her for being hyperorganized and tightly focused on meeting targets. In retrospect, Elaine proffered, those psychometric test results had probably helped her realize her own "true personality." However, Elaine promptly threw in a note of caution. What the test perhaps failed to appreciate, Elaine suggested, was that much like others she could be more than one person. There was, for instance, the "Elaine-at-home," who certainly could be both "scatter-brained and fizzy." But then there was also the "Elaine-at-work," who could be very organized, sensible, and "on the ball." Indeed, precisely one of the things that Craig's posts had taught her was that personality could also be "environmental": that is, it could change depending on your social context or which group of friends you happened to be hanging out with. Elaine believed that its definition was additionally dependent on the recognition of others. She reminded me, for instance, how astounded she had been in our previous group conversation when Iain told everyone that Elaine had a calm personality or at least a personality that had the desirable effect of calming her colleagues.

There was in such talk always an interesting tension between the idea of consistent difference in behavior as the defining aspect of an individual and the recognition that personality could be variable. This could be the case both in response to changing conditions and because of the unpredictable ways in which others defined or experienced your personality in the world. If everything did take place as if every human individual's personality were their totem, then there appeared times when a subject might experience that totem as more or less detached or autonomous from the individual being it apparently signified.

Indeed, as Elaine's musing illustrated, colleagues sometimes spoke as if the ways in which personalities interacted at work were beyond their control, a mediated encounter from which they remained aloof and sometimes totally unaware. Most obviously, the unintentional moral effects of personality on others—its ability, for instance, to calm them—could render personality as an external relationship to self. Or put another way, it could give the impression of self or signified being as a certain kind of patient in relationship to the doings or actions of personality. Such realizations could also be turned into a diagnostic. It was not uncommon, for instance, to hear colleagues identify "a clash of personality" as the reason for problems in the office or for the inefficient delivery of a campaign. As mentioned in the last chapter, Euan's biography, including his consistent failure to link his vegetarianism to moral choice, made others in the office sometimes feel awkward. At times colleagues even attributed that awkwardness to the effects of an odd or unsettling personality. Iain went further and claimed that their initially strained working relationship was because Euan's personality at first annoyed or aggravated him. It was only much later that Iain realized that this effect was not Euan's fault and that once they got beyond this clash, they could better appreciate each other for who they really were (as opposed to what their personalities were), enabling respect and a good working relationship to ensue. And of course, as we have seen, these interactions also took place within an infinitely wider field of nonhuman animal personality. But in this case, the relationship between totem and signified sentient being was even more contested and open to interpretation. For, of course, the perspective of the nonhuman signified being on its own externally observable totem was assumed to remain unknowable, or rather to be nonexistent.

* * *

So, paying attention to personality highlighted the commensurate and incommensurate natures of human and nonhuman sentient being. On one level, the fact that humans and other species apparently shared the same totem illustrated commonality and the power of the organization's claim to be all about animal sentience. On another level, it spotlighted an essential cleavage since the only species with any perspective on that totem was taken to be the human species. However, that distinction was itself complicated by the fact that human personality appeared able to evade or become alienated from what it signified. It could interact, for instance, without reference to either the first-person or second-person perspectives on that sentient being. This suggested a series of splits—for instance, between what the signified being did and how that action expressed itself through personality, or between a first-person perspective on that personality and the interpretation of others. All of these complications pointed to the fact that issues of virtue, if linked to the question of how one should live and what sort of person one wanted to be, were far from straightforward. Who am I? Who are you? In diverse ways, those inquiries take center stage in the chapters that follow.

Indeed, this short chapter has been an oblique introduction to such concerns. In what follows, I aim to trace in more depth how colleagues addressed themselves to these matters by taking a closer look at three expert positions within animal protection. In the next chapter, we turn to a figure with an iconic moral status both within the animal group and in the wider animal movement: the undercover investigator (see Song 2010, 43–45). As perhaps the clearest example of someone who has dedicated their life to the cause of animal protection, this figure stood out from the crowd of animal people. Ennobled by their level of commitment and by the sacrifices necessarily involved in this largely self-taught role or form of expertise, the investigator was also most closely identified with self-control and the disciplining of moral feeling. Their sideways stance, both toward the animating passions of colleagues and toward the self-definitions that accrue from moral patiency, made the investigator an unusual figure in the office. It further gave them an extraordinary insight into the first-person perspectives of those who were not animal people, including those involved in the systematic abuse or exploitation of animals. More than any other colleague, the investigator specialized in the cultivation of diverse personalities. This included close attention to the issue of those personalities' performance and effects as well as the professional possibilities of developing an external relationship to the personality ultimately taken to signify being.

For the two remaining expert positions, the relationship to personality is on the face of it less dramatic. Considering the equally iconic moral status of the rescuer, for instance, it appears that the totemic aspect of personality escapes problematization. The rescuer is who they are, and what they are is animal love. However, the apparent collapse of distance between personality and signified being sometimes creates problems elsewhere—for example, for those who must live with the rescuer. And the clarity of this role brings its own costs. Most noticeably, rescue involves a confrontation with evil, often indexed through casual acts of cruelty and its most recognized outcome, animal suffering, which risks exhausting the moral subject. But the rewards are also high, for the care work of rescue, more than any other role within animal protection, brings one into intimate contact with the personality of the rescued. Indeed, it opens the possibility of unique bonds of personality. In the case of our final expert figure, the policy director or lobbyist, we enter another field altogether. Operating in the terrain of interests, personality emerges alongside the capacity to compromise as a natural expression of expertise. But it also appears as a strategic resource that the policy director or lobbyist must learn to exploit in dealings with both government and multiple stakeholders or interest groups. Whether rendered as affability or more generally as a moderate disposition, personality is in some sense weaponized, put to good use in the service of protecting animals.

DISCIPLINES OF INVESTIGATION

What does it require to become an undercover investigator? What exactly is involved in the work of investigation, and why has this peculiar form of expertise become so closely associated with the moral project of animal protection? In this chapter, I seek to explore answers to these questions by leaning closely into the professional biography and outlook of Barry, who has been an investigator for over forty years. I begin by introducing you to Barry's distinctive voice. This extract is drawn from one of our many extended conversations and is entirely characteristic of the ways in which Barry narrated his life and practice and kept circling back to reflect on key moments in that history:

> It was my second day at the circus. We were at winter quarters in Andover, where they train the animals owned by a woman called Anne Meadows. She used to be a famous animal trainer. Anyway, I was cleaning out the tigers when the elephant keeper came over to me. He had a bit of material in his hand, and he was angry because one of the elephants had just eaten his jacket. I also noticed an iron bar in his hand, which was bent. So, after he told me what had happened, he started walking back to the elephant shed and I walked back with him. In the shed at that time there was one African female and one African bull elephant, and an Asian elephant. And it was the bull elephant that had eaten his jacket. He stood there showing me his jacket for a minute. Then he walked into

the enclosure with the iron bar, and he started laying into the animal, severely beating the animal. I was taken aback! I didn't expect that. My second day there and there's this guy beating the crap out of an elephant. And the elephant was chained, back and front leg. So, I started filming [with his body camera]. . . . In fact, I discovered that he did this every day, he beat all of the elephants. In the end we had to install covert cameras in the shed because we felt that this was going to be a prosecution. It wasn't the initial idea; the initial idea was just to obtain evidence of animals being used in the circus. But now I felt we had to set cameras up so I wasn't there, so he couldn't say that I was encouraging him to beat the elephants or that I was paying him to do it. He couldn't claim that he was doing it through bravado or anything like that. He was just doing it because he could. And Anne Meadows was later caught on camera too, beating the life out of a baby chimpanzee. She used to kick and punch and hit with a stick this baby chimpanzee. . . .

By this time, I think it was a year and a half [undercover in the circus world] and I was exhausted. I'd been Ozzy, which was a name that I had given myself, and I was just tired of being this person and playing this character. . . . But [exposing such people] is the most satisfying part of the job. Unfortunately, because of the way the system is, the way enforcement of the law is, 60 percent of the time these people get away with it. And it can be really upsetting, especially if you have been there and witnessed what they are doing.

Anne Meadows was a very brutal person towards people as well as towards the animals. But I played along. Ozzy was the village idiot, if you like, and she really did not know who I was. She had no idea, was not suspicious whatsoever. Then when she saw the film [footage] and the penny dropped, I was told by a police officer that, you know, you could have painted a picture of her face. She couldn't believe that all this time she had been filmed and infiltrated by us, me and my colleague. And all of her abuse was captured on film. That's the best moment ever, when you capture these people who think they can abuse animals and get away with it because they do it behind walls or they do in the countryside away from public view. And then suddenly they realize that in fact somebody had been filming them and everybody now knows about it, and everybody has seen what vile and disgusting people they are, and cowards as well. It is a cowardly act to abuse an innocent animal, particularly in the most brutal ways I quite often witness.

Undercover Hero

As Song (2010, 45) points out, the covert investigator who infiltrates, witnesses, and exposes human cruelty toward animals on behalf of others and who risks personal dangers in the process has been an essential figure in the emergence of the new wave of animal protection and animal rights movements since the 1980s. Addressing comment principally to developments in North America and to the example of the rise of PETA, Song explores how such investigative work merged with contemporary innovations in direct mail campaigning. This included the coordinated public release of photographed or videotaped footage of animal abuse collected by investigators and the emergence of a new campaign genre of observational diary or first-person reportage of those undercover actions, which combined led to a dramatic spike in both supporter numbers and the rate of donations (2010, 42). As a "surrogate witness of cruelty," Song argues, the covert investigator is necessarily anonymous to their readers but in that very anonymity serves to highlight the potential ubiquity of the "menace" of human-inflicted animal suffering in new ways, appearing in the process as a kind of neutral "instrument of monitoring" (45). For Song, the "dramatized intimacy" of the investigator's contact with that suffering can be juxtaposed with the "helplessly distanced" relationship of many supporters (44). In fact, Song views the latter relationship as partly an effect generated by the sorts of investigative materials that supporters receive and consume, originally through direct mail but also more recently through its presentation online and through new social media.

Both the strength and weakness of Song's account derives from the fact that it is largely the result of a discursive analysis of organizational literature. The investigator is much more a narrative figure that Song reads off those texts than a subject encountered directly. But as the opening reflections cited above illustrate, such figures of course have their own first-person perspectives on what they do and experience. Although these may overlap with aspects of the genre of observational diary or first-person reportage produced for campaigns, those unscripted perspectives necessarily offer an original outlook on the moral imagination of the animal movement and on the passions typically taken to motivate its participants. Indeed, it was the same peculiar role of investigator that granted Barry an uncommon kind of moral authority at the Edinburgh animal group, not just in the minds of the group's supporters but also among his colleagues.

As we will see, part of that authority was certainly connected to the role of heroic surrogate witnessing. Uniquely positioned as an intermediary between animal people and the sorts of suffering animals that were the chief object of their moral concern, Barry's up-close-and-personal relationship to the circus elephant being beaten with an iron bar or to any other animal being systematically abused was an essential part of his aura. No one else, not even those who devoted their nonworking hours to animal rescue (see chapter 7), encountered the scene or moment of abuse so immediately. Even though the investigator was not an anonymous figure or "nonperson" (Song 2010, 45) to his colleagues, Barry remained in this regard a mystery to them. But the moral status of the investigator was further tied up with another necessary aspect of that expert role. What struck other colleagues and supporters most forcibly was the fact that Barry's work required him to be up close and personal not just with the victims and processes of abuse but also with animal abusers. These included those "cowards" or "vile and disgusting people" responsible for the worst kinds of harm inflicted as well as those less objectionable collaborators in industrial or systematic forms of abuse that might be regarded as simply caught up in the violent culture of any institutional practice. Barry was then additionally an intermediary between animal people and the cruel or not-so-cruel persons that he had to covertly work alongside. Albeit undercover and always "playing a character," his knowledge of and proximity to such persons was taken to be extraordinary. In fact, these two intermediary positions were acknowledged to be what distinguished the investigator from others; neither colleagues nor supporters could countenance being able to do what Barry did. Even more than the power to expose human-inflicted cruelty, it was the immense self-sacrifice involved in such essential work that made the investigator a hero.

Barry was therefore an exemplary figure in the moral imagination and the only person at the animal group who had a public profile based on their role alone. As Song's account suggests it might, that profile revealed itself in the continuing prominence of the investigator in the group's campaigns and fundraising appeals. Despite the general organizational thrust to move away from a dependency on emphasizing suffering or human acts of cruelty, Barry's expertise remained central to the work of the organization, especially to policy and lobbying (see chapter 8). In fact, on occasion, the work of the investigations and field research officer (Barry's formal title) was the very basis for a fundraising call. These appeals drew directly on the motif of sacrifice and on the general assumption that the investigator fully embodied a set of admired virtues.

"Imagine day after day witnessing animals in distress or suffering. It's not something I would want to stomach, and thankfully I don't have to. But Jim does." So begins one of the animal group's periodic investigations appeals. In this case, the appeal invites supporters to make a range of donations so that "Jim," a pseudonym chosen for Barry, has the equipment to carry out his fieldwork. The call specifies, for instance, that "£8 will buy an Ordinance Survey OS Explorer Map, that will enable Jim to navigate safely in remote, hard-to-reach places and document and report exact grid references of any cases of animal suffering." Or, for the more generous donor, "£500 will buy covert spectacles with a digital audio and video camera." Indeed, the appeal letter connects the ask with the unique intermediary positions of the investigator, urging supporters not just to view Jim as the savior of suffering and abused animals but also to view themselves as vitally responsible for the sustainment of his role and person. "Without you, there would be no Jim," the letter continues, "and without Jim, countless animals would suffer at the hands of unscrupulous or misguided individuals." As well as stressing the sacrifices involved, it makes repeated reference to the virtue of bravery. "Often working in dangerous conditions," one passage reads, "Jim spends hours in the field gathering data that exposes real-life examples of animal suffering and neglect." The link between exposure and extraordinary powers of endurance was also a consistent theme, as were the virtues of commitment. Just as no one else, neither colleagues nor supporters, had the stomach to witness animals in distress on such a daily basis, so none of them, it was assumed, were as dedicated to the cause as the investigator.

Although he did not present his own actions in heroic terms, Barry did accept that he led an exceptional life and that his levels of commitment were unquestioned or comparatively absolute. Whenever we met, that fact was a consistent point of self-reflection, for Barry a mark of his inevitable distinction from others in the animal movement:

> I often wonder, if I could take my life back and live a different kind of life, go down alternative paths, where might it lead? And I think it would lead back to animals again in some form. You see, I dedicate my life to animals; every aspect of my life privately and at work is about animal protection. I am always thinking of new [investigative] projects and the best way to do things, the most effective way to live my life for the animals. And I do have, I don't know if the word "love" is right for me, but I do have a tremendous admiration for animals. I think they are amazing. I have some kind of affinity with animals as well; I can almost feel how the animal is suffering, you know.

As we will see, such levels of affinity were in Barry's eyes an asset to the work of investigation—part of what Barry regarded as his unique skill set—as well as a burden. While Barry acknowledged that a special affinity with animals was a common claim among animal people, for him the role of investigator meant that it was both newly purposed and originally challenged. Being able to "almost feel how the animal is suffering" took on an entirely different complexion when that suffering was not just an imagined or reported action but instead a process irregularly or systematically unfolding in front of you. Among other things, it reinforced the tragic dimensions of Barry's sacrifice: if each witnessing simultaneously drew out from Barry feelings *for* the suffering animal and what seemed like shared feelings *with* that suffering animal. This was not an experience, Barry implied, that ordinary animal people could fully comprehend. "What others who are close to me and who have a keen interest in animals don't understand," Barry stressed, "is that I wouldn't class myself as an animal lover." That remark echoes the hesitation expressed in the longer quote above; indeed, in some ways Barry held that his work alienated him from other colleagues and animal people in general.

So, by his own estimation, Barry's path through life and within the animal movement was quite unparalleled. "I do find," Barry once told me, "that a lot of people around me who are also involved in animal protection have come down a very different road." The observation was partly informed by a realization, in this case shared by colleagues, that a lifetime of undercover work necessarily marked Barry out. It was also partly shaped by his keen awareness of a difference in social background; unlike most of his colleagues, Barry came from a working-class family, he didn't attend university, and, as will emerge, most of Barry's considerable expertise as an investigator was self-taught. But perhaps the starkest difference remained in the degree of dedication, in Barry's strong sense of investigation as a vocation or calling. "Yeah, I could not imagine for one minute not working towards the protection of animals. I live and breathe it," Barry reaffirmed. "I couldn't do anything else. And I will never retire."

Investigative Expertise

While Song is largely concerned to draw out the rhetorical effects attached to the idea of the undercover investigator and to map the associated techniques by which movement building developed in organizations such as PETA, in this chapter I am principally focused on the expertise of investigation and the disciplines attached to its practice. For to a far greater extent than any other role in the animal group, this one presented as a discipline, both in the sense of rigorous ethical labor or self-mastery and in the sense of a task with a prescribed set of methods and exercised drills. Indeed, I was first struck by the professional seriousness with which Barry self-presented. For him, the investigative task was far less or only incidentally about fundraising or supporter mobilization and far more about the collection of evidence in the context of policy, campaign research, and sometimes legal or even criminal proceedings.

To a degree, that sense of discipline drew from a shared professional culture of investigation. Barry's habit of dividing his evidence-gathering activities between "overt" and "covert" operations, for instance, was part of a common language and practice. Barry estimated that there were at least seven full-time animal protection investigators in the UK. There were also several investigators Barry knew who did undercover and overt investigative work for conservation groups, in Scotland mostly centered on the observation or infiltration of illegal egg-collecting practices or the persecution of protected raptor species. In addition, the large welfare organizations such as the RSPCA or in Scotland the SSPCA employed a group of uniformed investigative officers, sometimes ex-policemen, who had some formal reporting powers in law. Although undercover investigators such as Barry at times worked closely with these uniformed welfare officers—for instance, passing incriminating evidence of the flouting of animal welfare regulations on to them (see Reed 2016a)—they otherwise operated quite separately.

Among the seven or so animal protection investigators in the UK, some were employed by single animal organizations, but others were self-employed and took up assignments by contract. Barry combined both arrangements, working for the Edinburgh animal group three weeks out of every five and taking contract jobs or conducting projects on his own account for the remainder of the time. Indeed, in the office, part of the mystery attached to Barry's work was a result of his relative absence. While certainly friendly with colleagues—present at team meetings and after-work social events, for example—Barry was also often shut away in the back room archiving unshared files of photographic and video evidence that he had collected or confidentially discussing the details of a project or upcoming court case with Maggie, the policy director. As we have already seen, some of Barry's past assignments involved teaming up with other investigators. Barry's long-term undercover operation at the circus of Anne Meadows (pseudonym used), for instance, began as a solo assignment with Barry turning up and simply asking for work. Barry managed to persuade the circus to take him on first as a ring boy and then as a beast man. However, once the extent of abuse became evident, the animal protection organization that contracted the undercover assignment (this operation was not initiated by the Edinburgh animal group) recommended that it become a two-person affair. So, after a while, a "girlfriend" also got work at the circus and lived on-site in the same camper van as Ozzy. On occasion, Barry had even worked in larger teams of investigators. This included contracted projects linked to the monitoring of hunts in England as well as projects where a group of investigators were hired to work together to track and observe animal transportation trucks moving across the Continent.

But in the main, Barry worked alone. Barry explained that apart from those hired assignments, he rarely saw or heard from other investigators. It was true that to an extent they shared practices in common, partly absorbed because of occasionally teaming up; nevertheless, for Barry, investigation was principally a solitary pursuit. Indeed, that was how Barry preferred it. Overall, Barry presented his expertise as learned on the job, as knowledge accrued and finessed from years of largely solo undercover experience and evidence collection. In his late forties, when we held our first conversations, Barry offered me an autobiographical sketch that had him falling into investigative work in stages. At each turn, he was motivated to carry on and innovate new working practices based on the nature of abuses that he had encountered. In fact, by the time we met, Barry seemed to have conducted undercover work in many of the most notorious sites of systemic animal abuse in the UK, from the perspective of the animal movement. These included experimental laboratories and their breeding stations, puppy farms, circuses, hunts, shooting estates, slaughterhouses, factory farms, and pet shops. However, as Barry told it, the first and formative experiences occurred when as a young man he got a series of jobs in zoos.

Born and raised in the North London district of Highbury Corner, Barry had instinctively known since primary school that he wanted to work with animals. Partly due to an early affectionate relationship with the family dog, that feeling grew through a boyhood obsession with dinosaurs that quickly mutated into a fascination with big cats and especially with tigers. Barry joined a zoological society library in London and started to study the natural history of those cats intensely. By the time Barry was at secondary school, he knew that he wanted to get to know tigers firsthand and that the obvious place to do so was in a zoo. The opportunity emerged when he was sixteen. Prompted by a discussion with his careers teacher, who pointed out that Barry could leave school straightaway if he found a suitable position, Barry successfully applied to work at a privately owned zoo in Warwickshire. It was a big responsibility. Almost immediately, Barry was placed in charge of looking after tigers, leopards, and chimpanzees. Although initially thrilled, Barry soon became disillusioned by the treatment he witnessed and by the cruelty exhibited by the zoo owner. To persuade chimpanzees to move from one cage to another to enable clean outs, for example, the owner would shoot any recalcitrant animal with an air pistol. At the time convinced that this experience was an aberration and that conditions would no doubt be much better in a public zoo, Barry determined to find a position elsewhere. In fact, Barry managed to get a keeper role at a very well-known metropolitan zoo, this time to look after mountain goats, bears, and sea lions. Once again, however, Barry soon noticed problems. The bears, for instance, seemed "all quite mad," swaying and pacing up and down, behaviors that Barry much later realized to be stereotypical of captive animals. Barry's unhappiness grew, especially after rumors spread among zoo staff about more active forms of ill treatment.

"And then I had a kind of light-bulb moment," Barry revealed, "where I realized that in fact I didn't want to work with animals anymore. I wanted to work for them, which is quite different." Barry therefore resigned as a keeper and took up rather monotonous employment in the invoice control division of a Central London department store. It was during this period that Barry happened across an advertisement on the London Underground for the Living without Cruelty exhibition at Kensington Town Hall (the same annual exhibition, organized by Animal Aid, that Eilidh had once visited at a crucial moment in her ethical biography [see chapter 1]). Among the stands Barry came across was one hosted by an organization called Zoo Check, later to become much better known as the Born Free Foundation. Barry began telling them of his experiences, and they encouraged him to submit an anonymous report to their magazine. Over the next few years, Barry did a lot of voluntary work for Zoo Check, at first helping in the accounts section but then increasingly doing fact-finding visits to various zoos and safari parks. Eventually, Barry joined the organization full time.

In this way, Barry stumbled more and more into a life of investigations. The original work with Zoo Check led to contacts with other animal groups and further invitations to conduct both undercover and more overt studies on their behalf. Part of the appeal of Barry to these organizations continued to lie in his hands-on experience as a zookeeper, his vital insider's knowledge. This allowed Barry relatively straightforward access to any facility that required animal management, since his work experience was also a useful CV. At that time, Barry told me, staff quite regularly moved between zoos and other kinds of institutions. Just as importantly, it enabled him to be a convincing colleague at these places. Indeed, Barry prided himself on being able to fit in with the sorts of persons, often from the same class or social background, that did the ordinary or mundane work there. Barry likewise prided himself on having that working experience with captive-held animals, manual and technical knowledge that he believed distinguished him from many other animal protection investigators.

Moving to Brighton, Barry started mixing with a circle of like-minded friends—for example, running a campaign to get the dolphins at Brighton Marina released. Because Barry had some interest and knowledge in using video cameras (quite unusual at that time), he also began experimenting with covert recording, which only increased his standing as an obvious candidate for undercover work. Around that time, in the late 1980s, Barry got his first undercover assignments at commercial laboratories. Initially, Barry found work as a technician at a local facility where they conducted experiments on monkeys. As well as documenting mistreatment, Barry noticed that primates were often arriving at the facility dead in their crates and that the baboons, for instance, were being wild caught. After successfully filming activities there for an antivivisection organization, Barry gained a position at another animal laboratory, this time in Yorkshire. His job was to clean out, feed, water, and ensure the day-to-day care of lab animals, including their monitoring during experimentation. This included tying monkeys down to Perspex chairs to administer gas inhalations, gavaging or force-feeding through oral tubes, and giving chemical injections into the stomach. Barry was employed there for over six months, and the covert footage he shot at the laboratory became part of a television current-affairs program that exposed practices in the UK industry and its relationship to the trade in primates. For a brief period, Barry found himself invited to give a series of public talks on the topic.

Although he continued to be contracted to conduct undercover work—the circus operation outlined at the beginning of the chapter occurred in the late 1990s—in the ensuring years, Barry's investigations also increasingly became self-generated. Barry developed an instinct for the kind of project that might get commissioned, an attitude that also led to a growing sense of his own investigative flair. "I always like to find something different," Barry explained to me, "to put my own spin on the investigation, to discover an angle that might take the interest of the public." In fact, one of the reasons that he joined the Edinburgh animal group in 2006 was that the terms of his contract allowed him to remain flexible. The other reason was that, like the organization, Barry was becoming more and more interested in wildlife crime and in particular the abuse of wild animals in the countryside, a joint focus of concern that converged around the management of grouse and pheasant and the wider activities of the Scottish shooting estate.

Indeed, during most of my time, Barry's investigative operations for the animal group centered on these estates and in particular their use of traps (see also Reed 2016a, 2017b). This was an interest prompted by the antisnaring campaign that across this period dominated a significant amount of colleagues' energy and time. It was in this context that I most closely got to understand Barry's working practice and some of the extraordinary disciplines of his investigative expertise.

* * *

One morning Barry and I convened in the animal group's boardroom, where Barry spread out some satellite image maps of the shooting estate that it had been agreed we were going to visit that day. Pointing at some strange-looking craters on the map, Barry informed me that we ought to check them out. Then Barry highlighted another, smaller square mark, which he suspected was a cage for trapping corvids such as crows, jackdaws, jays, and magpies. Barry thought that this was probably the same cage trap that he had come across on his last visit to the estate. Unfolding an ordinance survey map for the area, Barry next ran through the aspects that most concerned him. There was a series of geometrically planted clumps of trees and small woods, for instance, which he thought deserved a closer look. Barry explained to me that on grouse-shooting estates, the fringes of such plantations were often the most likely places to discover snares: wire nooses typically anchored to a tree or fence line that are designed to tighten on contact and can be legally laid by gamekeepers to trap predator species such as foxes. These places, Barry added, were also where one frequently found evidence of illegal raptor poisonings. The prospect of the latter had been his initial prompt to regularly visit this estate in the Scottish Borders, since Barry knew that a few years back, one of its gamekeepers had been successfully prosecuted for bird persecution. Once our deliberations were over, Barry checked that I had brought warm clothing—it was mid-October, and we would be tramping across upland slopes—before we set off to find the car.

Upon our arrival, just over an hour and half's drive from Edinburgh, Barry was at first a little disoriented. Since Barry's last visit, the estate had knocked down the plantation nearest to the road to put up three huge wind turbines, which had not appeared on his slightly dated satellite maps. After a quick readjustment, Barry suggested that we head down to a much smaller cluster of conifers on the path in front of us. Our status on this estate, Barry advised as we walked, was relatively secure since the Scottish Land Reform Act meant that members of the public had the right to roam across privately owned countryside (see Reed 2016b). For this reason, Barry classified the operation as an overt investigation. However, the Land Reform Act was less clear on the rationale for our visit; public access to such places was granted based on leisure or educational activities. So, Barry informed me that should we be stopped, we should self-present in this fashion. His favored attitude was that of a birdwatcher, an occupation that conveniently made sense of his general appearance, including the pairs of binoculars, stills, and video cameras that hung around his neck or attached to a pouch on his belt.

When we reached them, the cluster of conifers revealed nothing. Barry explained that since it was unfenced and hence open to grazing sheep, gamekeepers would be unlikely to lay snares. So, we moved off to the larger plantation on our immediate horizon. At the low drystone wall that marked the beginning of its boundary fence, Barry once again knelt to see whether any snares were attached. He inspected the ground for signs of animal tracks and quad bike marks. In Barry's experience, gamekeepers regularly set snares and dropped poisoned bait at these points. Barry also bent to examine the long grass, looking for signs of feathers and other bird remains that might indicate a predator kill—an action that he claimed mimicked the practice of gamekeepers, who often laid a line of snares in such locations. In addition, Barry took the opportunity to look through his binoculars to see whether any birds of prey were in flight and to begin the process of starting to identify any breaks in the wood line that might suggest a spot where a keeper had entered.

Indeed, as we hiked the conifer boundary, Barry periodically dove down below its branches to check for signs of gamekeeper activity. This included halting to sniff the air where he thought he might have picked up the stench of decomposing corpses, caused either by dead or dying trapped animals or by the presence of stink pits (dumps of rotting animal carcasses). The latter were often created inside plantations to entice predators into spaces laid with channels of snares. At one point, those inspections led to the discovery of a tangle of dead magpies and at another to a small, broken white egg that Barry thought probably belonged to a pheasant. He told me that such eggs were sometimes baited to attract foxes and other predator species, but on close examination we could not find any syringe mark. Nevertheless, Barry took photographs for his records. After climbing the fence and tramping across a slope to a further line of trees, we saw scattered feathers and entrails and then several patches where the long grass had been stamped down. Such marks, Barry advised, looked a lot like the familiar doughnut-shaped devastation that one found when a badger had been caught in a snare. During a previous conversation in the office, while showing me a selection of past images and footage from his field archive, Barry had outlined how trapped badgers, a nontarget species for snares, typically tried to escape by trying to dig themselves out of the situation, a behavior that invariably resulted in the noose tightening. But crouching down, Barry could see no sign of any snares and eventually determined that in this case sheep had probably caused the patches of stamped grass. A little further on, we started to notice the presence of blowflies—another common indicator of injured or decomposing animal bodies—and Barry once again sniffed the air.

The day continued in this fashion. For instance, we searched for plastic bags stuffed into the fence line, and Barry stopped to scour any gateposts for jars of carbofuran, the chemical substance typically used in raptor poisonings (since it was illegal, keepers rarely kept carbofuran in or near their properties). At a small stone bridge, Barry ducked under the arch to see if there were any Fenn traps, legal spring-loaded kill mechanisms typically set for rats, stoats, and weasels, all of which take grouse eggs. As with the inspection of snares, the point was to monitor whether the traps were being kept in good order and in accordance with the protocols attached to their license of use. Upon hearing the bark of dogs, Barry next led me into a deciduous wood beside a small pond that eventually emerged to open fields with the estate's grouse moor now sloping up in front of us. Scanning the horizon with his binoculars, Barry pointed out the obvious sign of recent burnings and then a freestanding wooden post, which he speculated might be an illegal pole trap for birds of prey. In fact, the search continued all the way along our return. At one gate, for instance, Barry stooped to examine the bottom of the wall line, moving a few stones aside to reveal a Fenn trap with a few eggs laid on the ground beside it as bait. Photographing the scene as evidence,

Barry highlighted the heavily rusted state of this trap, which he said clearly broke the licensing code. A little further on, we discovered another Fenn trap, this time littered with old stoat or rat bones, before finally heading away.

Although Barry was a little disappointed that he had not found any snares to show me, our trip gave a sense of what it was like to fully devote oneself to the role of investigator. His extraordinary levels of concentrated attention struck me. A complete novice to such practices, I could immediately see how much preparation and knowledge was involved. As well as needing to know the variety of traps laid on shooting estates and where to find them, to properly document evidence, one had also to understand the licensing regulations attached to each kind of trap. What constituted a breach of license on a spring-loaded Fenn trap, for instance, was quite distinct from a breach of license on a noose-tightening snare, which operated without bait, and likewise from a breach of license on a cage trap, which depended on the use of live decoy birds (see Reed 2016a). Welfare regulations pertained not just to the state of the trap but also to the regularity with which it was checked (in theory, a set snare was meant to be visited every twenty-four hours to ensure that trapped animals were not left suffering for too long) and to the method of dispatch or kill. Barry would require similar knowledge of licensing regulations for all the other kinds of institutions and places—zoos, laboratories, circuses, slaughterhouses, and so on—where he conducted investigations. Additionally, to correctly record what he found, Barry needed to know the wider law on animal welfare and on criminal responsibility. What should he photograph, for instance, if he discovered a pole trap? Illegal across the UK, these traps are designed to catch hawks and other raptors by placing a steel jaw trap atop a high wooden pole. Or how should he document and treat the scene if he did find a jar of carbofuran? As his opening reflections on undercover work at the circus illustrate, this often required Barry to think carefully about his own role as a witness and especially about the hazards of corroboration (see Reed 2016a). And all this forethought does not even include the preparation and knowledge demanded to collect evidence relevant to a particular campaign.

Our trip also highlighted the self-taught nature of investigation. Barry told me that when he began work on shooting estates, he knew very little about their practices. In fact, Barry insisted that much of his current attuned reading of the landscape, his almost forensic attention to a range of multisensorial clues, was a simple product of devoting long hours in this remote countryside and across different seasons. The obviously methodical dimension of his actions—slow, alert strides, interspersed with stops to look through binoculars, sniff the air, or bend down to carefully examine the ground, followed by a careful recording or GPS-integrated video notation of what he found—was therefore hard-won. While Barry regarded the operation as an overt investigation, he preferred to keep his activity on estates unnoticed to ensure he could work without disturbance. On occasion, this led to awkward and sometimes even heated encounters. A certain frisson of danger was added to the proceedings by the knowledge that gamekeepers nearly always carried shotguns. However, the solitary aspect of this investigation belied another crucial aspect of his method. As with his other projects, Barry prided himself on getting to know and anticipate the behavior of the persons in charge of the abuse he wanted to document. Although not working undercover and hence never in close quarters with gamekeepers, Barry nevertheless strove to understand and predict their actions by putting himself in their shoes. This was partly based on intuition and the few amiable conversations Barry did manage to have with gamekeepers over the years, but more essentially it relied on that search for clues, which Barry invariably figured as a reading back of what he took to be the gamekeepers' own methods of land management.

As well as making me feel that by comparison I noticed almost nothing about the welfare or animal protection implications of the environment around us, the visit reinforced my sense of the energy, time, and skilled commitment that lay behind Barry's determination to dedicate his life to animals. Indeed, Barry was very much aware of the costs and sacrifices involved. Once one developed a capacity for investigation, Barry explained, it was very difficult to ever fully relax or close your mind to the discipline of an evidence-gathering compulsion. On holiday, Barry liked to go climbing and kayaking; he also enjoyed walking in the countryside with his girlfriend and her dog. However, Barry reported that the investigative instinct often waylaid his enjoyment on these excursions. It also frustrated his partner, who would complain that on their walks together he went too slow or got too easily distracted by what he saw, heard, felt, or smelled in the landscape around them. As Barry pointed out on our drive back to Edinburgh, it was almost impossible to switch off. In fact, this journey was itself peppered with investigative insights shared along the way. At one moment Barry indicated, for instance, that we were passing through a valley where he knew fox hunts still took place illegally. At another moment Barry asked me if I knew that the roadkill we had passed was often actually dumped there by gamekeepers or farmers, as a handy way to disguise the evidence of their illegal or unlicensed persecution of badgers and raptors. While that instinct was containable when investigations were restricted to a facility or institution, it became much more difficult, Barry admitted, when one started making the countryside an investigative canvas.

If moments of exposure, when the cruel practices of those Barry observed or infiltrated were brought to light, might be read as opportunities to resolve or conclude that dilemma, then it is important to highlight that such moments were few and far between. It was very rare, for instance, that Barry's investigations led to a successful legal conclusion, in part because the covert nature of his work often compromised the admissibility of recorded evidence or his own witness statements in court (see Reed 2016a) but also because that work was often directed toward the collection of information for other purposes, such as specific campaigns or policy interventions. Likewise, the public exposure of systemic abuse was a rare occurrence, and when it did happen, his participation was usually disguised. As a rule, it was not good practice for investigators to expose or reveal themselves, even after an investigation ended. So, most of the time Barry remained a guarded and highly constrained figure. "You know, I sometimes feel alone," Barry admitted, "because obviously I can't discuss a lot of my work." In fact, many of those closest to Barry, both in the office and at home, had learned over the years to not even ask him about his activities. "So, you are by yourself," Barry explained, "and any emotions that you do have you have to keep to yourself." That solitude could be accompanied by more other-directed suspicions that were also part of the investigator's professional attitude: "Personally, I trust nobody," Barry once told me.

But the cost of being an investigator was not just about the inability to escape that investigative impulse and its accompanying suspicions. The real sacrifice, Barry insisted, lay in the necessarily altered relationship to the moral passions that typically motivated animal people. Feelings of both love toward animals and outrage at their human-inflicted suffering carried risks for both the overt and undercover investigator. Indeed, Barry located the true discipline of his craft in that ability to manage or control these emotions, an absolute necessity if Barry was to become an effective instrument of monitoring or gatherer of evidence. When Barry told me and other colleagues that he wasn't an animal lover or that he wasn't sure that love was the right word to describe his admiration for animals, Barry was also alluding to the fact that, for him, expressions of animal love were simply not a viable professional option. This was the case despite those emotions and the states of moral patiency that they animated remaining at a base level essential. For while Barry believed that the natural tendency of the animal person must be disciplined to become a competent investigator, he also believed that he could not do without it. If Barry entirely lost a sense of outrage, indignation, sympathy, or distress, for instance, then he knew that he would simply lack the will to keep going, to continue placing himself and his body—the Scottish weather meant that much of his snaring research was carried out in the wet and the cold—in these extreme situations. So, the sacrifice of investigation required a double shift. Like other animal people, Barry needed to keep a channel open to patiency, to an experience of all the passions that prompted an ethical stance, but at the same time to deny its purchase or to submit to a form of doing that actively and quite ruthlessly subsumed it.

The importance of this latter action was hard for others to understand. Barry told me that both colleagues and supporters often quizzed him about the consequences. "People always ask," Barry stated, "'how can you do the job that you do?'" Such a question usually had supplemental queries attached. "'Because we know that you are passionate about welfare and the protection of animals,'" they might next affirm. "'But your job, particularly when you do infiltrations, is to stand and watch animals being abused and die. Surely it must affect you in some way to allow that to continue?'" Privately, Barry conceded the point. "I've lost count how many times I've actually cried," Barry once told me before quickly adding a caveat: "But after the event, sometime down the line, after it's all over, reflecting back on what I saw, you know, what I was involved in." For Barry, that delay was imperative, a mark of his professionalism and self-control. "I'm not emotional enough to not be able to do the work," Barry added with emphasis and then continued more forcefully, "in fact, I cut myself off from that emotion." The cost of that disciplining was also not lost on Barry. "Maybe I have had issues in my own personal relationships because over the years I have learnt to do that," Barry admitted. But as Barry knew full well, the underlying implication in these inquiries was not just how he could stand by and watch human-inflicted animal suffering but also how he could participate in it. For even if he was not the person administering the violent intervention on an animal, his undercover role meant that Barry was never simply observing. To be effective, Barry had to be involved in some way in the institution's working practice, given a task in the processes that led to systemic abuse. That aspect of the sacrifice, more than any other, was almost impossible for his colleagues and fellow animal people to comprehend.

Barry's own response was to remain rigorously focused on the goals or moral duties of investigation and on the wider good (for animals) that a well-executed undercover operation, evidence-gathering trip, or exposé could produce. Indeed, it was the "lost opportunities" in past investigations, Barry regularly told me, that haunted him quite as much as the scenes of animal suffering that he had witnessed. Part of the reason Barry shared his memories of the undercover circus operation, for instance, was to illustrate the potential consequences of lapses in concentration. When Barry had first observed the circus hand beating the elephant with an iron bar, it was true that Barry started filming with his body camera, but the results were far from successful: "I got back to the caravan after the incident. But the camera hadn't worked. The off switch hadn't been switched on because I was too emotionally involved. And I actually cried because I had lost the opportunity because of it. I was really upset because I knew if this was it, I might never get that kind of footage again, and he would be allowed to continue what he was doing. So, I was upset for the animals. It was them I had let down." Although in this case the abuse carried on and Barry was therefore able to later capture it on video, other examples had less happy outcomes. Barry recalled episodes from undercover work in laboratories and zoos where the opportunity never arose again, and likewise episodes from his overt investigations on shooting estates. This included occasions when he had omitted to press the play button on his video camera when secretly observing a gamekeeper brutally dispatching nontarget trapped animals or when Barry had failed to measure the size of an illegal cage trap or to bag materials or animal tissues that he knew needed to be chemically tested to ensure an evidential standard. Each time the lesson was the same. "If you allow your emotions to go too far and get wrapped up in the situation," Barry repeated, "then you'll forget the reason why you are there, and you are there to obtain evidence only." That message was Barry's mantra. "It's the lost opportunities that I get upset about, and I think ultimately that's how I can cope with the work," Barry finally reflected.

Habituation Blues

For Barry, practices of habituation were another important aspect of the expertise and disciplines of investigation, which came with their own associated costs and sacrifices. Indeed, the processes by which animals were made to become used to something, so that, for example, they no longer feared it or found it distressing, was a central obsession of his work. From Barry's earliest days, as a young man absorbed with the idea of getting close to tigers and other big cats, Barry had spent long hours researching the methodology of habituation in the animal behavior field sciences. Seen as part of his skill set as investigator, that self-taught knowledge was something Barry sought to regularly operationalize. So was the peculiar relationship to nonhuman animals that habituation cultivated, something quite unlike the contact-based principles of animal love. In his investigative trips to shooting estates, for instance, Barry placed a high premium on stillness and careful movement not just because he concentrated on the search for clues and wished to avoid the notice of gamekeepers but also because he was highly mindful of the need to make himself unthreatening to the wild animals around him. That ambition became especially pressing when he anticipated or confronted the reality of individual foxes, badgers, deer, stoats, weasels, rats, crows, or birds of prey caught in traps. In fact, Barry invoked his conscious adoption of that methodology as a strategic resource in his legal and campaigning struggles, a move Barry was compelled to make in response to a common charge laid against him by gamekeepers and estate owners—that is, their much-repeated claims that the devastation and injuries Barry filmed and documented were the result of distress caused to trapped animals by the investigator's very presence at the scene (see Reed 2016a, 105). In his testimony, Barry was therefore always alert to the requirement to make clear the protocols he had followed, such as keeping a respectful distance from the trapped animal or ensuring that he remained out of sight and upwind.

Of course, practices of habituation were not merely choices Barry made from a concern for animal welfare or as a defense of the integrity of evidence gathered. They were also practices that Barry was obliged to carry out in his undercover roles, linked to tasks assigned to him in many of the facilities where he worked. In zoos, for instance, Barry had to participate in forms of training designed to habituate captive animals to the routines of their keepers and to other forms of intervention such as veterinary care. Barry had also been involved in attempts to ensure their habituation to environmental enrichment programs. Likewise, in testing laboratories, Barry had to conform to the strict procedures on handling, focused on rendering animals "deeply docile subjects" in preparation for experimentation (Lynch 1988, 289; see also Holmberg 2008; Friese 2013, 2019; Kirk 2016a, 2016b; Nelson 2018; Sharp 2018; Friese and Latimer 2019). As well as reward training designed to habituate laboratory animals to experimental methods, this included learning and reinforcing tightly prescribed behaviors—for instance, habituating mice to being picked up without causing them distress and hence disrupting test results through the repeated use of handling tunnels and various cupping techniques. Here, then, habituation served the welfare interests of animals but, more significantly (at least from the perspective of colleagues at the animal group), also the interests of those who used or abused them. In Barry's eyes, such practices were essential to the perpetuation of these institutions and to the systemic violence Barry witnessed because of infiltration.

As the variety of examples highlights, habituation was both diverse and task specific. In fact, as Candea (2010, 245; see also Candea 2013) illustrates, even within the practices of field scientists, that methodology is several sometimes-conflicting methodologies. Candea compares, for instance, the habituation practice of British field scientists at the Kalahari Meerkat Project, where Candea conducted ethnographic research, with that of certain primatologists advocating what they regard to be more radical approaches to doing habituation. While on the face of it the former aims to achieve a form of self-disappearance, to become from the perspective of the meerkats whose behavior they wish to observe a "part of the scenery" rather than a social agent within it, the latter seek to achieve habituation by actively building social relationships with the primates they study (Candea 2010, 245). There are different identified virtues attached to each prac-

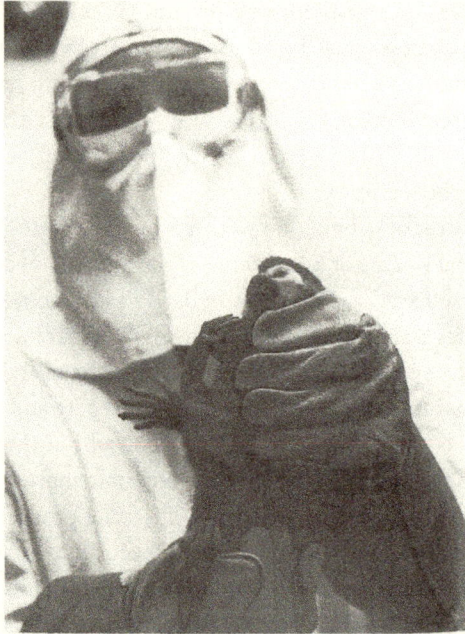

THE USE OF NON-HUMAN PRIMATES AS LABORATORY ANIMALS RAISES
SERIOUS ETHICAL QUESTIONS

FIGURE 6.   "The use of non-human primates as laboratory animals raises serious ethical questions." Image from 1991 annual report of animal group. Used by permission.

tice, but whether striving for detachment or closer integration with the social world of the animals or species concerned, the desired outcome, Candea argues, remains focused on the quality of data collection that results.

Although Barry intuitively leaned toward models of integration (in Barry's teenage years, he had enthusiastically consumed, like many other colleagues, the popular works of Jane Goodall), in most of his undercover work it was the virtues of detachment that dominated the proceedings. Indeed, whether having to adopt those habituation practices himself or having to witness their breach, that relationship occupied much of Barry's attention, especially as it pertained to animal handling.

To me, though, there was an equally important sense in which habituation mattered to Barry as a human-oriented strategic exercise. In fact, in many ways, that conflicted methodology was better represented in the tone and manner of Barry's undercover relationship to coworkers. It was certainly essential that Barry as a moral agent of animal protection entirely disappeared from the scene; the success of Barry's covert monitoring depended precisely on the ability to meld into the landscape of the facility or at least in being able to cease to draw attention to himself. Playing the idiot was one flamboyant way of doing this. However, Barry also used quieter strategies of inconspicuousness, most obviously through adopting a simple heads-down, getting-on-with-it approach to whatever mundane tasks his employers at an infiltrated facility assigned Barry. To redeploy a distinction that Candea (2010, 249) invokes to describe the habituation methodology of volunteer field scientists at the Kalahari Meerkat Project, this involved the cultivation of "inter-passive" rather than "interactive" sets of relationships. In the example explored by Candea, the emphasis lies on a quality of connection with animals that rests on willing or unwilling "pacts of inaction" between them (249). Treating the field as a lab, Candea tells us, these volunteers "refrain from certain active interventions, and the meerkats refrain from doing the first thing meerkats usually do when faced with human beings, namely, run away." While Barry's inaction did not signal a pact—by definition, Barry's coworkers at these facilities were simply not aware of his infiltration and therefore of the discipline involved in Barry not speaking out—it did reference a faith in the monitoring potential of passive relationships based, at least from Barry's perspective, on disengagement.

But Barry's habituating practice also included more actively integrating himself into the workplace culture of his coworkers. This involved not just being accepted at the facility or learning workplace linguistic codes but actually becoming liked, or as Barry once put it, "one of the gang." In this regard, collegiality, friendship, and other conventional interactive social relationships could become a legitimate strategy within any covert operation, just as they also sometimes were within the animal handling procedures that Barry learned. As Lynch (1988, 280) long ago pointed out, formal detachment measures in the testing laboratory linked to experimental preparation and the scientific rendering of the "analytic animal" began with and continued to rely on an informal empathetic orientation toward the "naturalistic animal" that, for instance, one needed to soothe and calm by stroking but also at times trick or con into submission through the reassurance of known and cultivated relationships with their handlers. Although others might view the transplantation of such interactive practice as a manipulation, Barry (and many animal activists he knew) preferred to identify it as another form of personal sacrifice linked to his role. Partly this was because it required becoming close to persons whom Barry might not like or whom he even actively detested, such as the "vile and disgusting people" he exposed at the circus. Yet just as importantly, it was also because the simulated act of friendship or collegiality sometimes produced genuine feelings of attachment. Indeed, the very success of Barry's attempts to make others habituated to him could sometimes result in those calculated relationships appearing as "real" affective or intersubjective

ties. The disillusionments that followed, like the inevitable severing of those relationships once the job was done, caused Barry some hurt and grief, as it did,
he acknowledged, for those he left behind. Barry's narration of past undercover
operations was peppered with shocks and regrets born of that kind of recurring
dilemma:

> For example, when I worked at a testing laboratory in Yorkshire, I noticed
> a girl in the canteen. I was single and quite attracted to her. I was build-
> ing up to talk to her, chat her up if you like. But then there came a time
> when we were gassing all of these monkeys. There were forty monkeys
> involved, and we had to kill them all in a day. You know, we would take
> two down at a time. They were injected with ketamine first to make them
> a bit sleepy. And with necropsy, you know what a necropsy suite is, right?
> [I shake my head] . . . Oh, necropsy is where they cut the lab animals
> up and take organs from them. In the necropsy suite, they have to take
> organs when there is no blood in the body, and so the way to kill an
> animal is to exsanguinate it, to bleed it to death. So, these monkeys were
> given this sedation and we took them down. We had known these mon-
> keys for weeks and weeks, so you got to know the characters of them.
> So, we took them down, and basically there is a table, it's like an autopsy
> kind of setup, and so the necropsy member of staff, they have a massive
> syringe, inject them directly to the heart. This is supposed to kill them,
> but before their heart stops beating, you have to hold them up by the back
> legs, cut the chest open, and bleed them to death. They are conscious,
> but they're sedated, so they would be aware of what's going on, but they
> would be unable to do anything. And who did this job? My friend [he
> laughs disconsolately], this girl I really liked!

This passage is notable in multiple ways. Although its denouement was clearly
meant to be horrifying, the nature of that horror carried at least a double dimension. In explaining to me what happened in a necropsy suite, Barry was describing a scene that he anticipated might disturb me and that he certainly knew
would disturb the imagination of animal people. Most pressing, especially for
those with an investment in the value of recognizing animal sentience, was the
image of these forty monkeys each sedated but nevertheless conscious throughout the procedure that led to their deaths. However, the fright that Barry felt
was directed elsewhere. For as the narrated details of this incident make obvious, necropsy was an already-familiar procedure to Barry, part of the broader
systemic violence of the lab that Barry was tasked to covertly document and that
he had trained himself to endure. In Barry's case, the true horror of the episode
lay in the shock of discovering who was delivering the final and in Barry's mind

most cruel and unfeeling element of the intervention: the girl that he fancied and wanted to ask out. Knowing that she was responsible for this task horrified Barry in a compound fashion—the issue of how she could do it merging with the issue of how Barry could possibly have liked her, and so to a period of intense self-questioning.

It is important to stress that this dilemma was not due to Barry's participation in the proceedings or to the participation of others in the handling team. Indeed, another striking aspect of the passage is Barry's relatively untroubled and consistent use of the first-person plural pronoun "we." There is a strong sense here of the investigator assuming a collective relationship with his coworkers, in part defined by a series of actions performed in common. "There came a time," Barry narrated, "when we were gassing all of these monkeys . . . we would take two down at a time," and so on. In fact, that pronoun remains in place even when Barry broke from an account of procedure and from the formal protocols of detachment around animal handling in the lab to reference the growing knowledge that he and other team members had acquired of the individual "characters" of these monkeys in the weeks leading up to necropsy. The succinct communication of that relationship (i.e., to each of the forty monkeys) could be read as having the narrative effect of individualizing and hence reinforcing the horror of their deaths; certainly, both Barry and the rest of the handling team found the task emotionally difficult. Yet the concluding lesson that Barry drew from his narration of this incident as a whole and especially from the shock of finding out what the girl he was attracted to did in the necropsy suite was once again the importance of self-discipline. Reflecting on it, Barry told me that the main problem had evidently been that at the time, his affections for the female colleague were "out of control." Barry's last words on the matter were clinically strategic. "You know," Barry mused, "if I was going to do it [i.e., start a romantic relationship], then I should only really be chatting her up to obtain information from her."

But the aspiration to control feelings or instrumentalize such interactive social relationships remained a challenge. Despite the necropsy incident, Barry reported that he continued to be a casualty of the relationships that he made in the cause of habituation. One friendship made soon after this event, at the very same facility, stayed with Barry as a continuing object lesson of what not to do. "There was one guy in the laboratory who I kind of bonded with," Barry told me. "A really nice guy, quite young, maybe twenty-eight or so, a farmer, although he'd come away from farming because working in a lab gave him more money," Barry elaborated. "He was very interested in the environment, wanted to get back to farming and set up an environmentally friendly farm"—and then Barry added with marked emphasis, "and he was uncomfortable about what he did." Barry described how he used to be invited to his coworker's house, how he even met his family and sometimes used to have Sunday dinner with them. Barry shared how his friend used to fix his car for him when it broke down. Barry paused. "And there was this monkey, a wild-caught macaque." Barry paused again. "Every time it was taken out of its cage, it put its feet and hands up against its face and just screamed." In fact, Barry told me that his friend, charged with handling this lab animal, gave the macaque a nickname that stuck. Everyone called the monkey Rape, he explained, because the jarring screaming sound that it regularly emitted sounded like that word, "Raaaape!" Since the animal was so hard to work with, handlers regularly got exasperated and at times became abusive. This was something Barry first noticed because he once happened to walk past his friend's station door. "He would shake it [the monkey], and he would punch it and slap it across the face," Barry outlined. Knowledge of that mistreatment deeply troubled Barry, but despite it, the two men remained friends until the day Barry abruptly left the laboratory.

In this case, the witnessed incident of abuse was somewhat different. Unlike the events in the necropsy suite, his friend's outrageous behavior toward the macaque was not part of the formal procedures of the facility. It was a clear breach of the protocols of animal handling and of the principles of habituation. Such uncontrolled violence disturbed Barry, yet it was also in some ways easier to understand. "I mean, what he did with Rape was absolutely unforgivable," Barry stated, "but generally he was OK with the other monkeys." In fact, it was their friendship that made Barry's sudden departure and the media exposé of the laboratory that subsequently followed difficult—so much so, that a few weeks after the investigation, Barry phoned his former coworker and apologized for deceiving him. "Because he was such a nice bloke," Barry explained, "I felt bad, I felt guilty for what I did," as if he owed him an explanation. I asked how his friend had responded. "Oh, I think he cried," Barry reflected, and then with considerable sympathy he added, "You know, I was there, and I was this character, and the next minute I wasn't." Barry felt that the phone call had helped the other man comprehend what had happened and where Barry, as a covert investigator, was coming from. "I can't remember my name at the time, because I have used so many different names over the years," Barry added, "but I believe he said to me, 'OK, so you were doing a job like I was doing a job.'" His ex–lab mate told Barry that he had now left the facility and gone back to college. "I really appreciate you phoning," were his friend's last words before they hung up.

Despite the evident gratitude, the two men never spoke again. Indeed, Barry now saw the phone call as a mistake or as a kind of rooky error. "I have realized I shouldn't feel guilty," Barry told me. Drawing on a familiar theme, Barry continued, "I've learnt over time that you are there to do a job and you shouldn't care about the people, or you shouldn't consider their feelings when you are investigating." When I suggested that his action could be interpreted as a welcome act of compassion, Barry swiftly interrupted me and returned us to the priority of his ethical mission and to the standards of professionalism it drew from him. Above all, Barry stressed that such actions were unsustainable for an investigator of animal abuse:

> After that job, I stopped thinking about why these people do these things, because if you think too much, you become crazy. And also, it can cloud your thoughts about why you are there. You are there for one reason. Don't worry about why the people do it, why they don't do it. You are not a psychologist. You are not there to judge. If it helps with the job, then yeah, try to work out why they do it, so you know when they are going to do it again, or something like that. But you know you are there for the job; you are there just to attain the evidence of what these people do.

This is much the same sentiment that we have heard before. Whether overtly or covertly gathering evidence, Barry believed that he had to learn to discipline not just his emotional responses to animal suffering (and to any positive intersubjective relationships that may develop between Barry and any lab animal that he might be put in charge of handling) but also his responses to those human relationships that were the artifacts of infiltration. As already indicated, both these detached stances required a further disciplining of Barry's response to the emotionally charged ethics of colleagues.

* * *

So, far more than any other figure at the animal group, Barry expressed a professional and personal concern to censor or regulate the kind of moral experiences that typically defined the ethical stance of animal people. I am referring to those experiences that ensue from what I have termed the moral patient position and the wider dynamics of moral patiency-cum-agency within animal protection (see chapter 1). Indeed, that effort, along with the unusual primacy placed on his strategic or calculative manner (a clearly agentive position that, as we have seen, when reduced to an investigative compulsion can itself reappear in the patient position), marked the moral exemplarity of the investigator in the eyes of others. In this regard, the investigator appears to be a rather more familiar subject, at least from the perspective of a tradition that centers attention on self-cultivated relationships by the exercise of care or restraint. Within anthropology's ethical turn, a lot of attention has fallen on the description of modes of subjectivation and the terms of self-interpretation and self-control by which moral subjects come to know and cultivate themselves. That literature generally assumes that we are dealing with subjects in the first person who may have conflicting values and suffer internal struggle yet nevertheless strive to be themselves or that version of themselves that most closely accords to the kind of ethical subject they wish or feel they ought to become. Although a focus on moral patiency—reported instances of being done to or acted on—may importantly resituate or even decenter aspects of that attention, it does not necessarily challenge the widespread assumption that moral experience consolidates a sense of who the subject is. Certainly, this was how colleagues generally understood the animating potential of those experiences.

However, a cautionary note highlighted by the anthropologist perhaps most closely associated with the legacy of the late work of Foucault points to an important exception. For there are some figures in moral regimes, Faubion (2013) argues, who appear to strive for exactly the opposite—namely, to want to eclipse or erase the self, to seek a moral state of "subjectless selfhood" or "selfless subjectivation" (2013, 288). Faubion offers the example of certain forms of Christian mystic. In Faubion's schema, such examples of "desubjectivation" have their value precisely in their anomalous status. For mystics are "maximally ethically ambiguous" (301), an extreme point that throws "the normativity required of the ethical practices of the mundane [world]. . . . in which the ethical Everyperson makes his or her home" into relief. Indeed, Faubion wants to suggest that the desubjectivating exercises of mystics enable them to serve as "systemic irritants" (304), to embody a capacity for "undoing" dominant principles of subjectivation and hence provide a notional space for the necessary work of "revision and readjustment and re-coordination" (302).

In some ways, I would argue that the figure of the covert investigator occupied an equivalent kind of maximally ethically ambiguous status for the animal movement. But at least in the case of Barry, that ambiguity was due not to a striving for self-erasure or the achievement of subjectless selfhood but rather to the sacrifice involved in needing to be or pretending to be someone else. Impersonation in and of itself necessarily complicates the picture of subjectivation, especially when it is performed in the service of ethical mission. Indeed, as we have seen, what especially marked Barry out in the opinion of friends, colleagues, and supporters was his commitment to impersonate subjects who were taken to be cruel and to be abusive toward animals. That ethical work then involved imposture of persons or behaviors that continually threatened to negate the conventional terms by which animal people related to the person they wished to become.

The peculiar status of the investigator can also be drawn out through a reconsideration of the totemic relationship that I discussed in the last chapter. I am referring to the observed relationship between a person or sentient being and the assumed-to-be signifier of that signified being: personality. To recap, colleagues at the animal group identified that relationship as central not just to the case for improved welfare conditions for nonhuman animals but also to an understanding of who they were, both from a perspective as themselves and from a perspective toward each other. What the category of personality in the science of animal sentience taught them to recognize or reemphasize was precisely the relative autonomy of that totem. This understanding was the result of reflections on the inevitable stretching involved in making a human-centric category fit for purpose in the study of the signified being of diverse species. Thus, Elaine came to a renewed awareness of the flexible dimensions of her own personality. This included a growing sense of the disparity between the personality she perceived and that experienced by colleagues. Partly inspired by the idea that animal personalities may be radically contingent on environment, it involved, for instance, a sense that Elaine's personality at home may be quite distinct from her personality at work. However, such reflections became of an entirely different order when one considered the imposture of the investigator. For while Elaine's personality might be subject to environmental change and divergently experienced in her own mind and the minds of others, it remained an undisputed signifier of her sentient presence, in the office and among colleagues.

By contrast, the characters or personalities adopted by Barry during covert operations were taken to be strategically cultivated and hence not true signifiers of Barry's signified being or person. In other words, Barry's relationship to personality confused rather than clarified the issue of who he was. This was the case not just from the perspective of those Barry infiltrated who took him to be Ozzy, for example, and judged that personality (i.e., a performance of the village idiot) as a signifier of the person before them. It was also the case for those who admired the investigator, including supporters who only knew him as Jim and helped fund the investigator's activities on that basis and colleagues who of course did know Barry yet didn't know Ozzy or the other personalities that they understood Barry became undercover. Finally, it was even the case for Barry, who knew himself as a dedicated and disciplined investigator working in the cause of animal protection but additionally knew that he left behind impressions of someone else, personalities habituated into activities and sets of relationships whose legacy, Barry admitted, he never fully escaped.

Circus Lies

Even though Barry expressed a growing disinterestedness in the motivations behind acts of cruelty toward animals or the reasons why ordinary people allow themselves to become involved in institutionalized systems of abuse, he was not immune to offering friends and colleagues a diagnostic assessment. Indeed, there was a kind of world-weary wisdom about these opinions. Combined with awareness of the investigator's mysterious participatory knowledge of abuse in action, this made those opinions compellingly authoritative in the eyes of many colleagues. Barry especially specialized in the analysis of certain mechanisms of human dissimulation. He professed an expertise in the terms of self-justification by which perpetrators of this cruelty convinced themselves and each other that these practices were acceptable or required. Barry even coined a term to capture the phenomenon as he had witnessed it during undercover operations. He spoke regularly of the need to detect and dissect "circus lies."

The term itself was a reference to the kind of covert work where Barry claimed that his understanding of the nature of that dissimulation properly crystallized. For, as Barry pointed out to me, the circus world was all about animal training, the mastery of which completely depended on a series of violent interventions. In fact, the more vicious and irregular violence Barry encountered there, such as the angry beating of a bull elephant or the punching of a baby chimpanzee, was in his opinion possible only because other forms of violent action were normatively integrated into training methods. At the winter quarters where Barry conducted his long-term undercover operation, the circus also ran a company that provided animals for film and television entertainment. It was here that Barry witnessed some of the worst training violence. Barry reported observing, for instance, that to get monkeys to stand up on their back legs when they appeared in shows, the animals' limbs were intentionally prolapsed, stuck up, and then sewn up in the right position by staff. At that time, violent measures were also integral to the training of larger wild animals in the circus, the whip of the lion tamer being just the most obvious and public example. Barry made the same point about animal handling in laboratories. In his experience, docility in lab animals was not just instilled through procedures of habituation; it was also often achieved through more ad hoc techniques of disorientation, sometimes violent in nature (claims that correspond with the early observations of Lynch [1988, 281], who reported witnessing a range of rougher measures being used to "soothe" a lab animal into a docile state in preparation for experimentation, such as swinging a rat in the air by its tail to induce dizziness).[1] But in terms of matters of dissimulation, the crucial point was that circus trainers or laboratory animal handlers justified such measures as unavoidable and necessary procedures.

Wherever Barry did investigative work, the same kind of thing occurred. Thus, Barry started noticing the moment and labeling the kind of lies particular to each facility or working practice. During our conversations, Barry would often refer to them:

I believe that comes to another circus lie. I believe that's a lie that they've got to learn to live with. In my experience of working in laboratories, I have had technicians and researchers speak to me. Sometimes if I feel it's safe, I play devil's advocate, and I have had them say to me, "Steve" (Steven was a name that I used in one place), "I wish I didn't have to use these animals. They give me the results, but I do it because there's nothing else. There are no other models to test on." Because to keep on researching, you have to produce so many [academic] papers to keep on going. And the way that the laboratory world is set up, in the majority of this world, animals are there to be experimented on, and animals are seen as the method to progress with the experimentation, the testing. But I know that there are other methods, and there could be more if more time and effort was invested in that. But, and it's not just one researcher that's said this to me, it's the only method that they can use to get their papers written and move on to another stage. And if they don't get these papers published by carrying out this research on animals, then they don't get the money coming in to carry on.

In such reflections, the quality of the circus lie identified often depended on an observed assertion among those Barry infiltrated that there was no alternative to doing what they did to animals. Barry's comments above swithered between a depiction of lab technicians and researchers as pawns caught in a system not of their own making and a depiction in which they acted in a self-aware fashion, deeply regretting the use of animals but still believing it was essential. Frequently, the terms of the lie noticed by the investigator closely matched the wider terms of dispute around a given issue, in this example drawing on long-standing arguments made in the antivivisection campaigns of the animal group and wider animal protection movement. For, of course, Barry's ability to define the claims of lab technicians and researchers as lies rested on a confidence that realizable alternatives to animal experimentation did exist. So, what the investigator presented as evidence of dissimulation could be reread as just another way of confirming entrenched positions. The idea that opposing interest groups representing scientific research or field sports or zoos or industrial agriculture regularly told the public lies about both the nature of their activities and the reasons for them was a common assumption. But at least as far as Barry was concerned, the more striking fact was that in his experience these people seemed to believe their lies.

Indeed, Barry was quite genuine in his desire to comprehend how people lived with what from his perspective were clearly wrong and unjustifiable actions. "Over the years, from early on I started to try and understand, get behind these people's minds, why they did what they did," Barry explained. "Because I'm not saying that everybody I worked with were animal abusers or were malicious," Barry added. In fact, usually he wanted to claim quite the opposite. "Yes, they did hideous things to animals, but they were very pleasant people. You could have a beer with them in the pub, for instance, and not guess for a moment what they did." As we have seen, that observation came directly from the knowledge gained through undercover work. It was the kind of insight that once again marked Barry out from colleagues and other animal people, who might be less willing to distinguish the perpetuator of animal cruelty from their deeds (see chapter 7).

What struck Barry was the capacity of ordinary people to sustain their lies or to live with extraordinary contradiction. Here Barry referred not just to the contradictions that he identified in their specific circus lies but also to the contradictions he witnessed between their working practice and the rest of their life. Most obviously, his reference point was the relationship to the lab pal he got to know and like in Yorkshire. But Barry reported that across the years there were plenty of other examples. During our field trip to one shooting estate, Barry told me that of all the conversations with gamekeepers that he had had, there was one that he will always remember. In a very friendly exchange, this man emphatically stated that he could never shoot his own grouse, those he reared and managed. Yet, almost in the same breath, the gamekeeper announced to him that he could quite happily shoot the game birds of another keeper. "People are just weird!" Barry concluded. By way of further verification, Barry proceeded to recall the actions of a woman that he once knew at another laboratory. She was responsible for experiments carried out on beagles. Every day, this technician would receive several beagles, gas them, and force tubes down their throats. But what bemused and fascinated the investigator was the fact that at the same time, the lab technician would regularly ask her employers for a half day off work to be able to take her household dog to the vet. Such "disconnects," as Barry called them, were important since they continued to prove that most people were not straightforwardly bad. "She wasn't a monstrous person," Barry insisted; "what she and others do is monstrous, but they are not monsters."

Of course, Barry's remarks look like a concession only from the perspective of animal people and the broader negative attitude toward those who work in such facilities typically found within the animal movement. As many commentators observe, it is perfectly possible to identify moral regimes that can accommodate these apparent contradictions (see Sharp 2018; Svendsen 2021). Returning to Lynch's famous essay on laboratory culture, technicians and researchers do not just have to resort to the circus lie. In Lynch's account, they can also refer to a positive framework that considers the transformation of the "naturalistic animal"

into the abstracted "analytic animal" of scientific procedure as a moral achievement (1988, 266). Likewise, the idea of combining care with killing, including
heartfelt grief for any untimely death of animals ultimately raised to be slaugh-
tered, has a long history within other British animal handling practices such as
farming (see Law 2010). It is certainly true that in popular imagination, dogs are
usually exempted from the more unfriendly or exploitable categories of animal
classification in Britain; the thought of beagles as laboratory animals would and
does regularly trouble mainstream opinion. But, in general, the act of classification itself continues to be accepted as a legitimate way to make distinctions and
to absolve the activities linked to those distinctions from the charge of contradiction. Rather than worry, for instance, over how certain mice or rabbits can be
loved as pets, while others can be treated as experimental subjects or rendered
immediately killable as pests (typical sets of concerns among animal people), it
seems that the majority remain untroubled or simply do not recognize the terms
of that moral dilemma.

Barry, like other animal activists, refused to accept that indifference. Barry
never strove to empathize with moral traditions that might be associated with
infiltrated working practices. However, an insistence on the humanity of many of
those Barry covertly worked with did mark him out. Indeed, there was a strong
sense in which Barry understood contradiction to be an essential condition of a
normative moral life, not just for the mainstream but also for animal people (see
chapter 9). More than that, Barry sometimes seemed to express regret at what the
efforts to purify his own life of contradiction (and of circus lies) cost him. There
was nostalgia over lost pastimes, periods of his investigative prehistory when
Barry too once lived untroubled, alongside others.

As a teenager growing up in Highbury Corner, Barry told me, he had once
reveled in the opportunity to go coarse fishing. This was a sport Barry's father
had taught him. It was a "workingman's activity," cheap to pursue and a common
way for many boys and adult men in his neighborhood to get out into nature.
Barry would spend hours sitting by London canals and local rivers trying to catch
perch. In fact, in those years, Barry participated in and won many fishing com-
petitions. Despite a developing awareness of human-inflicted animal suffering,
Barry's passion for the sport continued. It was not until the RSPCA's Medway
Report came out in 1980, demonstrating in Barry's words that "having a hook
stuck in the mouth is as painful for them [fish] as for humans," that he became
convinced angling was cruel and with great reluctance resolved to give it up. A
few years later, Barry even accepted an assignment to go undercover with groups
of anglers to expose that cruel practice. The operation concluded with the publication of a short film, whose final scene showed Barry symbolically and with

some emotion smashing up his fishing cups. However, Barry told me that he still loved fishing and that he continued to miss it. As with so many other aspects of Barry's investigative role, having a high regard and respect for animals left him feeling ever more isolated. Barry reported a sense of alienation not just from other people and very often from the animals whose mistreatment he witnessed but also from a version of himself.

EVIL PEOPLE (AND THE BONDS OF RESCUE)

Of course, all colleagues wanted to see animals protected from harm. They were committed to improving welfare standards and more broadly to advocating for the autonomous rights of individual animals to live a life free from pain and free to express their positive sentient characteristics or normal behaviors. The animal group was fully committed to the Five Freedoms.[1] But it is worth reiterating a point frequently made by historians of the Anglo-American animal protection movement and illustrated across the previous chapters. These aspirations for animals were always also tied to aspirations for humanity and to civilizing dreams of moral improvement. Indeed, I was consistently struck by the fact that the animal group and those who worked for it were almost solely concerned with protection issues that emerged because of documented human action. It was harm done to animals by humans that mattered rather than the harm suffered in any generic sense. Where individual animals suffered without signs of human causality or where they clearly suffered at the hands of other nonhuman animals, colleagues usually imagined that it was not their role to intervene (see Reed 2017b). To make such intra- or interspecies situations a legitimate welfare or protection concern, they would need to demonstrate that humans were in some sense ultimately responsible for the conditions of that interaction. Individual suffering due to a drought or accelerated predator threats, for instance, would have to be shown to be the outcome of intensified human activity. In some ways, the growing recognition of anthropogenic climate change turned all animal suffering into potential animal protection issues. However, in the moral imagination of colleagues or animal people, it was still the direct and immediate consequences of our negligence or abuse and in particular evidence of human cruelty toward animals that provided the challenge.

Signs of that cruelty, indexed in the maimed or injured animal body and mind, invariably led colleagues to worry over the state of humanity. This included worrying over its future development. Individuals regularly observed or made predictions about a psychic crisis in the world. In diverse ways, they concerned themselves with the harm done to humanity by the harm we continued to inflict on animals. Barry, the animal group's investigator, offered a typical example of such anxious reflection:

> I've always felt, from the first time I was aware that people were cruel to animals, I've always had this strong feeling that it's wrong and it shouldn't happen. These animals are defenseless, there wasn't and still isn't, in my opinion, enough people doing anything about it. And as I've grown and thought about it more, I've come to understand that, you know, if we continue to allow ourselves to treat animals, especially animals who can feel pain and suffer, the way we do, then I think we will have a lot of trouble through our lives developing into more civilized people. And to use a cheesy quote, as Gandhi said, you know, you can judge a nation by the way they treat their animals. And, unfortunately, that is true. I mean, for instance, in this country, a country of animal lovers, this country is supposed to be one of the most civilized countries in the world, and we still allow animals to be bred, animals to be born, to go through a lot of suffering during that process of growing to adulthood. Specifically, only so somebody can take the pleasure in killing them again. To me, that is completely bizarre, that's savage. For a country that's supposed to be civilized! You know, I try to put myself in the mind of maybe an alien, a civilized alien, let's say, who comes to this planet, stands and watches this whole industry of these animals being purposely bred so a human being can then kill them, for their pleasure. And that's OK to do; in fact, that's legal to do. To me, that's incredible. And I think, as long as we allow people to legally take enjoyment or pleasure in killing animals, then I think we are not going to progress.

This is a very different invocation of alien presence from the one introduced at the beginning of the book. It is not a celebration of the sentient superpowers of diverse animal species; recall that the animal group's teaser film urges viewers to forget little green men and instead focus on or make contact with the amazing alien life that already surrounds them (see the prologue). Instead, we have an attempt to adopt the perspective of a quasi-human "civilized alien," this time not of this planet. In fact, in the quoted passage, Barry attempts to get into the mind of someone watching the treatment of nonhuman animals for the first time. This includes observing the industrial-scale exploitation, the extraordinary license given to cruel and perverse forms of killing. Although it is noteworthy how quickly reflections shift back to the first-person perspective of the investigator, the implication hangs that, like Barry, these civilized aliens would be shocked at what they found. Barry assumes that they would depart deeply pessimistic about the chances of humanity ever moving forward.

That this pessimism appeared grounded not just in the witnessing of mass industrial systems of animal handling and slaughter but also in the fact that the humans observed took pleasure in their killing should not surprise. As Song (2010, 39) emphasizes, the stress in the moral condemnation typically offered by the animal movement often rested on a claim that opponents appeared to "kill for fun." Indeed, Song identifies that claim as "central to the charge of cruelty" within animal protection (2010, 40). What apparently worried animal activists most was not the utility behind human forms of exploitation or even the sheer number of animals suffering and abused as a result but rather the cruel intentions sometimes displayed. Furthermore, as Barry's musings also illustrate, they worried about the cruelty exhibited toward animals by specific peoples. The passage's invocation of concerns about civilization and its progress switches between reference to our common humanity and the humanity of nations. It was especially incomprehensible to Barry that his nation, "supposed to be one of the most civilized countries in the world," should continue to allow people to kill for pleasure. Barry's principal example in our conversation was foxhunting and the annual grouse and pheasant shoots that took place on Scottish sporting estates.

Indeed, the mocking tone in Barry's voice when he repeated the popular cliché that Britain is "a country of animal lovers" was very familiar from my conversations with other colleagues. So was the implied comparison between his country and others. As we will see, focusing on intentionally cruel behavior often led colleagues to make observations about cultures of cruelty. These could be perceived as operating within or between countries, linked to certain activities such as field sports or certain industries such as dairy farming or the pet trade, or equally sometimes be assigned to a set of national characteristics or to the cruel culture said to distinguish a given region of the world. Turning their attention toward human cruelty then could produce a kind of negative sociology of behavior or anthropology of perverse or cruel societies.

In saying this, I don't want to overemphasize the prevalence of such an outlook on humanity. As we have seen, part of what distinguished the animal group, especially during the period when Eilidh was CEO, was a commitment to stressing not just positive expressions of animal sentience but also the positive ethical potential in all human beings. Part of the drive Eilidh led to engage the main-

UNCIVILISED BEHAVIOUR

FIGURE 7. "Uncivilised Behaviour." Image from 1995 annual report of animal group. Used by permission.

stream involved an active attempt to end the public impression of animal groups as "the antis," which included a struggle to appear nonjudgmental (see chapter 3). While many of the CEO's colleagues remained skeptical about aspects of this project, in their everyday lives they too were concerned to avoid preaching and particularly to resist some of the harsher assessments of human behavior that they recognized were available to them in the wider animal movement. No one wanted to claim, for instance, that kin, partners, or friends might be cruel or bad (see chapter 4). Even Barry, who alone witnessed some of the most disturbing practices of animal exploitation up close and personal, resisted that move. People might do monstrous things, Barry advised us, but on the whole, they were not monsters (see chapter 6). However, it is also important to register that the investigator's statement was made in full knowledge that colleagues or animal people did sometimes label others as monstrous. On occasion, colleagues expressed the belief that there really was "evil scum out there," to quote one of them. In fact, as Barry also well knew, expressions of the most heartfelt animal love could sometimes be accompanied by expressions of the most immediate or tangible sense of evil. This was another reason why the investigator abjured from embracing moral emotion.

In this chapter, I explore the stance of colleagues toward the category of evil and those moments when they claimed to feel or experience the presence of evil in the world or to encounter evil people. I do so especially by looking at the context in which they express their animal love perhaps most unconditionally—that is, through the extraordinary levels of care many of them show toward rescued animals. Although this care was not part of their formal working lives, as already discussed, ethical practice linked to rescue was central to their wider moral imagination (see chapter 2). Many of them volunteered at animal shelters or sanctuaries on the weekend or during their holidays, and all their companion animals were rescue dogs or cats. Likewise, those who left the animal group often did so to join or take a sabbatical volunteering in animal rescue centers. The only thing that Barry could imagine doing when he finally quit investigation would be working in an animal sanctuary. In fact, despite the animal group being a generalist animal protection organization, there were occasions in its history when it flirted with the idea of amalgamating with a rescue organization. Across the period I worked there, it maintained close ties with a center that specialized in rescuing injured and orphaned Scottish wildlife. A representative from that sanctuary sat on its board of trustees, and among the group's patrons were a few people who started and ran their own operations. This included someone who ran a shelter for Staffordshire bull terriers, a much-demonized breed in the UK, and someone who coordinated an online group devoted to rescuing and rehoming mistreated or abandoned cats and dogs from Eastern Europe. Rescue, then, was in the blood, and around and about the working day of the animal people that I knew best.

Problems with Evil

In the summer of 2010, soon after I joined the animal group, a story hit the headlines and grabbed public attention. Commented on and much debated around the proverbial office water cooler, on countless internet forums and social networking sites, the story remained in the British news for several months. It was subsequently picked up and followed by news agencies around the world. First promoted by the tabloid newspaper *The Sun*, the story centered on the report of an act of animal cruelty. One evening in an unremarkable neighborhood in the cathedral city of Coventry, an "unmarried bank clerk" was walking home from work (Parker 2010). As that person passed a row of pebble-dashed houses, they noticed a tabby cat sitting on a low wall. The clerk stopped, and the tabby raised itself and moved toward the figure to receive some strokes. After patting the cat in an apparently absent-minded fashion, the clerk looked over their shoulder to observe a green refuse bin. Then, with only a slight pause, they casually lifted the bin lid up, grabbed the tabby by the scruff of the neck and threw the cat in. The

FIGURE 8. Cat in the Bin. Comic strip. Artwork by Ryan Hamill. Used by permission.

bank clerk closed the lid. Briefly exercising fingers as if to shake off any residual cat hair, the clerk hesitated a moment before proceeding down the street.

But what made this incident especially newsworthy was the fact it was caught on the CCTV camera attached to the front of the house where it happened. When, after fifteen hours, the family eventually found their tabby cat mewing and covered in her own mess, the camera's owners checked what had been recorded and released it online. Very quickly replayed thousands of times on YouTube, it was this footage that journalists at *The Sun* picked up. Journalists at other British tabloid newspapers like the *Daily Mail* also closely renarrated the bank clerk's recorded actions, often frame by frame (see Ellicott 2010). Indeed, *The Sun*'s headline was informed by the announcement that the newspaper had managed to track down the culprit, named as Mary Bale and identified as from the local area (Parker 2010). The story soon gathered momentum. This was partly helped by an interview that Bale perhaps unadvisedly gave to the newspaper a few days later, where the bank clerk appeared not only unrepentant but also entirely bemused by any public outrage. "I really don't see what everyone is getting so excited about—it's just a cat," Bale was quoted as saying (Parker 2010). Those words only exacerbated the media interest and escalating online abuse that followed. The bank clerk's name became mud, calls began for Bale to be fired from the bank, and death threats ensued. Rather more prosaically, the Royal Society for the Prevention of Cruelty to Animals (RSPCA) pursued a prosecution. This eventually led to the issuing of a £250 fine. Ordered to pay costs and a victim surcharge, Bale was additionally banned from keeping pets for the next five years.

Quite early on, the issue of evil took center stage. By this point popularly known as the "Cat Bin Lady," Bale's actions were condemned as a "stroke of pure malice" in at least one newspaper headline (Ellicott 2010), and Bale was repeatedly condemned as an "evil bitch" on social networking sites. Extraordinarily, thousands signed Facebook pages entitled "Mary Bale is worse than Hitler," "Death Penalty for Mary Bale," and "Mary Bale is EVIL." Several tabloid newspapers, themselves never shy of applying the term "evil" to either the actions or persons described in its pages, mischievously stoked the proceedings, but it was undoubtedly the publicly accessible footage of the incident that allowed the abusive commentary to spiral. After a week or so, the footage on YouTube had received hundreds of thousands of views. People could watch Bale's cruel behavior countless times and analyze and discuss it in endless detail.

It was clear, though, that what to many viewers made the recorded action of Bale especially disturbing and hence "evil" was the matter-of-factness of the behavior exhibited. This was seemingly reinforced by the attitude implied in Bale's quoted remarks. Not only was there no obvious provocation for Bale's cruelty, but the clerk also had no history of dispute in relation to the cat or its owners. There also appeared very little forethought. In the CCTV footage, we see Bale for less than a minute before the incident, but there is no indication of any plan to stop. It was further evident that part of the interest of the story, linked to the rush to personify Bale as evil, lay in the clerk's ordinary appearance—that is, as a middle-aged, "grey-haired" (Ellicott 2010), and bespectacled figure who worked in a bank. So, this was not the usual suspect for opportunistic petty acts of animal cruelty in the street, in either popular imagination or media representations. As several observers in the British broadsheet newspapers commented, if a local male youth or group of youths had carried out the action, it would have attracted far less attention. While there was certainly something to say about the gendered dimensions of Bale's public demonization and in particular the repeated emphasis placed in the tabloid coverage on Bale's status as "spinster," as if being older and unmarried enhanced the story's sinister dimensions, it was that ordinariness that seemed to matter most. Bale was precisely the kind of person who went unnoticed, someone you might pass by without giving them a second thought.

In the offices of the animal group, colleagues responded to the breaking story with a mixture of emotions. At first disgusted by Bale's actions, they quickly became suspicious and then concerned by the nature of the coverage and aspects of the public reaction. Eilidh particularly worried about the consequences for the reputation of the animal group and the wider animal protection movement. For early in the story's history, the most extreme forms of abuse directed toward Bale were identified in the media as coming from enraged "animal lovers" and advocates of animal rights. In fact, the analysis by reporters and columnists in tabloid newspapers soon equally split between attacks on Bale and counterattacks on those animal activists seen to be channeling their anger and hate through social media. After their initial disgust, colleagues also became worried by the disproportionate attention this incident appeared to be gathering. It was, as many of them pointed out to me, hardly the most serious case of animal abuse that they had encountered. Not only did the cat survive comparatively unharmed by the ordeal; the violence of Bale's actions was also comparatively muted. Bale didn't, for instance, hit the cat or seek to otherwise maim and injure it. Bale's actions might have been baffling. (Bale appeared to partly share that sense: "I don't know what came over me, but I thought it would be funny to put it in the bin," the bank clerk told the same newspaper [Parker 2010]. Bale's mother also told journalists that in fact Bale "loves cats" [Parker 2010].) But the story was hardly worthy of so much attention or outrage. In fact, colleagues soon began to compare the coverage negatively against the limited media or public scrutiny focused on a whole range of much more substantial examples of cruelty toward animals. This included those examples regularly covered by the animal group, made the target of their organizational campaigns, and forensically brought to light through the investigative work of Barry. They particularly complained that the demonization of Bale eclipsed the institutional and routinized nature of animal abuse in Scottish or British society, to which the public largely remained indifferent.

But there was in turn something quite normative about colleagues' response—that is, the expression of suspicion or skepticism. As Csordas (2013) highlights, naming someone or some action as evil tends to draw out certain kinds of distrust from others, including from anthropologists. Few of the latter, Csordas observes, wish to take seriously the utility of the category of evil, either as a category of description or as a category of analysis that might have purchase within the anthropological study of moralities (see also Parkin 1985). Almost as soon as the presence of evil is recognized, countercommentaries about the problems or dangers attached to its invocation tend to arise (Csordas 2013, 525). The instinct of many observers is to immediately challenge the term's use—for instance, to want to demythologize evil or to reject its metaphysical connotations (2013, 525–26). This invariably includes stressing what risks getting obscured by its mention. In this move, emphasis lies in a loss of understanding, including the loss of contextual or situational knowledge about the behavior of given people and an overlooking of systemic actions assumed to be underlying it. Talking about evil therefore is often taken to lead to a sacrifice of objective precision. Csordas discusses the common view that reported instances of evil ought to be unpacked—for example, broken down into apparently more neutral "material" categories of abuse such as slavery, torture, murder, rape, or genocide (2013, 526).

Csordas also discusses the widespread assumption that evil simplifies or denies the complexities of historical causation and of human psychology.

As well as being echoed in the suspicions expressed by colleagues, this skepticism about evil manifested itself in much of the media response, especially from within the broadsheet newspapers. Their headlines typically reported the Cat Bin Lady story with a heavy dose of irony and sober reflection. Moral cautions about the dangers of invoking evil mixed with amused commentary on the "weirder truths of British life" revealed by the public reaction (Bell 2010) and in turn mixed with a series of factual corrections.

"Is Mary Bale the most evil woman in Britain?" one piece starts by asking its readers, before going on to immediately answer the question with a "firm no" (Bell 2010). In addition to casting doubt on the wisdom of portraying anyone as evil, the article seeks to provide context for Bale's action by exploring what prompts acts of cruelty toward animals. A psychologist is quoted, for example, to explain that "people who are cruel to animals usually feel a lack of power in their own lives. . . . They may feel helpless, lack authority and want to exercise control over something or someone who cannot retaliate" (Bell 2010). Bale's action is presented as potentially symptomatic of something else and therefore deserving of sympathetic understanding. "She may have been abused herself," the article continues to cite the expert: "She needs help" (Bell 2010). Another retrospective piece entitled "The Trial of Mary Bale" adds further individually specific psychological context (Hyland 2011). We are informed, for instance, that Bale had just returned from visiting her critically ill father in the hospital and that at the time she might have been suffering from depression. The columnist then shifts to a metacommentary to talk about bigger issues identified at the heart of this story, such as the spread of uncensored abuse on social networking sites or the problems associated with the common use of CCTV cameras in residential neighborhoods and its links to a broader surveillance culture. Overall, the effects of these rhetorical moves are to consistently downplay evil, both as a characteristic attributable to persons and their behavior and as a tangible presence or force that one might encounter or experience in the world.

The resonance between the skepticism toward evil expressed in such newspapers and that informing the attitudes of colleagues was not accidental. For it was precisely these kinds of newspapers and media sources that colleagues generally read and whose views they took seriously. The preferred broadsheet for most of them was the left-leaning, socially liberal paper *The Guardian*, which at the time also carried similar reports and opinion pieces about the Cat Bin Lady story. Indeed, those sympathetic psychological interpretations, of the sort that strove to contextualize Bale's actions, were entirely familiar to colleagues. As we have seen,

Eilidh's drive to get the animal group to engage the mainstream heavily invested in just such forms of understanding (see chapter 3). For example, Eilidh favored and strongly pushed the idea that cruelty was to be reappreciated as an expression of alienation, that animal abusers were likely past victims of abuse themselves, and that abuse toward animals was a likely sign of future abuse toward fellow humans. While not all colleagues were interested in redeeming Bale in this fashion—that is, from the tabloid and public accusation of inexplicable cruelty—they did all acknowledge that Bale's actions needed to be partly read as indicative of wider societal problems. Likewise, they concurred that the bank clerk and the issue of what Bale personified were in general terms a distraction from the serious business of exposing and petitioning against industrial levels of human-inflicted animal suffering.

But colleagues could not straightforwardly divorce themselves or their working practices from the Cat Bin Lady story or others like it. Part of the task of the animal group was to get their message across. This included persuading journalists, tabloid and broadsheet, to publish the group's viewpoint with quotes attributed on any relevant animal stories that they were running. Even though Eilidh wanted to reposition the animal group away from the traditional animal protection focus on suffering, this invariably meant having to respond to news stories of this nature. Indeed, since tabloid-fueled demonizing reports of individual acts of cruelty toward animals often received the bigger headlines and wider coverage, this was rather inescapable. For the animal group, then, the problem of evil was twofold. To skeptical-minded colleagues, it was not just mythologizing and distracting but also a form of narrating animal abuse that they needed to learn to negotiate and exploit out of professional necessity.

The issue cropped up regularly. It featured, for instance, in the movement-building work of David and Iain. At first, they tried to seed the positive sentience values of the relaunched organization and mobilize support across social networking sites on this basis—for instance, by soliciting responses to amazing animal stories that Craig posted or by encouraging the group's followers and friends to post accounts of their animal-friendly kind acts. However, the results were poor. All their attempts to experiment with search engine optimization or to strategically hashtag in the hope of catching the attention of high-profile, usually celebrity Twitter users likewise produced minimal response. It was only when their strategies were redirected toward stories of animal cruelty, either begun on social networking sites or initiated by newspapers and then taken up by social media, that some improvement occurred. Riding on the back of Twitter storms, like that generated by the Cat Bin Lady furor, notably increased their traffic.

However, it principally fell to Euan, in his role as communications officer (see chapter 4), to negotiate with journalists and to find ways to take advantage of animal-related breaking news. Sometimes this task was prompted by the solicitation of journalists seeking a quote from "the antis" in relation to a particular animal cruelty story. In such cases, the approach was often unwelcome. It became a situation that Euan needed to manage, either by seeking to reshape the expectations of journalists and persuading them to use a quote that better represented the interests and public image that the animal group wanted to cultivate or by creatively batting away the requests. On one occasion, for instance, the group was asked to provide a quote in response to the news that Bernard Matthews had died. A bête noire of the animal protection movement, Matthews was a celebrity farmer who pioneered the introduction of industrial-level turkey production in Britain and became famous in the 1980s for appearing in a series of television commercials promoting the Bernard Matthews company meat product. As Euan immediately realized, the tabloid journalist wanted a hateful quote: "There was an element of him [the journalist] expecting us to go, 'Whoa, he's dead! Yipee!' You know, he wanted finger-wagging outrage, a kind of 'ding-dong, the witch is dead' angle to his story." Euan enjoyed disappointing the journalist: "It was the day Bernard died, so all we could really do was say we might have disagreed about his farming processes, but we wished to send condolences to his family." Needless to say, the quote wasn't used.

But at other times, a big story emerged that Euan knew he had to try to exploit, particularly when the story appeared to spark or capture a public mood. For Euan, the Cat Bin Lady story was a perfect example:

> Sometimes you just hit the zeitgeist in a way you just can't predict. It just so happens that you talk about an issue, and the next day or the day after, it becomes a huge issue. And if we look at things like when we launched [the rebranded organization], we were really talking about people becoming less kind to animals. And that coincided roughly with the story of a woman putting a cat in a bin, which went everywhere. And that immediately gave everybody, "Oh yeah, oh, that's what animal cruelty is, I get it now." Such moments don't happen that often, because everything is so fragmented now; there are much fewer shared media experiences [Euan references the diversification in British media due to the rise of cable television, the internet, and new social media]. There are very few times when we come together as a culture anymore. So, you tend to find that the news agenda is dominated by those few things that everyone will know about. And this was definitely one of them.

Despite reservations about the demonization of Bale and the misleading consequences of personifying Bale's actions as evil, for a period the story became a useful narrative template for communicating the animal group's new ethical mission. Indeed, in the radio, television, and newspaper interviews that Eilidh and Euan conducted at the time of the relaunch, it was a constant reference point. Making the connection was strategic, Euan advised, because Bale's actions made the moral concerns of the group potentially legible to everyone, to the mainstream "culture" as a whole. Doing so required a good deal of translation work and carried certain risks. As Euan well knew, the aim behind the organization's new mission was not really to discuss how people were becoming less kind to animals. It was rather to actively promote animal protection through the positive reinforcement of the virtue of kindness, precisely without direct reference to cruelty or the movement's traditional focus on graphic cases of human-inflicted animal suffering. In this regard, the details of the Cat Bin Lady story were usefully subdued. All the media interest might fall on the question of Bale's evil status, but the actual cruelty involved was not too distressing. In fact, the suffering of the cat was hardly center stage. Likewise, the ordinariness of Bale, which in the public imagination often made Bale's behavior more sinister, was in some ways conveniently pitched. Bale could stand for either a recognizable version of the mainstream that the animal group wished to engage or alternatively an illustration or warning of what could happen to the most ordinary of people—that is, if kindness and an animal-friendly attitude remained uncultivated.

But colleagues' skeptical or strategic stance toward the invocation of evil was not consistently held. As already mentioned, there were times when colleagues themselves recognized the presence of evil in the world or imagined that they in turn were grappling with evil forces. In that context, Mary Bale and Bale's actions were poor exemplars. Instead, evil was illustrated by violent and extreme acts of cruelty and by people who inflicted harm on animals in a serial fashion with vicious and unfeeling forethought. Colleagues did sometimes invoke the category of evil in the office, especially when riled up by shocking news of abuse or during heated discussions about already well-documented cruel practices such as those linked to blood sports or religious forms of non-stun slaughter. However, evil was most consistently invoked in the context of animal rescue. This was an outside-of-office commitment that reconnected colleagues to the much-valued experience of working with animals rather than just for animals or on their behalf. In what remains of this chapter, I seek to explore the nature of that doubled encounter, both with rescued animals and via them also with the evil people taken to remorselessly persecute those animals and to inflict terrifying levels of pain and suffering.

Healing Light

I always saw Cassie outside of the office. Indeed, we usually met at Cassie's home, a bungalow-style property with a garden that sloped down to a small river. Largely populated by Edinburgh commuters and middle-class retirees, Cassie's part of the village was a quiet spot. Only nine miles from the city center but beyond its encircling bypass, it almost felt like you were in the countryside. Cassie shared the house with her husband and with a host of cats and dogs that Cassie had rescued and either was fostering or had adopted herself. A tour of their garden revealed several enclosures, one of which was a secure outside pen for those of her cats that were blind; this was a specialized rescue interest of Cassie. One of the other enclosures was for rescued rabbits, and a third was for rescued hedgehogs. But our conversations mostly took place inside the bungalow's large conservatory. The air in there was often thick with released cat or dog hair. Amid household furniture interspersed with food bowls, sleeping baskets, and various-sized scratching posts, we often sat and whiled away a few hours:

> All of us are deep people. That's one commonality. We feel things passionately.

> How do you mean?

> Well, I know within myself that I would do anything for an animal. I would starve before they would starve. I would go into a river and rescue them. I am one of these people, you know, that will stop to get a slug or snail off the pavement or weep over a squished spider. We will do anything for animals, and equally we get totally torn up about what other people do.
> *Haggis! Haggis! Don't!*
> [Cassie breaks off the conversation to tell off one of her blind cats for playfully pouncing on another.]
> He's a rotten horror! He ambushes everybody all the time.

> So, you would do anything?

> We say that we are hardwired, and that just means everything you are is for animals. Right from the start, that's all we do. We think of animals. We love all animals. We would do our best for any animal, anywhere in the world.
> [Cassie pauses to get up and check on the location of another cat called Feisty, whom I had noticed earlier because of an evidently poor condition, which included a constantly running nose and tendency to go around and around in circles.]
> Oh, good. He's sitting quietly on the chair. *Look, darling, hi there, hi there, yes.* [Cassie strokes Feisty, who responds with a little meow].
> That's as loud as he speaks. He can walk. He's just very, very wobbly. You know, I think his legs have got to the point that he is not coping any-

more. He still jumps, but his back legs give way. *Hi peeps, Feisty darling.* I might have to bring him up for a cuddle. He tends to get a bit unsettled when he needs to be fed. *Are you hungry? Oh, you need your nose wiped.* He has just got constant respiratory problems from getting hit by a car.

[Feisty emits a string of little meows as Cassie cuddles and strokes him.]

Hey, hey, hey. I better go and feed him.

[We pause to allow Cassie to prepare the food, which needs to be liquidized and administered patiently with a syringe.]

None of you come to this later in life, then? There's no transforming moment?

No, it's just there. It's not something we learn. It's like a calling. You know, it's all we want to do, and it's all we'll always do. You don't give up on it, and you don't walk away from it, and you feel it intensely, and you just know it, and then you find it in other people.

The "we" Cassie referred to in this conversation was at one level animal people in general but also and far more pressingly those among them who, like Cassie, self-defined as dedicated animal rescuers. In fact, the collective invoked by Cassie was more specific again. Cassie meant to directly acknowledge the fellow members of her own rescue group, a transnational network committed to identifying, saving, and caring for abused cats and dogs from Romania and other parts of eastern and southern Europe. Initiated by Cassie and overseen from her laptop computer in the conservatory, the online group brought together local rescuers from across that region (after a while extended to include Egypt) with prospective sponsors in the UK, northern Europe, and sometimes as far afield as the United States. As well as funding the initial costs of rescue, which included often expensive veterinary bills and the immediate expense of feeding and sheltering rescued animals in situ, these supporters facilitated the complicated process of international rehoming. For after being saved and given sanctuary by a local rescuer, the cat or dog in question was typically collected and driven back across the continent to new homes or fosterers. Indeed, it was through this very process that Cassie's first rescued cat, the blind but mischievous Haggis, arrived at her door. Like Feisty, Haggis came from Romania, as did two of the dogs. Another dog came from a local Edinburgh animal shelter, and at the time of this visit, there was also a blind cat and a one-eyed cat from Bosnia and two blind cats from Cyprus. Much like herself, Cassie told me, the members of this online rescue group were "hardwired" to love and help animals, a "calling" that they straightaway recognized in each other.

Animal rescuers, then, identified certain qualities in common. As well as intense feeling, registered in the emotion of love and the ensuing hours devoted to caring for and rehabilitating rescued animals and financing that rescue, this manifested in expressions of grief. As Cassie emphasized, rescuers were "deep people." They would do their best for all animals and mourn the loss of any. Equally what they shared was a sense of this passion for animals as come unbidden. For them, rescue was almost a compulsion, an instinct that they could never remember being without. Of course, their practices also conformed to a wider ethics of animal rescue. Indeed, Cassie saw her home as a place of sanctuary. Cassie explicitly drew on many of the principles behind the animal sanctuary movement. Among them, as Abrell (2016a) describes in a study of US-based animal sanctuaries, was a common emphasis placed on creating spaces outside of property regimes and dominant forms of human control, where every animal's right to life was respected (2016a, 43). So Cassie assumed, as the tradition of sanctuary also implies, that the caring environment she and others in the online group provided for rescued animals was a qualitative improvement in their lives. It was a removal from sites of peril or torture and from experiences of unendurable human-inflicted suffering (217). While it might still be a form of captivity and it might still depend on transfers of legal ownership to protect the new status of these animals as rescued (314), Cassie insisted that her work and the work of the rescue group was a defiant experiment. Cassie absolutely invested in sanctuary-familiar forms of "care-based interspecies relationships" (59) and ideas of "interspecies community" (197).

However, my motivation for first visiting Cassie's home was not driven by knowledge of this project. Instead, I had gone to see Cassie out of an interest in meeting one of the animal group's chief patrons, a position to which Cassie had recently been appointed not so much because of her rescue work but rather because of Cassie's other profile as a published writer of fantasy stories for young adults. That initial visit occurred at the prompting of Eilidh. The CEO had suggested that Cassie might be worthwhile to meet, an introduction easily facilitated by the fact that the two women were close friends. Indeed, although Cassie was not a colleague in the strict sense of the term, I soon came to realize that she represented an important constituency or voice among colleagues or within the organization, as of course within the wider animal movement. Distinguished from those who simply got their companion animals from rescue centers and otherwise focused on their role as ethical consumers, dedicated animal rescuers like Cassie were completely committed to the practice of care-based interspecies relationships and tended to organize their lives around it. They presented an outlook and zeal for rescue and for the caring relationships that ensued. Many of them also claimed healing powers or a special connection with animals or had that reputation assigned to them. When Eilidh first mentioned that I should talk with Cassie, for instance, the recommendation trailed off into a mumbled reference to her friend perhaps having something of the "animal whisperer" about her.[2]

Among colleagues more strictly defined, one could also take the example of Elaine, the personal assistant to Eilidh during much of the time I worked in the office. Elaine had adopted several rescued cats and dogs as companion animals over the years. A little more unusually, Elaine had always tried to select older animals or those with underlying health problems, as she knew they were more likely to remain unclaimed. For a period, this had included the adoption of a few greyhounds, since she had once known a woman on the animal group's old board of trustees who specialized in rescuing and rehoming this breed of racing dog. But Elaine explained to me that she also liked to rescue those "little strays" that sometimes appeared in her garden or even on occasion tried to sneak through the cat flap of the back door. "Sometimes, they just come to me," Elaine revealed. As well as volunteering on weekends at a friend's rescue center where she fed and cleaned out the pens of dogs, cats, ferrets, and ex-battery chickens, Elaine devoted much thought to making her home and garden friendly to a wide array of wild animals. "My garden is completely chemical free," Elaine told me. "I put food out for the birds and make sure the feeders are well placed, you know, so the cats can't reach them." Her list of caring actions extended to smaller wild creatures too. "I also grow plants to attract butterflies," Elaine continued, "and I have a big area of long grass because the bees like sheltering there when it rains." She added, "I've got bits of wood rotting for insects too, which also helps the birds." Much like Cassie, Elaine placed considerable importance on the care of local spiders, especially household ones. There was, for instance, Elaine proudly told me, a big spider in the kitchen that she affectionately called Bertie as well as a couple of "corkers" that shared Elaine's bathroom.

In addition to trying to retrieve and release the birds and mice that her cats occasionally caught, Elaine continued to take in and look after any injured wild animals. Elaine rescued not only fledglings and nest-fallen chicks that she found in her garden or the surrounding fields of her West Lothian home but also, again like Cassie, fly-blown hedgehogs. Elaine also rescued any wounded badgers, foxes, or other victims of local traffic that she came across in the lanes. "I speak to them, you know, try and send them white, healing light, stuff like that," Elaine offered with an embarrassed chuckle. The same caring attentiveness informed Elaine's movements in the city. Elaine explained that she had more than once been late for work because of an emergency rescue. This included the time, Elaine recalled, when she found a wounded crow in the middle of the road just outside her office and had to rush the bird off to a vet. Indeed, most working days in Edinburgh, she encountered situations that prompted an impulse to rescue. "Only the other day," Elaine continued, "I went out at lunchtime and came across a guy begging on the street with a dog holding his cap in her mouth." Bending down for a pat, Elaine immediately noticed that the dog didn't wag her tail or even look up. Elaine worried that its jaw was sore from the repetition of this trained action and made a note to return later to check whether the dog was given regular rest intervals.

Both Elaine and Cassie stressed that the capacity to love animals went hand in hand with special or instinctive abilities to read their needs and wants. Elaine, for instance, told me that one of her rescued dogs had occasional fits and that she quickly found she could tell when he was about to go into one. "Other people would just think your dog is coming through to the kitchen, to stand there and look at you," Elaine explained, "but I knew he was going to begin fitting." It was the same with her cats: "One had high blood pressure, and I knew when he wasn't right." That sense was partly about learning to pick up behavioral signals, but Elaine advised that it was also about opening oneself up to the possibility of interspecies communication. "I knew when he was about to have a fit," Elaine elaborated, "because he would come and tell me." Cassie went further. "It's about an instinctive touch," Cassie explained, "being tuned in at an empathetic level to the animal in front of you. . . . Apart from the fact I literally talk to animals, it's sharing their body language, learning from them, and understanding them as they understand you." Eilidh's suggestion that her friend might be an animal whisperer derived from observing Cassie in practice, particularly from witnessing the ways in which Cassie drew out the trust of rescued animals previously too traumatized to accept human company. As Cassie acknowledged, her special strength was "stopping the hurt, making them loved."

But as all the other dedicated rescuers among colleagues at the animal group insisted, the greatest reward of rescue was the love and companionship that these animals gave in return. In fact, Cassie and Elaine sometimes assigned healing powers or whispering skills to certain rescued animals. Elaine, for instance, told me that her cats and dogs immediately understood when she was not at her best or was feeling unwell. "I mean, sometimes I may lie in my bed a bit longer than normal," Elaine illustrated, "and you can bet I'd have the dog and three cats up there checking on me." Elaine chuckled. "You know, asking, 'Are you OK?' . . . They can pick up if you are not right." Likewise, Cassie reported that her cats and dogs always knew when she had another bad migraine, and long before her husband did. Cassie had become even more aware of this fact during a recent period of more serious illness. "Now take Smudge," Cassie said, pointing at a big white blind cat sitting on a radiator that had arrived at Cassie's house from Romania since my last visit:

Smudge used to sing to me when I was ill, or, you know, upset. . . . I used to lie at night with her curled up next to me, and she made me feel safe, at peace with the world. I think she recharged my batteries. You know, she was with me everywhere, didn't matter what I did. If I went out [to the hospital], I came back to her. I would check she was still there, like touching base, and know that everything is going to be all right. I was really hurting, and she would come and sing to me, and she was with me wherever.

In such reflections, Cassie came close to suggesting that she had been rescued or saved by Smudge, a sentiment I heard repeated many times.

Rescuers, then, were deep people not just because they felt things passionately or shared experience of an impulsive love for all animals but also because their care-based interspecies relationships taught them that these affections could be reciprocated. Saving animals frequently led to intense kinds of bonds with specific animals and on occasion to defining relationships. However, that connection and the love that flowed from being able to read and anticipate the needs of rescued animals (and sometimes to sense that the rescued animal was able to read and anticipate your needs) was only one side of what united them. The passions stirred and affirmed by rescue also included rage and expressions of hatred toward their animals' tormentors. As well as introducing me to the extraordinary levels of patience and empathetic care required to successfully rehabilitate and heal traumatized rescues, it was Cassie who acquainted me with the struggle against evil. This included the assumption that animal rescue necessarily meant confronting the realities of what evil people do.

Evil Cultures

In a sideways review of the current anthropological literature typically assumed to address the ethical, Csordas (2013, 525) highlights the inclination to focus analytical and descriptive attention on "actors who recognise moral challenges and want to make the morally best choice." This includes authors that address themselves to the everyday struggles of the ethical subject, to the issues of freedom and human agency, and to the kinds of ethical work or regimes of self-management consciously performed. However, Csordas suggests that what this literature often ignores is the formative place of both the idea and experience of evil. This is a serious blind spot in the discipline's ethical turn, Csordas argues, since "in a sense if it wasn't for evil morality would be moot" (2013, 525). Whether figured as an external force that negatively acts on moralities or as something intrinsic to those moralities, the reported presence of evil reorients the shape of moral challenge and the nature of moral subjecthood described. It also throws up different sorts of questions. "Is it possible to be/do evil and not know it?" Csordas asks, for instance. Csordas also suggests that there may in fact be some value in taking seriously the elaboration of evil as a moral or existential category (2013, 526). For Csordas, part of the interest in doing so lies in acknowledging and exploring evil's evident descriptive capacity, which Csordas identifies as uniquely able to convincingly organize or draw diverse forms of abuse together.

If Csordas challenges anthropologists to hold their nerve and engage with the moral category of evil directly, there was a sense in which dedicated animal rescuers associated with the animal group threw down an equivalent challenge to their colleagues. Confront the ugliness around you, they implicitly and sometimes explicitly beseeched. This invitation addressed a twofold audience. On the one hand, it issued a challenge to those who supported the animal group's move to engage the mainstream and the accompanying organizational shift in emphasis away from animal suffering. On the other hand, it also challenged those who remained skeptical of that project and continued to stress the importance of publicly spotlighting suffering and confronting abuse but who still refused to recognize cruelty's evil dimensions—perhaps because those colleagues insisted on focusing on its systemic or industrial scale rather than the individual act itself. That distinction could also be drawn within the rescue movement; farm animal sanctuaries, for instance, tended to focus on the systemic cruelty within factory farming, unlike dog and cat shelters, where the individual story of abuse was often foregrounded (Abrell 2016a). For Cassie, that invitation was addressed to animal people in general. However, during our many conversations, its recurring reference point was Eilidh, the friend Cassie first met at university when both women were running separate animal activist groups. Indeed, Cassie's personal relationship to the organization's CEO was a common frame for such reflection. The observed differences in temperament and outlook between the two friends were frequently used as a basis to measure the difference between dedicated animal rescuers and other kinds of activists within the animal movement. First and foremost, that difference lay in the rescuer's willingness to accept the true horror of human psychology and society, which included the full force of evil as it played out in the extraordinary and terrifying range of our apparently endless capacity for intentional acts of cruelty toward animals.

Cassie's understanding of that horror directly emerged out of her growing involvement in animal rescue and from the experience of founding and running a transnational rescue network. As the coordinator of a Facebook group, Cassie received daily reports from rescuers across eastern and southern Europe documenting the abuses they witnessed or heard about in their local neighborhoods and the ensuing suffering inflicted on cats or dogs targeted for rescue. Indeed, each time I visited Cassie, we would sit down together at her laptop computer to scroll through the latest reports and review the cases Cassie was currently focused on for fundraising purposes. Cassie narrated the details of each case to me with a mixture of despair, anger, fear, and revulsion:

> Look, that's Bosnia [Cassie indicates the page now opened on her screen]. Our rescuers had a family of dogs living in cardboard boxes that they were looking after in the park. Then this guy came with a hammer and just killed the lot of them. That sort of stuff happens all the time. . . . And here's a dog that got beaten up with an iron bar and paralyzed. We rescued it, but this nutter went into the local rescuer's house and attacked the dog again. We got it on video camera, but they wouldn't take any action against the person who did it. . . . One of the horrible things they do is get a nylon sack and just put them all in and tie it up and dump them. Then they all die of starvation trying to break out. There's a lot of that kind of horrible stuff. If my dog strayed [into a neighbor's property] over there, they would just get their legs cut off and be left to die. That's what they do. They have a completely different mindset.
>
> This is Romania, and one of this lot [Cassie turns away from the screen to point to a blind cat sitting on one of the cat trees in her conservatory] was there. Well, their average answer [to dealing with a cat that is blind] is to chop its legs off and throw it down a hole. That's what they do. They are very, very violent. . . . [We scroll through a series of other pages.] He was stabbed and chained in the mud. . . . She was beaten up, dumped on the side of the road with a broken leg and acute internal injuries. Well, you can see how bloody hungry that dog is! A lot are left to starve. That's what we see every God-damn day, walking skeletons! Cars have hit a lot of them too, quite deliberately so. . . . The neighbors beat up the dogs and poison them, and they get left on the road to die. Some of the dogs we picked up from winter were frozen to the road. They are still alive, but they've been hit. People never stop.
>
> He was very badly injured, thrown in a minefield and forced to crawl into a river, which is where we found him. Amazingly, he recovered. We got him to a fosterer in the UK. . . . [Scroll down to another page.] This one was beaten by dogcatchers. She is going to have to have a leg amputated. . . . When they get shot—a lot of them are shot for fun there—their back legs don't work, and they end up with appalling drag injuries. [Scroll again.] That's a shelter. It's pretty grim. Nobody gives a toss in Serbia.

That's Spain, which is dreadful. Spain is horrendous. Spain and Greece are just a nightmare! You get photographs of dogs left on skips. They've been beaten up with a hammer, tied and left to die. . . . Worse than that, in Turkey and Greece dogs are getting raped. And that's quite common too. Generally, they get killed afterwards. . . . There you are! [Cassie found the page she was looking for]. "Abused and raped by children" [the title of the page reads]. This is actually Egypt. I mean, it's just bloody awful [Cassie plays me a brief video clip taken by a local rescuer, which shows some kids in Alexandria being chased off, as the sound of a dog yelping can be heard in the background]. The dog is not able to stand and is totally distressed, but all these people are walking on by. . . . What you are not seeing here is the torture stuff that comes up all the time. We are getting lots of animals set on fire or kicked to death. It's just endless!

Cassie was careful to stress that cases of extreme cruelty also occurred at home. Cassie was referring not to the muted violence in the reported actions of figures like Mary Bale but to something more full-throttle or unambiguously evil. "I mean, there are utter scumballs over here too," Cassie insisted, referring me to a recent newspaper headline in Scotland. It was about a man in Fife convicted for tying a dog that bit him to a tree, stabbing the dog, and then dousing the dog in petrol and setting the dying animal alight. However, Cassie's online rescue work had drawn her to a realization that in some places such cruel actions were endemic. While in Britain the mainstream was largely indifferent to animal welfare, elsewhere in the world it appeared that extreme cruelty toward animals often defined the mainstream. So, it was not just the "endless" abusive behavior of individual evil people that concerned Cassie's rescue group but also the abusive behavior exhibited by evil societies. This awareness developed as a simultaneously external and internal perspective on different national cultures in the various countries where rescues took place. As Cassie knew well, she might be criticized for judging these societies from a small village on the outskirts of Edinburgh, but Cassie insisted that others in turn needed to acknowledge that her judgment was informed by the reports regularly fed from rescuers on the ground—that is, from people who came from those countries and belonged to those societies. It was their desperation, their daily struggles, the incomprehension, and often verbal and physical insult that these local rescuers suffered doing their work that convinced Cassie that they and her rescue group in general were "working against cultures that loathe animals."

Online rescue work, then, quickly led Cassie and other members of her group to adopt a comparative perspective on the varying degrees of evil to be found in different societies or subcultures. Indeed, in Cassie's case, running the Facebook

group was a stimulus to explore wider examples, either through her own search-
ing of the internet or via the promptings of group members. On occasion, this
resulted in a dramatic transformation in her understanding of the realities of a
known culture of persecution; at other times it led to the introduction of new,
previously unknown horrors.

Cassie told me that for the longest time she had experienced a special love for
wolves, animals that she had felt a close affinity toward since childhood. Cassie
wasn't sure why exactly; there had been no native population of wolves in Britain for several centuries. In fact, that feeling had developed further as she grew
older and began to appreciate the species' persecuted status, especially in the
United States. Cassie had donated monies to North American wolf-protection
groups, which tried to combat the hunting and farming lobby's claims that wolves
were a threat or nuisance to livestock. However, it was the online viewing of the
annual aerial shootings of wolves in Alaska and footage of wolf hunts in Idaho,
hyperlinks both shared by fellow rescue members, that strengthened Cassie's
passion for the species and her commitment to help campaign against public
misunderstanding. Even more importantly, the internet provided access to the
cruel behavior and mentality of hunters. Cassie showed me images and videos of
trappers doing "unspeakable things" to wolves caught in leg traps—for instance,
standing to pose and laugh beside a caught but still living animal rather than
shooting it immediately to ensure a more humane death. "They are absolutely
ghastly, bloody people," Cassie exclaimed heatedly; "they think it's perfectly OK
to leave wolves in a trap to die or to let them chew their legs off trying to escape."

Among the long list of other horrors regularly viewed and shared was the
skinning, baking, or boiling alive of animals for the fur trade. Cassie had viewed
several video clips of such live skinning, reported as mostly taking place in parts
of China, the Philippines, South Korea, and Indonesia. This had included footage
of a dog being bound and hung by a hook before having the skin literally peeled
from its body like a sock, a scene that terrified Cassie but that she could not stop
rewatching. Although this cruelty was driven by utility rather than killing for
fun—its practitioners claimed that live animals were simply easier to skin than
dead ones—Cassie and her rescue group kept returning to this example. "We try
and understand what it is that they don't have that they could do that to animals,"
Cassie explained: "I mean, we look at them as completely soulless, because we
don't understand how anyone can skin a cat or dog alive; we just don't." Here,
the presence of evil, indexed in actions such as live skinning, prompted a reflection on what was absent in the cruel subject to enable this practice to take place.
Evil, then, sometimes occurred because human subjects lacked human qualities;
the indifference of the skinner to the suffering of the cat or dog suggested that
cruelty was practiced in an almost zombielike fashion. But it could also be identi-

fied through the more positive or active presence of evil qualities. Typically, this was measured by the degree of unnecessary violence in a cruel action, by signs of a sadistic element to abuse or evidence that inflicting harm and suffering on animals was a source of pleasure. Once again, the sense here was of evil as a force not just positively embodied in human actions but also capable of infecting or even defining the depraved shape of human society or culture.[3]

Such differentiating observations (e.g., "Spain is horrendous" or "Nobody gives a toss in Serbia") were often underpinned by a workaday theory of how evil generally emerged in society. This usually rested on an interpretation of "culture" as a phenomenon that negatively impacts the instinctive expression of animal love. "So, we believe really strongly that people have a natural bond with animals when they are young," Cassie explained, "and that all this cruelty follows only once they become involved in a certain culture." As she stressed, referring to the reported rape of street dogs by children as well as adults across southern Europe and North Africa, "It's something that that they learn from somebody else." Extreme cruelty toward animals, then, was an encultured activity, born of alienation from natural sympathies. "For us," Cassie concluded, "that bond was never severed, but for so many other people it is."

As Cassie readily admitted, this explanation resonated strongly with aspects of the Edinburgh animal group's new vision to engage the mainstream. In fact, Cassie's words were accompanied by an acknowledgment that she admired what Eilidh was trying to do, in particular the organization's plans for an early-years education program (see chapter 3). It was one of the chief reasons Cassie agreed to become a patron. Cassie, like her friend Eilidh, was totally committed to the idea of working with children. It was not just the presence of a tiny minority of animal-loving local rescuers in Eastern Europe that gave her rescue group grounds for hope but also the promise invested in cultivating future relationships with those rescuers' sons and daughters and by extension with a new, young generation across the region. However, that endorsement came with a heavy caveat. For unlike Eilidh, Cassie had considerably less faith in the notion that adults, especially those within national cultures or subcultures of extreme cruelty, could be reformed or reconnected with that natural bond. "We believe that it's something you can't change," Cassie once told me, "that people who are violent towards animals will always be violent towards animals." The experience of rescue taught them that there was no redemption for the grown-up perpetrator of evil; to think otherwise was simply naive.

In Cassie's eyes, such naivete could be linked back to the inability of many animal people, including colleagues at the animal group, to engage in a sustained way with the realities of evil. "Now, I am somebody who will look at it [i.e., online images, footage, and reports of extreme forms of abuse and animal suffering],"

Cassie pointed out, "but Eilidh, if you said anything to her, anything vaguely horrible, she couldn't handle it. She doesn't want to know." Although many colleagues understood that horrors are going on and could talk about cruelty and animal suffering in general terms, they did not want to see the cruelty and suffering for themselves or know its terrifying details and sordid, devastating immediate outcomes. "Well, that's a very common reaction," Cassie stated, "even amongst people that love animals. They can't cope with it, and they don't want to cope with it." This didn't mean, Cassie hurriedly conceded, that Eilidh and others like her friend didn't positively respond to what they did choose to know or hear about. Cassie highlighted, for example, that she knew Eilidh didn't allow meat in the house, a principled position that Cassie admired and contrasted with her own "reluctant vegetarianism" and with the fact that Cassie continued not only to buy meat on occasion but also to cook it for her husband and extended family. Likewise, Cassie was very grateful that Eilidh had recently agreed to adopt a tri-paw cat that the Facebook group had rescued from Cairo. Indeed, such acts demonstrated that colleagues and dedicated animal rescuers for the most part existed in relationships of encompassment; animal rescuers were, after all, also animal people. There was also a complementarity to their positions. In the case of this friendship, one life devoted to rescue balanced out against another life perhaps more consistently committed to campaigning and consumer activism.

Yet from Cassie's perspective and the perspective of her rescue group, an irreconcilable difference remained. "We know that we have to see the really horrible stuff because we've got to know what we are dealing with," Cassie explained. "We believe that if you don't see it, you don't know it, and you can't change it or fight it." There was here a vivid sense of the dedicated rescuer as the frontline soldier of animal protection, of rescue as a battleground where the virtuous forces of protection struggled head-to-head against the forces of animal cruelty, this time personified in evil people. Viewed in these terms, the animal group's project to engage the public without reference to brutality or the endless cruelty witnessed by rescuers was an exercise that had its value but that could never supplant the cold, hard facts of the real struggle. Cassie, like her rescue group members, sometimes dreamed of an alternative engagement with the mainstream. "I just like to think," Cassie once told me, "that we will link up more and more until we can start forcing them [the mainstream] to listen." Instead of gradually building empathy or supporting an animal-friendly capacity, this would involve shocking the public into self-awareness of the evil around them. In her more strident moments, Cassie envisaged compelling people to watch footage of the thousands of animals annually skinned alive, making them listen to detailed reports about stray cats and dogs thrown into mobile incinerators or injected with bleach or bagged and smashed to death with hammers or mutilated and set on fire.

But it was just as likely that the battle completely eclipsed any concern for the mainstream. The general pessimism of Cassie and her rescue group occasionally led them to re-present acts of extreme cruelty as more akin to natural behaviors and to figure the cruel subject as simply always so disposed. Such a shift brought the rescuer and the perpetrator of evil into a relationship of antagonistic equivalence. Instead of a distinction between an instinctive animal love and an encultured cruelty, they sometimes spoke of two sets of passions both mutually attuned to their respective moral source. "I think collectively we think that people who are cruel to animals are that way inclined, in the same way that we are not," Cassie once proffered. "They are not neutral; they will deliberately hurt animals because that's the trait they have got." Cassie paused for emphasis. "So, they have the opposite [trait] of us." In this strange ethical affinity, where rescuers and abusers shared opposite versions of the same patient-cum-agent status, it was perhaps the state of neutrality or indifference that emerged as the truly mysterious human condition.

Indeed, for Cassie there remained something chilling about the idea of an unimpassioned majority, a public who failed to feel deeply for animals (or to loathe animals with evil intent) and who in all probability would continue to do so. It made a more combative or polarized vision of good versus evil, rescuer versus perpetrator of extreme cruelty, an altogether more compelling proposition. As Cassie conceded, hatred of that perpetrator or evil scum often surfaced as the rescuer's more cultivated companion emotion, almost as strong and overpowering as animal love itself.[4] Its expression certainly drew them back into relationships of antagonistic equivalence, an irony that was not lost on Cassie. In fact, the puzzle of intense love operating alongside intense hatred was a frequent source of reflection:[5]

> See, we love anything, from earwigs to elephants. And yet there is a paradox there. Nearly all of us think that the only way we are ever going to stop people that hurt animals is an eye for an eye. So, try to make that one out if you can! We revere life, but [when we hear of abuses] the overwhelming reaction is to do the same thing back to the perpetrator.
>
> *Charcoal, neither of us wants a kiss* [her largest rescue dog comes for a nuzzle, offering paws up on both our laps]. *Now, now, now.*
>
> Whether that means setting fire to him [the perpetrator] or tying him up and throwing him into a river or torturing him to death or skinning him alive. You will get people, including me, who just think they will never change, and we wish they would rot in hell basically. That's a very explicit contradiction, isn't it? You know, there was this woman [a story recently reported in the British newspapers] who put her kitten in

the microwave because it scratched her, then took it out and said as if in surprise, "Oh yeah, it was screaming!" And it's just like that every day.

Did you fall off, Sweetheart? [Cassie responds to the sound of a light thud and the realization that one of her blind cats had toppled from a chair].

She must have tiptoed over the edge.

Sweetheart, that must have hurt, clanging down like that [Cassie rises to offer strokes and comfort]. *Yes, that was rotten, wasn't it.*

So, if I could stick them on the moon, all these people who hurt animals, and just leave them to die, that's what I would do. And that's what all other animal lovers would do too. We want them off the face of the planet.

At the beginning of this chapter, I invoked an alien perspective suggested by Barry to throw a critical gaze on what he regarded as the uncivilized behaviors of the human species. Here, Cassie reverses the direction of space travel. Instead of civilized aliens visiting the planet and witnessing with disapproval the cruel outcomes of our practices and pursuits, Cassie recommends the earth's natural satellite as a dumping ground for our planet's evil perpetrators—that is, "all these people who hurt animals." The move appears benign when compared to the other options offered by Cassie and the members of her rescue group. In their eye-for-an-eye morality, the abuser or torturer of animals should be harmed in like measure. So, the people who torch dogs or cats should in turn be set alight. Those who tie animals up and throw them into a river should themselves get bundled and tossed into the water, left to drown or starve to death. And those who skin animals alive should be hung from hooks and forced to endure the agonies of having their skin peeled away. The impulse to retaliate was central to the moral imagination of many of the rescuers I knew, but as Cassie reminded me, it remained just that, a set of violent thoughts and desires rather than actions. In the end, animal rescuers enacted love and care rather than revenge, while evil perpetrators were largely left unpunished.

Over the Rainbow Bridge?

In more forgiving moments, Cassie offered a vision of evil or extreme cruelty toward animals as a symptom of a shared fate. "I think if we carry on the way we do [toward animals]," Cassie once reflected, "we will just become soulless." In this inclusive iteration, that condition of lack threatened to define humanity in general. "A soul for me is something that is innate in humanity, it's the positive power in our life, how we relate to other people, how we relate to animals,"

Cassie elaborated, "but it's also what we are losing, which is why we are doing all the [bad] things we do." That creeping soullessness could overtake all of us, Cassie warned, even the dedicated animal rescuer. The thought gave her pause. But when Cassie continued, her words were more rather than less heated, and the inclusive pronoun had been dropped: "The people who are soulless are the ones involved in torturing animals." Cassie paused again, this time for emphasis: "I think they are ugly and self-centered and destructive."

Rendered as a form of combat, rescue work could require a certain battle-hardened disposition. It drew out virtues of heroic self-sacrifice but also left behind trauma and causalities. Even the toughest rescuers admitted that facing "the really horrible stuff" eventually took its toll. Sitting in front of her computer hour after hour, often late into the night, reading the latest reports of violent abuse as they came in, left Cassie emotionally drained:

> That's a real problem. I mean, I was talking to Eilidh about it. And she said, "You can't carry on doing this, because you will get totally sucked into it." You know, it's just one horror after another. My husband gets pissed off with me, because I sometimes need to scream. I can just end up weeping half the day. He's told me to stop going online. . . . In fact, he thinks I have, that I am now on the computer all the time only because I am writing [Cassie's sly expression lets me know that this is an impression she hasn't tried to correct]. . . . So, the problem is most of us are frazzled. The stuff we see is really, really nasty. Nobody wants to see it apart from us, and quite a few of us burn out. I mean [Cassie names one member of the rescue group who runs its petitions page], she is just on complete burnout now. She just can't deal with it anymore. It's amazing how many rescuers end up like this. Their marriages are often clanking too. There are an awful lot of rescuers who have got problems at home.

Part of the issue, as Cassie diagnosed it, was precisely the extraordinary capacity of rescuers to feel deeply. Unlike others, they could not simply stop engaging with the realities of evil. They could not switch their sympathies off or disengage to manage compassion fatigue in the manner that Cassie heard medical practitioners were trained to do. "That doesn't seem to work for us," Cassie told me, "because we respond to every damn picture we are shown. Mentally, we can't shut it out; we can't stop thinking about it." Again, it was the distinctive quality of animal love that defined the burden or costs of the struggle. "I cannot love a person I have not met," Cassie explained, "but I can love an animal on the other side of the world." Among the consequences of that peculiar affection was a massive escalation of the potential number of animals not just to be loved but also to be grieved over—an escalation that became almost unmanageable once their rescue work converted to the scale of the internet. "Every time we lose one, even though it's an animal we have not met or probably will never meet, we just cry buckets," Cassie said, "and that's every day." Committed to a life of almost unremitting grief, to the obligation to witness one horror after another, Cassie and her fellow rescuers appeared to sacrifice the prospect of ever recovering equilibrium or composure.

When I next saw Cassie, she informed me that her husband had left her. He was, Cassie explained, exasperated by the emotional energy she spent on rescue. In the end, his patience had finally worn down. For both, the tipping point came when Cassie lost Smudge, the blind cat that sang and comforted her throughout her period of serious illness. This loss had been especially hard to bear due to the nature of the death. After being rescued from Romania, successfully healed, and given sanctuary from harm, Smudge had not passed away peacefully. Instead, Smudge had died in sudden, violent circumstances, killed by a rescued dog that Cassie was temporarily looking after. Struck down by immense guilt at this failure of protection, Cassie suffered a deep and prolonged bout of grieving that neither her husband nor her family nor many of her friends really understood. Through that experience, Cassie came to a series of realizations. First, it became clear to Cassie that she had "ended up married to someone who doesn't feel for animals at all." Second, she had lost an irreplaceable companion. "I loved Smudge more than anyone else in my life," Cassie began to discern; "she was just a soulmate, one of the special ones."

This grief totally eclipsed the day-to-day grief she had previously experienced as a rescuer—that is, the grief felt for those faraway and unmet animals lost at the hands of evil perpetrators of cruelty. It also eclipsed any grief previously felt for other animals that had died in Cassie's charge, as well as the grief felt for departed family or friends. In fact, for a period of over four months, the death of Smudge totally eclipsed all other relationships or rendered relationship into a singularized experience. There was just Cassie and Smudge or Smudge and Cassie—two sentient beings who had been soulmates, who at some level understood each other completely, but who were now rendered asunder. "A lot of people believe in the Rainbow Bridge," Cassie once explained to me, clicking a link to a website that described the concept. It involves faith in a space "just this side of heaven" where humans and their closest animal companions will once again become reunited (https://www.rainbowsbridge.com/poem.htm; see also https://petloss .com/). Indeed, due to the high attrition rate of their rescue work, Cassie's group had added its own "Rainbow page" to their Facebook pages. When they lost an animal, someone usually posted a statement announcing that this dog or cat has crossed the Rainbow Bridge. "But, see, when I lost Smudge and they go, 'Oh, she's in a better place,' I think no, she isn't, because the best place she could be is with me and the best place I could be is with her." Fellow rescuers tried to reassure her. A few in the group were "animal communicators," able to get in touch with dead companion animals, and they claimed that they could see Smudge and that they knew Smudge forgave her. But Cassie could not see Smudge. Members of the group told Cassie that this was because her grief got in the way. However, Cassie remained skeptical. "I don't believe in the Rainbow Bridge, really," Cassie tearfully confessed. "If I did, I would just want to go to her now. But all I know is she has gone. I can't feel her, and I can't find her. And I have no idea where she is."

Describing herself as "spiritual" by nature but at the same time suspicious of established religions, especially Christianity, Cassie struggled with the whole tone of this afterlife notion. Cassie could see the appeal of collectively imagining such a place. "When an animal dies that has been especially close to someone here that pet goes to Rainbow Bridge," the website we read together told us. "There are meadows and hills for all of our special friends, so they can run and play together. There is plenty of food, water and sunshine and our friends are warm and comfortable." The sense of sanctuary continues. "All the animals who have been ill and old are restored to health and vigor. Those who are hurt or maimed are made whole and strong again." Likewise, Cassie could see the comfort in its peculiar promise of eventual get-together: "Suddenly he begins to run from the group, flying over the green grass, his legs carrying him faster and faster. You have been spotted and when you and your special friend finally meet, you cling together in joyous reunion never to be parted again" (https://www .rainbowsbridge.com/poem.htm). But to Cassie it was a fantasy—a necessary one perhaps, nevertheless still unreal.

What Cassie did know to be true was the enormous hole in her life left by Smudge's passing. "The minute I saw her dead, I just knew that was my life changed forever. I loved her so much. It's like there is no here and now for me," Cassie reflected. As well as the ongoing pain of separation—a fellow rescuer in England helped fund a series of sessions with a trauma counselor who told Cassie her grief was still too raw to be able to help her—Cassie knew the unabated pangs of guilt. But Cassie also knew the love and support of other animal people. Eilidh remained a good friend throughout this period, but it was the other members of the rescue group that made the difference. In the face of increasingly unsympathetic and baffled responses from family and from friends who were not animal people,[6] Cassie took comfort in the messages of condolence received. Rescuers from her network across Eastern Europe as well as members of the rescue group from Britain and North America bombarded Cassie with letters, cards, and lockets. These included pictures of Smudge that a few of them had drawn from the photograph Cassie uploaded on their Facebook pages. They also shared or reshared their own experiences of loss and grief. Indeed, as the months passed and the pain became more manageable, Cassie recommitted herself to the rescue group with renewed zeal. Guilt at the violent nature of Smudge's death might not have gone away, but at least it served as a reminder of the moral chasm between Cassie and those she and her fellow rescuers struggled against. The indifference of the mainstream might be hurtful, but it was evil people that they needed to combat, Cassie reminded herself—that is, those who injured or killed animals without remorse and who reveled in the pain and suffering that they inflicted.

8

BEING MODERATE IN A WORLD OF INTERESTS

In this chapter, I focus more squarely on the virtue of being moderate. Although introduced toward the beginning of the book and closely tied to a long-standing ethos of the animal group (see chapter 2), that virtue has somewhat slipped from the center of my narrative attention. At times it was obscured in all the high drama of certain forms of iconic moral patiency-cum-agency—for instance, the heroic self-sacrifices of the covert investigator or the epic struggles of the rescuer. Being moderate, though, that quieter, more tempered expression of a moral life, remained essential. Indeed, in many ways it crystallized not just what the animal group stood for in the wider context of the animal movement but also how colleagues continued to typically view themselves—that is, as "moderate activists" working to improve the welfare of animals through campaigning interventions largely in the sphere of lawmaking, policy, and regulation.

So, a moderate stance, attitude, or orientation seemed to define the way that colleagues regarded their activism in the office. But the issue of moderateness also played out in the rest of their lives. It found expression in the common desire not to preach to people (see chapter 3) as well as in the numerous usually unspoken acts that shaped those lives and resulted in outcomes such as a willingness to marry or live with someone who ate meat or who did not share a passion for animals. The virtue of being moderate in part signaled a commitment to tolerance, both in personal relations and in the relations that shaped the organization's involvement in the legislative process. This included an ability to understand and accommodate different viewpoints and lifestyles. Indeed, in certain ways, the new vision's ambition to engage the mainstream was simply a recalibration of the tradition of a moderate disposition. The latter included a historical attraction to like-minded counterparts across division—the willingness to identify and work with "moderate scientists," for instance (see chapter 2), or with moderate figures in the sector of farming or in government.

To some, the idea of a moderate activist might sound like an oxymoron. The cause should surely be everything, or at least uncompromised by too many accommodations. As colleagues well knew, this kind of judgment would not be hard to find within the animal movement. Typical stress in the term "activist" falls on strong action, an agentive stance that sits awkwardly perhaps alongside such a modifier. The suggestion of doing something moderately risks making a commitment or conviction appear halfhearted. It also sits uncomfortably within the "politics of prefigurement" (Polletta 2002; Escobar 2004; della Porta and Tarrow 2005; Juris 2008; Maeckelbergh 2009, 2011; Graeber 2002, 2009; Razsa 2015; Alexandrakis 2016; Krøijer 2015a; Laszczkowski 2019), a current focus of much anthropology of activism, which in significant part has developed through a dialogue with insights drawn from the discipline's "ethical turn" or its renewed attention to the subjective dimensions of moral life—that is, with the tendency toward socially encompassing political or moral visions, where the future to be prefigured by the activist's present actions maps onto relatively small-scale subjects such as like-minded individual activists or activist collectives and movements, and where those collectives or movements are often nonetheless unproblematically considered metonymic of the future to come. Again, the idea that one might prefigure a future society by enacting new forms of community, behavior, organization, or social relationship in moderate fashion jars with the sense of zeal (but see Ginsborg 2020). It also jars with the prefigurist value typically placed on completism and the integrity of desired political reality. Nevertheless, it is the very unglamorous, almost apologetic attribute of the moniker "moderate activist" that I want to argue gives it substance and interest. While open to derision from some quarters, in other quarters it speaks in a meaningful sense of a life well lived.

As a virtue, being moderate is usually associated with an undertaking to avoid excessive behavior or polarizing opinion. It can be tied with qualities such as mildness and reasonableness, and it necessitates a pragmatic outlook or a willingness to compromise. There are close connections to the Aristotelian-inflected principle "that what counts as virtuous in any given situation is the mean between two extremes of possible action" (Zigon 2008, 24–25). Indeed, more generally the virtue suggests an orientation toward the average. Being moderate, then, seems to suggest a more serious attention to the question of scale; it is not just about disposition but also about a more worldly sort of positionality (see Samanani 2021). Indeed, part of the art of being moderate involves knowing where that average position lies in any given issue or concern. Being moderate can also indicate average levels of intensity, including in the exercise of moral feeling. As we have seen, for some colleagues this could necessitate self-control, the regulation of natural passions in the service of organizational goals and hence animal protection. However, the virtue of being moderate does not have to always rest on disciplining actions. In addition to being a result of active struggle, an even temperament can itself appear as a natural condition of the person or of personality. Indeed, in this chapter, we will explore that tension by considering further the attributed personality of Maggie, the animal group's policy director and the figure perhaps most closely associated with the quality of moderateness. This includes a closer examination of Maggie's role as policy adviser and lobbyist and of the legislative field where Maggie led the interventions of the animal group and where I argue that the virtue of being moderate was particularly prized.

Of course, the animal group's long-standing ethos was equally recognition of the fact that animal groups did not fully own the debate on animal welfare. Engagement with the processes of lawmaking and regulation most forcibly brought that reality home, for all sides involved in consultative or petitioning proceedings could express moral concern for the welfare of animals. However skeptical colleagues might be about such expressions of concern, they had to admit that government and other lobbying groups also spoke the language of welfare. For example, Maggie held regular meetings with civil servants in the Animal Health and Welfare Division of the Scottish government. Likewise, opposing teams of lobbyists regularly referenced their conformity to industry-specific standards—that is, to the whole raft of welfare provisions governing the use and treatment of animals in various sectors of the economy. And more positively, they could also stress a commitment to improvement through, for instance, pledges to raise welfare levels above the industry standard or to innovate various enrichment programs. In fact, at the parliamentary-sponsored forums that I attended, spokespersons for industry bodies such as the National Farmers Union or professional groups such as the Scottish Gamekeepers Association sometimes insisted that they were the real guardians of animal welfare. To have a stake in the legislative process, then, necessarily involved participating in a field where welfare (if not protection) was a normative yardstick for discussion and debate, albeit interpreted and measured very differently. In a sense, it drew out comparison and encouraged dialogue and an incremental outlook on welfare improvements that itself favored an orientation toward the average. So, from certain perspectives, involvement in that process predefined the members of any cooperating animal group as moderate activists.

Although an emphasis on being moderate could be read as a historical effect or consequence of aligning the animal group with a regulatory concern for welfare, colleagues consistently argued that it was primarily a matter of disposition. As already implied, the invocation of the virtue drew attention to the notion of not just personality but also character. This was the case both in the sense of the distinctive types of persons led to work at the animal group and in the sense of an individual's steady moral qualities or cultivated disposition. Yet, as well as moral character, that disposition could be cultivated for strategic purposes. Indeed, across this chapter, the fluctuating nature of moderateness highlights not just who someone was or how they wanted to live but also the issue of appearances and effects—that is, to what ends a moderate disposition could be the means. This element of performativity, which understood the need to be moderate in a knowing fashion, so to speak, opened the possibility of considering the virtue in a threefold manner : as at once itself, as an object of concerted effort in the interest of self-improvement, and as a resource to be deployed to influence (human) others in the ultimate interest of protecting (nonhuman) animal others. The same element of performativity also enabled moral character to be presented as a multiple.

Heart and Head

Soon after I joined the animal group, the BBC commissioned a documentary film about Scottish shooting estates and their possible links, widely reported at the time, to the killing of rare and protected raptor species such as the golden eagle. The suspicion existed that gamekeepers were targeting raptors as part of wider predator-control practices to reduce the threat to populations of grouse or pheasants. As well as interviewing representatives of the shooting industry, the filmmakers spoke with conservation organizations working to preserve endangered birds of prey in Scotland. They also spoke with Barry and Maggie. Indeed, the planned documentary was perceived as a great opportunity to promote the investigative and campaigning work of the animal group and additionally to spotlight the organization's broader role in lobbying for better regulation and welfare improvements around predator control.

The film itself, when finally aired on television, opened with footage of the broadcaster standing next to Barry, who was disguised by a woolen hat and had his face blurred to protect his identity. Both men were described as visiting an unnamed shooting estate. The camera follows as the pair, under Barry's expert guidance, surreptitiously move across the grouse moor to inspect a range of traps. That journey ends with the discovery of the decomposing bodies of two hen harriers, assumed to have been killed and discarded out of sight by a local gamekeeper. At this point, the film cuts away to present a general history of grouse shooting in Scotland, which then segues into a conversation with a spokesperson for the industry body of shooting estates.

Unfortunately, the long interview that the broadcaster also conducted with Maggie did not make it into the released version of the film. However, a sense of that contribution was captured in a blog post about the BBC documentary that the policy director subsequently published. Its key lines were reproduced across the group's other social media platforms and further surfaced in various carefully scripted policy submissions, press releases, and briefing statements:

> When wild animals are illegally trapped and killed, just like domestic animals, *there's only one number that matters*—the number one.
>
> Each of these animals is an individual. When any animal suffers and dies after consuming carbofuran . . . or dehydrates in an unchecked trap or struggles as a snare cuts deeper into its neck, abdomen or leg, or has its body peppered with shotgun pellets, the only number that matters, then and there, is one. If we consider ourselves a humane society, the suffering of that single individual, regardless of its species, should matter to us just as much as any other.

Deliberately crafted by Maggie as a series of sound bites, this kind of statement neatly communicates the classic animal protection message and the group's essential perspective on predator control. In fact, the policy director's words were constructed to operate in contradistinction to the sorts of arguments that Maggie expected representatives of the shooting lobby to present in the televised documentary. These were chiefly economic statistics about the importance of shooting estates to the Scottish economy. The lines were also constructed to act as a counterpoint to the sorts of arguments Maggie anticipated from the voices speaking on behalf of the interests of conservation, the fair-weather friend or ally of the animal group on this issue. Knowing that both would more than likely engage in a "numbers game"—for example, trading economic statistics against the population figures for endangered species—the policy director quite deliberately offered an alternative mathematics precisely grounded in the protectionist calculation that "there's only one number that matters."

By Maggie's admission, this message was a straightforwardly emotional appeal, intended to register as a direct overture to the audience's capacity for cross-species sympathy. In fact, she regularly made similar kinds of appeals in her scheduled conversations with politicians, senior civil servants, and officials in the Animal Health and Welfare divisions of government, on occasion at Westminster but far more commonly at Holyrood. Yet, as any competent policy director and lobbyist working in those legislative environments understood, such appeals were never enough. Broad, impassioned moral statements about what constituted a humane society could certainly be heard, but Maggie knew that policy submissions on animal welfare concerns needed to be properly presented in evidential form. Argument and scrutiny would ultimately center on sets of statistics and technical details. This included a heavy reliance on experts (all sides routinely commissioned "friendly" scientists to write consultative reports as part of their policy submissions) as well as competing organizational interpretations of very specific welfare concerns.

Remaining with the example of predator control, lawmakers would typically hear evidence on and debate issues around a series of fixed reference points. How many hours could a trapped fox survive before dehydration became a real concern? Did the stop on a snare effectively work to prevent the noose tightening too much and hence causing death by strangulation? Could minimum loop sizes ever resolve the problems of bycatch, such as the capture of deer by the foot? And could any of these adjustments deal with the realities of animal struggle, which might result, for instance, in a snare wire slowly cutting through the neck of a trapped fox? These were the kinds of questions that mattered most in the legislative sphere. And despite the relaunch of the animal group and its new commitment to reorient attention toward an engagement with the mainstream, those sorts of questions and how to best answer them continued to dominate organizational time and attention.

So, while no politician nor any petitioning group participating in the consultative stages of lawmaking wished to appear passionless or without moral feeling, the expectation existed that those emotional appeals would be kept in check. Policy directors and lobbyists working in that environment understood that this required being able to combine displays of "heart and head." In fact, that commonly uttered idiom neatly expressed the essential quality that all actors in the policy domain wished to claim for themselves. I heard it normatively invoked from all sides. Talk of heart and head surfaced in the submissions of interest groups giving evidence to lawmakers as well as in the deliberations of members of Parliament sitting on committees responsible for scrutinizing bills and in the summing-up of government positions by ministers. A consistent emphasis fell on the virtues of moderating moral language, typically assumed as informed by powerful emotions of a personal or partisan nature, as well as on considering such utterances against the facts or through sober reflection on a decision's practical consequences. What we might call the ethics of heart and head, then, took center stage, a crucial moral grammar of the legislative process.

Of course, the idiom has a much wider purchase, not just in discussion and debate within the animal movement but also in diverse forms of popular discourse.[1] In a UK context, it is almost impossible to avoid its invocation. One can find the idiom expressed, for instance, in endless column inches devoted to lifestyle choices or popular psychology. "Should you follow your heart or your head?" appears as a regular question in straplines on social media, in newspapers, and in magazines, as does advice on whether readers should "let your heart rule your head." In such cases, the idiom is mostly devoted to musing on romantic choices or to reflections on the wisdom of a whole raft of consumer or investment decisions. Here the heart usually stands for impulsive or spontaneous decision-making and more broadly for choices often made against one's better judgment, whereas the head typically stands for considered opinion or for caution, a careful weighing of the pros and cons of any action, and close attention to the risks involved. When presented in this fashion, the idiom tends to suggest a subject torn between two poles—on one hand, that of a patient who suffers and feels compelled to act on a certain basis, and on the other hand, that of a conscious agent battling to act with reason or forethought and to take everything into account. Alternatively, the same subject could appear more on the sidelines of such internal struggles, as a narrator or observer of the push and pull of heart and head. In these scenarios, the individual's role seemed merely to decide which source to listen to or trust.

Perhaps of more direct relevance to the domain of lawmaking, the idiom of heart and head was also often central to the terms of public and political debate, especially around major national decisions. During the contentious 2014 referendum on Scottish independence, for instance, I lost touch of the number of times the idiom made a rhetorical appearance. Both sides in the independence debate regularly claimed to be speaking or campaigning with heart and head or accused each other of failing to do so. For the No campaign, this often meant trying to convince the Scottish public that remaining in the UK could also be a heartfelt matter—that, for instance, their campaign was not just about fears of the practical consequences of independence but also based on real emotional attachment to the idea of the UK or of Britishness. For the Yes campaign, the problem was typically figured in reverse. Politicians and pundits usually assumed that the Scottish public already had pride in and an emotional attachment to the idea of Scotland; what typically needed to be proven was the evidential case for a sustainable nation-state. Members of the public, including colleagues at the animal group, would regularly express their quandaries over how to vote in just this fashion. For example, one commonly heard the wavering voter on vox pop declare that "my heart says yes, but my head says no." Such statements usually expressed the sentimental draw or appeal of independence, coupled against the voter's concerns about what were perceived to be the economic and political costs of withdrawing from the UK.

Once again, the struggle between heart and head could operate in different registers. On the one hand, voters could present the issue as a straightforward tension between what the heart desires and what the head counsels or warns against. This in turn could be collapsed into a cruder distinction between emotion and reason. But on the other hand, the struggle could be presented as beyond the conscious mastery of the subject, with both the urgings of heart and head presented as being as mysterious as each other. I can recall one colleague, for instance, rather adamantly telling me that she was going to vote no and in the process outlining a series of concerns about economic security and the future opportunities for her kids in an independent Scotland. However, a few minutes later, that resolve seemed to break down entirely. "I just can't understand why I can't say yes," the colleague confessed with obvious frustration. Here, the privileging of head at one moment felt like a disciplined strength and the next felt like a personal failing. It was no coincidence, perhaps, that a common narrative among those campaigning for independence leaned on that notion—that is, the assumption that Scots had in the past suffered from a crisis of confidence or had had their belief in themselves as an independent people or viable nation undermined by the experience of Union (of course, as the lively past and present of unionism in Scotland testify, emotional attachment to the idea of Union could also be strong [see Webster 2020]; the terms of the referendum, though, tended to negate that emphasis). The implication in this diagnosis was that listening to the heart could be a process of self-renewal or discovery, of affirming who "you" were but also who "we" really were and hence could become. In this framing, appeals to the head could obscure as much as they clarified moral purpose. Yet, as already mentioned, the charge could just as convincingly be worked in reverse. From the perspective of No campaigners, listening to the head could enable the demystification of nationalist passions and rhetoric, for example. And letting the head rule the heart could ensure moral probity, including virtues such as prudence and fiscal responsibility.[2]

This was the case not just in the context of the independence debate. One heard very similar albeit differently targeted references to the idiom of heart and head in the polarizing UK-wide national conversations around the Brexit referendum. But to return to the domain of lawmaking and the policy work of Maggie, it appeared that here it took on new, added purchase. *Heart and head. Heart and head.* The policy director deployed and redeployed that idiom continually to communicate the animal group's policy positions and message.

In terms of the BBC documentary that I began with, Maggie's unaired contribution was certainly not limited to an appeal to the heart. Despite an initial distancing from the numbers game and an insistence on a radical protectionist measurement of true costs centered on the individual suffering animal, Maggie soon proceeded to throw counterstatistics back at those anticipated from the animal group's opponents. "Here's another number: £240 million," the policy director at one point announced, informing prospective viewers that this was the amount the shooting industry and the Scottish government typically claimed that the sport annually contributed to the Scottish economy. Maggie then attempted to dismantle or at least to unsettle the security of that calculation. She pointed out, for instance, that this figure was based on data collected on behalf of the shooting industry way back in 2004 and that more recent estimates by a government-funded public body put the amount, this time inclusive of the massively popular field sport of angling, nearer to £136 million. A typical statement followed this reassessment. "[Our group] would not dismiss any significant contribution to the Scottish economy that keeps people in jobs," Maggie told the broadcaster, "even if we would prefer wildlife tourism to be nonconsumptive. But if we are to play the numbers game, let's have the whole story." The quite deliberate reasonableness of this position, grounded in a willingness to concede the importance of keeping people in jobs *even* where that requirement threatened to clash with animal protection moral priorities, was a perfect illustration of the kind of heart-and-head stance that the animal group liked to finesse.

For it simultaneously spoke to an awareness of the value placed on being moderate, including within government and lawmaking circles, and to an awareness of the need not to play into the hands of the animal group's opponents. Maggie knew, for instance, that the shooting lobby liked to portray animal protection concerns about predator control as too emotive—that is, without regard for the practical considerations of land management or the hard facts of economic prosperity and job security. Standing as an obvious counterweight to industry charges of the overly impassioned and hence unreasonable animal activist, Maggie's statement demonstrated, once again, the group's ongoing desire to appear as a "sensible" participant in the legislative process. It showed willingness to abide by the terms of engagement within lawmaking. But it also reflected a general acceptance that while the animal group might continue to campaign for big changes in welfare policy, the legislative struggle generally came down to small wins in welfare standards precisely achieved through taking part in that process. The contrast here was not just to animal groups that refused to engage but also to those campaigning positions dominated by rational first principles or philosophical arguments such as the language of animal rights. For Maggie knew that from the perspective of government, that language risked appearing exclusionary or immoderate.

Just a Morningside Lady

One of the newer colleagues at the animal group once told me that when she first applied for her fundraising post, Maggie was on the panel that interviewed her.

At the time a little nervous, not just about what she would be asked and whether she would get the job but also about what kind of workplace she would be joining if she did, the candidate for the fundraising post sought some reassurances. In particular, she wanted it confirmed that the organization was not "one of those more extreme animal rights groups." It was Maggie who put the candidate's mind at rest. Leaning forward, with a smile that appeared to comprehend the possible degrees of anxiety such a concern could generate, the policy director dispelled the awkward moment by making a jocular self-reference. "Look, I'm just a Morningside Lady," Maggie said, before she and the other panel members launched into a more formal response about the group's new vision, its desire for inclusivity, and the general moderate nature of its stance.

Maggie's quip had very specific connotations to Scottish ears, especially for those like the interviewing candidate born and bred in the capital city. Indeed, it was typically well judged. Much referenced in the media and in ordinary conversation, the notion of the Morningside Lady has a vivid place in the popular imagination. As the term suggests, this legendary figure hails from the South Edinburgh suburb of that name, an area with a long reputation for gentility and nonshowiness and for a long period of slightly faded affluence. More particularly, Morningside was once famous for its high preponderance of older middle-class women, assumed to be either unmarried or widowed, an impression further bolstered by the national success of a series of picture-book stories for children entitled *Morningside Maisie*. As well as being depicted as of a certain age, the Morningside Lady presents as someone who dresses conservatively, who likes to lunch with other ladies, and who, when she ventures outside of that suburb, tends to shop at Jenners, until quite recently the city center's traditional department store. While the picture books follow the adventures of a mischievous kitten that treads the local pavements with her shopping trolley bag in tow, in more conventional parlance the expression is marked out as a model of propriety. In alternative guises, the Morningside Lady can be mocked for her airs and graces. Often referenced in the Scottish media—for instance, in opinion pieces or in journalistic character sketches of public figures—it can be used as a kind of shorthand, "She's certainly no Morningside Lady" being a familiar variant of that expression. But as well as communicating a sense of either class respectability or affectation, the figure can be admired for her fearsomeness—that is, as an older woman who speaks her mind in a no-nonsense sort of manner.

In the case of the job interview, the reference succinctly signaled several things. It was at once a way of saying, "Look, I am an ordinary person like you," and simultaneously a way of communicating that "I am a very distinctive kind of person, who you wouldn't necessarily expect to find in an organization such as

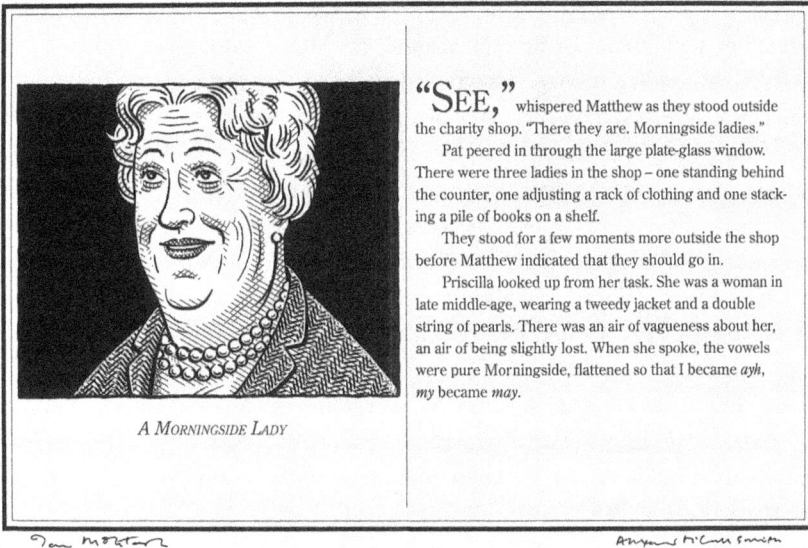

"SEE," whispered Matthew as they stood outside the charity shop. "There they are. Morningside ladies."

Pat peered in through the large plate-glass window. There were three ladies in the shop – one standing behind the counter, one adjusting a rack of clothing and one stacking a pile of books on a shelf.

They stood for a few moments more outside the shop before Matthew indicated that they should go in.

Priscilla looked up from her task. She was a woman in late middle-age, wearing a tweedy jacket and a double string of pearls. There was an air of vagueness about her, an air of being slightly lost. When she spoke, the vowels were pure Morningside, flattened so that I became *ayh*, *my* became *may*.

A MORNINGSIDE LADY

FIGURE 9. "A Morningside Lady." Artwork by Iain McIntosh with text by Alexander McCall Smith. Used by permission.

this and who certainly wouldn't suffer being told what to think or how to act." In fact, part of the fun of the quip lay in the ways in which Maggie did and did not meet these expectations. While her soft Edinburgh accent spoke of an identifiably middle-class background, Maggie was hardly the airs-and-graces type. Instead, her smart and efficient two-button jackets and business skirts or trousers suggested a busy professional woman. Among colleagues, the policy director sometimes liked to emphasize how much older than them she was. But at the same time, Maggie was self-evidently not of an age or disposition to restrict her movements to the cafés or eateries of a suburb such as Morningside or to occasional trips into town to browse the floors of Jenners. It was possible, though, to imagine that you might find Maggie in either place on a day off or a loose weekend. Not exactly fearsome, Maggie was nevertheless impressively well informed, with a detailed knowledge of her policy brief. She was also very much no-nonsense in demeanor. Other colleagues would happily take my invitation to extrapolate on the philosophy or terms of debate within the animal movement—for instance, to discourse on classic liberationist arguments about the moral equivalence of slavery or the oppression of women to the historical treatment of animals. Maggie, though, would invariably cut short such ruminations or dismiss them altogether as "pub-talk." If it was irrelevant to the task in hand, which was nearly always pressing, then the policy director tended to refuse an engagement.

But Maggie's quip was also a private joke, shared with colleagues on the job panel. For they knew that in fact Maggie was Morningside born and bred. As well as living in that suburb, Maggie had attended its famous fee-paying school, which was where she had met her husband and where they had subsequently sent their children. In a comparatively small city, in which an astonishing 15–18 percent of parents place their sons or daughters in such fees-paying schools (the UK national average is around 6 percent), that meant Maggie was comparatively well connected. It also meant that she was well versed in the etiquette of Edinburgh's professional classes. This included those who continued to dominate the lawmaking process (senior civil servants, lobbyists, legal experts, etc.) and those who contributed specific forms of expertise around animal welfare issues, such as vets and academic scientists. The same class fluency assisted the policy director in dealings with the animal group's opponents, such as the representatives of the shooting industry. Maggie knew something of the world they inhabited and the terms of social engagement. This was partly through her upbringing but also because of diligent research. The policy director made a point, for instance, of regularly reading both industry and countryside lifestyle magazines. Maggie further understood what made the lives of civil servants and politicians easier and what might antagonize them unnecessarily.

So, while Maggie knew how to speak her mind, the preference was always to do so quietly yet firmly, and with the facts to hand. Literally speaking, Maggie was a Morningside Lady. However, it was the virtues of reasonableness, pragmatism, adaptability, and competence that mattered most, as well as the importance placed on being personable. For these were the qualities that Maggie principally took from that background or utilized in her policy work. To be a favored partner in government consultation exercises and parliamentary committee bill scrutiny, for example, Maggie knew that the animal group needed to appear both trustworthy and reliable. As well as playing the game of lawmaking, this involved working toward what was possible rather than what the heart or sometimes the head demanded. It also meant ensuring that the organization and its mobilized supporters respected the time and pressures that other actors in the legislative process were under. The policy director was always very careful, for instance, to avoid making vexatious Freedom of Information (FOI) requests. Maggie knew that these requests could exasperate overworked civil servants, since the FOI requests invariably bogged them down in lengthy bureaucratic trawls through paper and electronic documents. Likewise, Maggie tried to prevent campaign actions that invited supporters to "bomb" MSPs (i.e., Members of the Scottish Parliament) with constant petitioning demands. While other animal groups and campaigning organizations did partake in such actions, Maggie understood that these email bombs risked clogging up email servers and hence alienating or irritating the politicians that they targeted. Once again, it was reasonableness that was the watchword, a virtue that Maggie naturally leaned toward but that she also believed was a tactical strength. In her opinion, it was far more effective in getting achievable legislative results than the all-guns-blazing attitude of some of the other animal groups that Maggie encountered.

Maggie's own biography was a testament to this heart-and-head approach. Indeed, the policy director presented as a moral subject that came to animal protection via a series of professional and personal circumstances. These enabled a careful education in the pertinent issues and hence an informed and evolving self-awareness. Unlike the patient-led biographical narratives of many colleagues and animal people, which typically began with intense childhood emotional contacts or bonds with specific animal others, Maggie tended to narrate a gradual enlightenment that fed or enriched an always-discrete sense of moral passion. When I began working with the organization, she had been policy director for nearly four years, and before that time she had worked for a longer period at the SSPCA. However, Maggie's driving moral concerns had originally focused elsewhere. Like Eilidh (the group's CEO), her professional background and interests initially centered on children's welfare issues. But once Scotland gained its own Parliament in 1999 and animal welfare became a devolved issue, Maggie found herself moving more and more into that policy sphere. By 2004, she felt confident enough to try going freelance as an animal welfare policy expert. This involved doing consultative work for the umbrella group of local authorities in Scotland, including in and around the complex regulatory issues linked to the 2006 Animal Health and Welfare Act. This continued until a conversation with Craig, whom she knew through their shared policy work, led to an invitation to apply for the newly vacant role of director.

Since that time, Maggie had kept developing her welfare knowledge and improving her broader policy expertise. Between ongoing family commitments and a busy social life as an active and enthusiastic member of a local choir and an amateur dramatics society, Maggie had managed, for instance, to complete an evening degree in law. This move was designed to further enhance her fluency in the legislative process. Indeed, it was the expanding nature of both Maggie's expertise and her policy brief that led her to make an incremental series of more self-conscious ethical decisions. For example, on discovering more about the widespread inhumane rearing systems for pigs, and especially after learning more about the animal's levels of intelligence, Maggie determined to give up eating pork. Similarly, conducting policy work on poultry farming eventually led Maggie to stop eating chicken meat, though for some time she kept eating lamb precisely because through her role she understood that the rearing system for sheep was far less intensive and hence more humane. Largely worked out in this piecemeal fashion, Maggie slowly became vegetarian.

In narrating each decision made, Maggie was careful to highlight that there was nothing unconsidered in what she decided but also that she was not really that different from other "ordinary" people that she knew. While the family had always had rescue dogs and cats and Maggie had found and taken in a stray cat soon after she married, the policy director was resistant to the idea of any special or inspirational attachment. "I'd always had, as I said, great fondness for individual animals, been a bit tenderhearted," Maggie told me, before hastily adding, "like lots of people are." It was true that Maggie's animal protection profile, both professionally and personally, had at times led her to reflect on past moral discomforts. However, in her biographical telling, these never really resulted in the clarification of an ethical journey or the discovery of an animating moral source:

> I remember being at a dinner party and meeting a very nice, very charming couple, and they were, they had an [shooting] estate in the north of Scotland, and there was this passion for shooting. So much of the conversation was shooting, and I was thinking, I really hate this! [Maggie chuckles]. And I began to think there's obviously a divergence of interest here. I mean, my dad used to fish, for instance. I liked going fishing with him. But then feeling the fish pulling on the line, I didn't want to catch it. So, it's [Maggie's ethical development is] made up of incidents, circumstances, like going into that world [the professional world of animal protection and policy work]. And then, of course, from there I had to learn a whole lot of stuff that I didn't know happened. Like farming practices.

Turning to vegetarianism out of an informed concern for the suffering of industrially farmed animals rather than out of what she termed a "pure" animal rights advocate belief that "we have no right to kill animals," Maggie kept returning to the fact that she could understand alternative choices. "I do genuinely believe [in the moral value of the individual animal's life], but I am fairly objective about others who don't share that view," Maggie said. "I would recognize that other people can square it with themselves, with their conscience, and that it's probably OK." Maggie paused and then introduced another caveat: "Until, of course, you get into egregious suffering, you know; that's where we wag fingers and say, 'Just don't do it!'"

As suggested, Maggie's attitude was partly informed by skepticism toward the all-worked-out or purist stance, whether expressed in moral argument or through policy positions (see chapter 9). Speaking in response to my query about the group's standpoint on animal experimentation, for example, Maggie once told me, "We don't take what you might call the abolitionist position, which is a purer animal protection approach; some people would say it's a more ethical one." When I inquired why not, Maggie replied, "Because it's not credible. You know, we've got to be effective. There is no point our taking a purist ethical stance if nobody will listen to us." The policy director then explained that while a ban remained an organizational "aspiration," the actual policy position of the animal group emphasized the need to keep pressing for the development of humane alternatives, in this case through a lobbying focus on any possible review of the relevant European directive. Doing so required a good deal of patience, Maggie emphasized, including an acceptance of the pace that lawmaking imposed. It also required discretion. Maggie once explained to me that her expertise was far less about full and open criticisms of government (or even of any opponents) and far more about learning how to gradually "chip away at their position over time." The policy director admitted, "It's a bit slow, and it's a bit frustrating, but there are other people who don't share our views, and they are very influential as well." In the context of animal experimentation, Maggie's specific reference was to lobbying groups linked to the pharmaceutical industry and to academic medical research. "So, it's more how can we effectively achieve that change," Maggie emphasized again, "rather than being so pure and ethical and achieving nothing for the animals." This latter point was essential. Maggie's moderate disposition was always grounded in a conviction that the utterance or articulation of moral positions did not necessarily have ethical outcomes.

A Balance of Interests

So, Maggie's specific heart-and-head stance may have reflected what she perceived to be her ordinary ethics, but as already mentioned, it was also very much a professional necessity. Indeed, the policy director's moderateness was an essential resource in the animal group's perennial struggle to present itself as the reasonable voice of animal protection. As we have seen, this included a capacity to appear evidence-based and measured in both its positions and its lines of argumentation (i.e., neither too emotional nor too partisan) and hence also capable of compromise. As well as projecting that impression to those whose opinion the group wanted to influence, Maggie continually witnessed that virtue of moderateness being played back to her in the lawmaking process (see Smith and Holmwood 2013). But in addition to the jostling to claim a moderate stance, the policy field was defined by another linked perspective. At its center lay a further assumption that any specific policy issue needed to be interpreted as interest-laden and that its resolution would inevitably emerge through a balancing of those interests.

As the policy director knew, this was an orienting perspective not just in government but also in the wider media environment (of press releases, briefing statements, and journalistic quotes) in which that policy work took place. For instance, Maggie assumed that in all probability it was an institutional commitment to achieving balance that had resulted in her contribution being bumped from the documentary on raptor persecution. The BBC's charter obligations require its program makers to ensure that viewers hear from a representative sample of different voices on any given issue. Once the editors of the documentary had decided to include the long opening footage with Barry, Maggie appreciated that it was always going to be hard to give space to another animal protection perspective, especially from within the same organization. The final program did indeed carefully move back and forth between interests or viewpoints. As well as conversing with a representative from the association of shooting estates, it presented an interview with a spokesperson from the bird conversation group RSPB, then with a gamekeeper and a shooter, and finally with someone from Scottish Natural Heritage, the public body responsible for managing and monitoring natural environments. Only after that display of balance did the interviewing broadcaster provide a set of personal reflections framed as a summing-up of all that he had heard and seen.

But in lawmaking circles, the commonly uttered reference to a "balance of interests" was most closely connected to another term typically used to designate an archetypal actor in the policy field: the "stakeholder." Literally defined as anyone with a legitimate interest in a particular issue, this was the favored category by which the Scottish government throughout the period of my fieldwork addressed and engaged with lobbying parties in the legislative process. That practice was itself an inheritance from predevolution moves at Westminster in the mid-1990s, where, in the name of enhancing democratic government, the inclusion of stakeholders in decision-making procedures became a normative standard, especially under the Labour administration of Tony Blair (see Du Gay 2005). This move was reemphasized in the foundation of the Scottish Parliament with the creation of a separate Public Petitions Committee (PPC). It sat alongside the usual array of parliamentary committees scrutinizing government-initiated bills, but unlike the other committees, the PPC could hear directly from petitioning individuals or groups with suggestions for new legislation. Certain caveats existed about who could assert an interest in any given issue. However, in general, those interests were simply self-declared and accepted once a stake had been claimed. Even in standard legislative committees, stakeholders were involved at almost every stage. In drawing up consultation exercises in preparation for bill development, for instance, civil servants routinely advertised for responses from stakeholders. And members of Parliament invited and sometimes selected "relevant stakeholders" to submit evidence to various committee stage hearings. While, formally speaking, a stakeholder could be any member of the public, in practice consultation responses and submissions to parliamentary committees came from recognized "interest groups." This occurred either via their formal contribution or via the mass of individual contributions that those lobbying groups encouraged their supporters to submit on their behalf. This included campaigning or advocacy organizations like the animal group but of course also industry associations and professional bodies.

Being forced to consider itself as a stakeholder or interest group had implications for the organization Maggie represented. Some of those implications were welcome, others of them far less so. While the animal group absolutely wanted to claim that it had a stake in all animal protection issues, it was less comfortable with the wider logic of the stakeholder process. Specifically, colleagues were troubled by the presumption that other interest groups, including those that one might hold responsible for the systematic exploitation of animals, had an equally legitimate stake in the proceedings. That concern was exacerbated by the suggestion that at one level all interests were equivalent, a principle ingrained in the eligibility criteria for consultation exercise participation. For, of course, none of them could accept the notion that their moral concern or "interest" in the welfare of animals was in some fashion comparable to, say, the "economic interests" of their perceived opponents (for a broader exploration of the mobile duality of the category of interest in moderate animal activism and several other professionalized contexts in the UK, see Candea et al. 2024). More broadly, few colleagues could identify with the dispassionate rendering of their moral purpose as an interest or with a description of the animal group's morally charged activities as the strategic action of just another interested party.

Nevertheless, this was one of the consequences of the stakeholder ideal that Maggie particularly had to address. It was a cost of choosing to engage in the process of policy petitioning as it was laid down and implemented by lawmakers. Regarded more positively, that flattening perspective, which rendered radically incommensurate sets of actors into equivalent sets of interested parties, was precisely what had allowed the animal group in recent years far greater access to and potential influence within lawmaking. Of course, some of the more troubling aspects of that stakeholder perspective were partly due to the peculiar ways in which the concept of interest worked—particularly the way it was able to contain a sense of an expression of concern (i.e., including moral concern or an attitude of something being important) alongside a sense of preference or self-interest. In fact, as Du Gay (2007, 66–67) describes, after Hirschman (1977), the history of the concept of interest suggests an initial emergence as a solution to failing English (and Scottish) polities. A range of seventeenth- and eighteenth-century thinkers posited the commercial spirit of self-interest and accompanying self-interested conduct as a solution to more violent or warring passions and as a strengthening of the evidently weak human motivator, reason. Rendered as a cooling or mild impulse, interest then got revalued as a "third way" between those passions and reason, as a disposition with a moderating effect. One might even read a similar intervening role for interest in that ethics of heart and head that I previously explored. For once recalibrated as a matter of balancing interests, the struggles of heart and head somewhat dissipated.

Indeed, colleagues did sometimes invoke the language of interests—more formally so, through occasions when they echoed the utilitarian-inspired logic of certain animal liberation arguments, such as those put forward by Peter Singer, but perhaps more pressingly so when colleagues invoked that language in the context of ordinary interaction. Recall Maggie's words in the passage quoted earlier, where the policy director mused on the personal emotions stirred by an awkward dinner party. Liking the couple but hating their enthusiastic talk about shooting, Maggie described the latter as a "passion" clearly at odds with Maggie's concern for animal protection. But Maggie avoided confrontation (and perhaps the sense that she ought to say something) by then diagnosing the situation as a straightforward "divergence of interest." The reference here was to the idea of topics of interest, unfortunately shared and only politely tolerated, yet also to an awareness that certain activities or pastimes can drive such interests. In fact, the invocation of a divergence of interest led directly on to Maggie's reflections about accompanying her father on his fishing trips.

However, for Maggie and the animal group, interest was first and foremost the dominant language of government. For them, it was most personified in the stance of ministers, those who represented the views of government across the passage of any bill through the Scottish Parliament and who were formally responsible for introducing new legislation. Indeed, both before the debating chamber and in front of parliamentary committees, government ministers typically liked to present themselves as those striving to dutifully weigh competing interests—that is, as the figures in control of striking a judicious balance. A regular desire existed to present the minister as above interests or without any specific stake in each issue, or alternatively to represent the balancing act finally made as a decision taken "in the national interest" rather than the particular interest of any stakeholder or interest-laden group. In effect, national interest, at least in the context of animal protection, usually referred to the perceived economic interest of the nation, of which the government assumed guardianship.

The role of ministers as relatively independent strikers of balance was especially exaggerated in the case of legislative issues related to animal protection. This was in large part, as Maggie pointed out to me, because those issues were not a government priority. Indeed, for the Scottish National Party (SNP), who governed either in coalition or by absolute majority across the period of my fieldwork, animal protection was not even a manifesto commitment, or certainly not a core one. There was then little internal pressure within government to make such legislation happen. Ministers often literally had no strong view on such matters or could afford to present themselves as without a stake or interest. That posture of independence was reflected in the way in which the main legislation of this period—the Wildlife and Natural Environment Bill—was introduced. Civil servants from the Animal Health and Welfare Division communicated that this bill was a "tidying-up exercise," that it brought together and cohered diverse strands of past legislation rather than embodied a new philosophy or approach toward environment or animal protection, such as might be enshrined in an electoral promise. In response to questioning from MSPs on the scrutinizing parliamentary committee, the responsible minister insisted that the bill did not contain a "vision." It was instead a "bill about management and regulation . . . not designed to be other than what it is."

This description of the legislation as a largely technical venture closely connected to the way the minister presented her role. At the first-stage chamber debate on the bill, for example, the minister's opening statement included a commitment that throughout its passage the government's "watchwords" would be "balance" and "compromise." Likewise, during the Q&A sessions of the various stages of committee hearing, the minister consistently referred committee members to the need to listen to "different voices in this." As well as balancing the concerns of interest groups, the minister regularly highlighted the need to recognize and balance different kinds of often-conflicting interests at play within the management and regulation of the natural environment. At times this almost sounded like a plea to directly assign competing interests to various wild animals. At one point, for instance, the minister called on the committee to recognize the tension "between the protection of raptors and the protection of the small birds they may harm." Indeed, such calls were intended to highlight the need for balance even within the terms of interest of those interested parties. In this case, that meant a need to compromise or strike a balance between the interests of predators and the interests of prey. This was something, the minister concerned implied, that still needed to be addressed by those conservation stakeholders petitioning for the enhanced protection of raptors.

Responding specifically to a set of questions related to clauses in the bill on snaring (the primary petitioning target of the animal group), the minister once again offered herself as the neutral arbiter and pragmatist in chief:

> Nobody likes to have to think about this kind of animal management, but the truth is we kill animals all the time. The only people who can take that moral high ground, I suppose, would be those who are vegetarian. There may be some here today, but my guess is not many. In which case, we start with the premise that killing animals is not in and of itself something we would be completely opposed to, obviously. So, the question then becomes, well, trying to manage things in a way that balances all the interests. . . . And I think once you become accustomed to the way the countryside is managed, it's harder to take the kind of very purist, hard line about snaring. I appreciate there are strong views on all sides on this.

Plotting a path between "strong views" and willing to confront hard "truths," the minister presented what to Maggie was a recognizable form of reasonableness. This included a familiar skepticism about purist stances, played back to refer to the animal group's own policy position, as well as a familiar call for adaptability. Indeed, although the statement was clearly a riposte to the petition of the animal group and of other stakeholders calling for the committee to pass a clause banning the use of snares, the minister was careful to acknowledge that the concern would not be settled by any enacted legislation. "I well appreciate," the minister added, "it will never go away, because there are particular groups for whom this will continue to be a campaigning position and will always be brought back." This might have been a rather negative nod to animal protection ambitions, but it nevertheless reinstated a ministerial awareness of competing interests and of her self-assigned task to find balance or locate the basis for compromise between them.

Reduced by government in this fashion, one might think that the job of a policy director was to rearticulate the interests of their interest group in as loud and fluent a manner as possible. However, as already mentioned, full participation in that process required something else—that is, a willingness and aptitude to abide by the conventions of lawmaking, which included the myth that interests were spoken directly and balanced in a transparent, open manner. For Maggie, being a competent policy director involved simultaneously affirming the formal terms of the stakeholder process and circumventing them in a way that was itself taken to be conventional in such policy work. For instance, those in the know understood that the official bill consultation exercise, open to all who wished to stake a claim, was often preceded by a preconsultation consultation exercise. This was where the civil servants of a government division responsible for developing legislative plans invited a limited number of favored or "trusted" interest groups from all sides of an issue to discuss what they thought should go into a bill. It was often these same interest groups or stakeholders who later received invitations to appear before parliamentary committees scrutinizing that bill. That was precisely the kind of insider stakeholder that Maggie wanted the animal group to be. Indeed, as well as refiguring ethics as part of an interest-laden field, the pragmatist approach that dominated policy work stressed that utterances, including verbal expressions of moral interest, were invariably multisided. There was a distinction not just between what actors said and what they meant or aspired to achieve but also between what they said and whose views or interests they really represented.

During the passage of the Wildlife and Natural Environment Bill, I often sat and watched the proceedings in the parliamentary chamber or committee room with Maggie. When we couldn't make a session due to calendar clashes, she and I used to watch it back together in the office, through the televised recordings archived on the Scottish Parliament website. Maggie's reading of these sessions was always illuminating. The policy director liked to offer further interest-laden contexts for the statements and opinions proffered by each member of the parliamentary committee. It was through Maggie that I learned, for example, that the Scottish Liberal Democratic member was in fact sympathetic to the idea of a ban on snares. But, like the SNP politicians sitting on the same committee, that member was unlikely to vote for a ban, since Maggie knew that the Scottish Liberal Democratic member was ultimately not prepared to vote against the official party line or policy position. Likewise, Maggie explained to me that one committee member had a rural constituency and that this meant it was not in their interests to be seen to back a ban. However, Maggie added, there was another Scottish Labour member on the committee known to be interested in or "keen on birds," and so Maggie anticipated that this member might get behind the campaign of the animal group.

The policy director also paid close attention to any specific claims made during a hearing or chamber debate, making a note whenever someone said something that Maggie thought was factually incorrect or contestable. I remember Maggie shaking her head, for example, at a claim made by the minister that the SSPCA were "on board" or in agreement with the government's proposed snaring regulations. Leaning over while the minister continued answering questions, Maggie informed me that the advising senior civil servant sitting beside the minister was known to be a field sports enthusiast and hence very close to the shooting industry. These kinds of observations fed into Maggie's ongoing interpretation of what was said and claimed. For instance, when the minister argued that the interests of raptors needed to be balanced against the interests of "small birds," Maggie pointed out to me that everyone knew this was a familiar argu-ment of the shooting lobby. The interests of small birds conveniently coincided with the interests of game birds, or rather the interests of those shooting estates that reared or managed grouse and pheasants so that they could be eventually shot by paying guests. In the ministers' words, therefore, Maggie couldn't help but hear the shadow of typical industry lobbying.

Of course, this understanding of the minister's words was also informed by Maggie's own lobbying practice and by the general assumption in the field of policy work that further interests nearly always lay behind the expression of any singular interest. This included the strategic influence that Maggie in turn aimed to exert on lawmakers and occasionally saw realized. As we watched the parlia-mentary sessions together, it soon became clear that Maggie was also listening for any references to the scientifically commissioned report on snaring formally submitted by the animal group as part of the bill consultation exercise. But addi-tionally, Maggie was listening for evidence that questions asked or responses given, in either committee or chamber, might be informed by her own private briefings with MSPs, government ministers, or civil servants. The policy director was very pleased, for example, when the Scottish Labour member who was keen on birds called for SSPCA inspectors to be allowed to enter shooting estates to collect evidence of wildlife crimes such as raptor poisoning, for this was a sug-gestion partly sponsored by the animal group. Maggie's satisfaction increased further when another committee member name-checked the animal group in a discussion about vicarious liability. This principle, which the group lobbied to be enshrined in the bill, would have made shooting-estate owners liable for illegal actions of their employees, such as the poisoning of raptors by gamekeepers.

Indeed, Maggie not only regularly fed policy positions to politicians and civil servants; the policy director quite often drafted the petitions or debating points spoken on the floor of the chamber by sympathetic MSPs. During the passage of the bill, for instance, Maggie drafted a range of parliamentary questions, which MSPs from various political parties agreed to put to the chamber or submit in writing to the minister (parliamentary protocol required that ministers or their civil servants respond to such letters in a detailed fashion). These included ques-tions designed to tease more information out from government or to get min-isters to clarify previous claims or sometimes even to try to trip them up ahead of a next-stage committee hearing or chamber debate. For example, I can recall an unattributed question lodged by one MSP that asked the government for its current estimation of the economic value to Scotland of the shooting industry. The question was phrased in such a way that required a denial or confirmation of old industry figures previously cited by the relevant minister and, if the latter, a full explanation justifying those figures' continued use. Likewise, Maggie got a

Scottish Green Party MSP to formally ask the minister to demonstrate how the government knew that the SSPCA was satisfied with the welfare content of the proposed snaring regulations. Another question submitted on behalf of the animal group asked for the estimated effects of a snare ban on the economic value of shooting and farming, with a subsidiary question asking how the minister came by those figures. In this case, Maggie told me that she was fairly confident that the government had no research findings on that matter. So, the minister would need to either prevaricate or commission such work. Both outcomes served the interests of the animal group.

Getting a question answered or hearing a version of your policy positions read back to you, whether acknowledged or not, was of course highly gratifying. But it was not really a source of surprise. As Maggie and all other policy directors and lobbyists understood, lawmakers relied heavily on the research and information provided by interest groups. Even as government ministers or lawmakers presented their role as a balancing of interests, they very often depended on those interests for the concerns that they expressed, just as those interest groups depended on them to get their interests heard. In this scenario, the achievement of balance was precarious. This was less because competing interests were hard to reconcile and more because the final balance struck always risked appearing laden with specific interests. That vision of interests operating all the way down might have a moderating effect. It could enable, for instance, disputing or polarized interest groups to participate together in the legislative process based on a shared stake in the same issue. But it made the likelihood of moral progress or breakthrough hard to envisage. Interests might be negotiated. One might sometimes get quite unexpected deals of mutual convenience made between stakeholders that enabled this or that clause in a bill to be passed and go forward to the next stage. However, the parties concerned ultimately remained defined and separated by their interests. In the next section, I want to explore how that gap might be bridged. This includes a closer consideration of the ways in which the animal group went about trying to persuade or influence lawmakers, as well as a consideration of how they sometimes sought to convert lawmakers, to get them to adopt the broader moral outlook of animal protection.

Appeals to the Heart

In terms of the ethics of heart and head, this chapter has so far largely focused on the attempts of the policy director to moderate the impassioned and antagonistic stance of animal activism. Maggie presented this as a project of both professional necessity and personal ethical inclination. However, we have heard little of what role moral passion as well as forms of what we have been terming moral patiency might play in the policy work of moderate animal activism. Is it merely that aspect of animal protection that needs to be pragmatically kept in check? Or might what colleagues sometimes termed "appeals to the heart" have a more direct and recognized place in the slow tactical struggle to influence policy, including the deliberations of members of Parliament, government ministers, and their civil servants? The focus now falls on the efforts of Maggie and the animal group to prick the conscience of lawmakers or to instill in them an animal-friendly change of mindset.

These questions necessarily return us to the issue of animal suffering and particularly to the power of images that capture that suffering—in other words, to the old tried and tested techniques for exerting moral influence. It is true that Eilidh, the animal group's CEO, had introduced an organization-wide ban on the display of such images across its printed materials, website, and social media platforms; colleagues replaced them with images that depicted more positive expressions of sentience such as animals at play, in courtship, or exerting natural talents (see chapter 3). However, an exemption remained in place when it came to the policy work of Maggie. Throughout this period, Maggie's private meetings with members of Parliament, government ministers, and civil servants continued to rely heavily on the extensive use of filmed footage and photographs. This included both premortem and postmortem images that graphically illustrated the extent and consequences of various levels of injury and harm. Indeed, it was the perceived capacity of these images to dramatically impact the emotions of lawmakers that made their exclusive use in the controlled circumstances of policy work acceptable. This was extraordinary not just because it clashed with the animal group's new ethos of positivity but also because Maggie knew that the images that she shared fell far short of the evidential form expected of policy submissions (see Reed 2017b).[3] These images were consistently shown to lawmakers even though this action risked playing up to the reputation of animal groups as overly emotive or immoderate stakeholders.

One morning, I came into the office to discover Maggie and Barry discussing a recent meeting that the policy director had had with a member of the parliamentary committee that was scrutinizing the Wildlife and Natural Environment Bill. Maggie had organized a series of individual meetings with all the MSPs sitting on the committee to brief them on the animal group's position on snaring and to lobby for their support. That last session, which lasted forty minutes, had been with the Scottish Liberal Democrat committee member who was also the party's rural affairs spokesperson. Maggie explained that as well as handing the MSP a quite substantial written briefing on the matter, she had passed over public opinion polling statistics, commissioned by the animal group, which apparently showed that 82 percent of Scottish Liberal Democrat voters in Scotland favored an outright ban on snares. This lodging of evidence alongside the call for the MSP to consider the Scottish Liberal Democratic Party's self-interest in the matter was followed by a reiteration of Maggie's main lobbying point—that is, that snares may be an affordable form of predator control, but at the same time they were very clearly indiscriminate and inhumane kinds of traps. Maggie told us that it was at this point that she got out a folder of "really horrible pictures." The pictures were quickly followed by a video presentation based on investigative footage taken by Barry on a shooting estate (the vast bulk of the photographic images used by Maggie also came from the investigations archive). That footage showed a badger caught in a snare, which Barry had subsequently attempted to release with the help of an inspector from the SSPCA. Obviously trapped there for days, the badger was in evident distress and very poor condition. But it was not until Barry and the SSPCA inspector turned the animal over that they realized the snare wire was in fact caught around its abdomen. As Maggie narrated for the MSP, the internal organs were all exposed, and so the inspector decided that they would have to euthanize the badger on the spot. "And this happens all the time, every day, and all over Scotland," Maggie told the politician.

Such footage and images were shown only to try to provoke a reaction. The aim was to hit the lawmaker's heart and in doing so awaken their "instinctive sympathy" for the suffering of animals. "Sometimes we will just go in and start doing that pitch, and [the lawmaker] will say, 'Oh, I completely agree with you; that's terrible,'" Maggie explained. "You know [they will say to me], 'I am absolutely opposed to that; I will do everything to stop it.'" But at other times, the desired response would not occur at all. Like Eilidh, Maggie believed that the capacity for empathy depended on how someone was brought up and on experiences in their early years (see chapter 3). Yet equally it could be due to internal struggles between heart and head. When Maggie showed the photographs and footage to the Scottish Liberal Democrat committee member and rural affairs spokesperson, the policy director watched the member closely. "He didn't say anything," Maggie reported, "but his face said quite a lot. I don't think he thought it was fine to see what he was seeing." However, despite that interpretation and the more evidential briefing and interest-directed pleas offered in the rest of Maggie's submission, the MSP remained resistant to the group's case for a snare ban. Maggie reported that the Scottish Liberal Democrat MSP told her that he instead preferred to support other clauses in the bill, such as those outlining tighter regulation and the introduction of a code of practice for the trap's use. "So, we didn't achieve that change of mindset that we were seeking," Maggie admitted. An element of disappointment could clearly be heard in Maggie's words, but principally the policy director expressed a professional resignation. "There are a lot of other interests speaking in his ear at the same time," Maggie conceded, before observing that the committee member was probably just trying to weigh the lobbied claims made by competing stakeholders.

Returning to the attempt to optimize the combination of appeals to heart and head, Maggie then reflected on the strategic lessons to be gleaned from the interview with the Scottish Liberal Democrat. Their conversation had highlighted, for instance, that the animal group needed to develop one or two further lines of reasoned persuasion, areas "where we have to get arguments that they just can't combat." Maggie particularly zoomed in on the issue of alternative means of predator control. For Maggie realized from her conversations with members of Parliament that they needed to present more concrete options for humane control measures, including some evidence of the efficacy of such measures. In addition, there was a need for further nuances in economic argument. Ideally, Maggie wanted statistics that demonstrated that the shooting industry was not completely reliant on snaring for its continued existence. Both needs, Maggie recognized, could require commissioning another consultative report from academic scientists. If that was too expensive, it would require Maggie devoting hours of her already stretched time to further research. But it is worth noting that in this postmeeting debriefing there was never a suggestion that appeals to the heart might be ineffective or counterproductive. All the estimated readjustments were rather focused on strengthening the evidential case that accompanied such appeals or on tinkering with the techniques for changing mindsets.

On occasion, images and video footage of animal suffering would also be presented at larger, closed screenings. This was especially common during the annual conference season of the main political parties. For years, the animal group had paid to be a conference exhibitor at such events. The group typically hosted a stall alongside other interest groups, organized photo-calls with supportive members of Parliament, and sometimes cohosted fringe meetings for conference delegates. Although the screenings were "hard-hitting and not very new vision," Maggie insisted that they were right and appropriate for that context. This was because party conferences were a gathering for "decision makers" as well as for those wider party members and branches that supported or elected them. And those people "needed to know what was really going on." In Maggie's opinion, this was especially the case with the membership of the party in government. Indeed, my own sense of the extraordinary capacity of these images of animal suffering to elicit extra-evidential sympathy properly emerged at an SNP annual conference in Perth—more specifically, during a 2010 fringe meeting that Maggie arranged with the policy director of the League against Cruel Sports on the legislative issue of snaring.

"The only animals that can actually blush with shame are human beings," one invited speaker at the fringe meeting announced to the audience of SNP delegates. "And I think the minister must have been blushing with shame when she came out and supported this bill as it stands." Behind the speaker, a slideshow of Barry's photographs and video recordings of animals trapped in snares played in a continuous loop. This packed fringe meeting took place in a hired room at the old hotel just around the corner from the modern conference center. Laid out with four rows of seats and a large table at the back with sandwiches and drinks, the venue was lined with shelves of hardback books and, rather incongruously for the occasion, with a variety of glass cabinets each displaying different mounted and stuffed wild animals. As a past SNP party candidate for the Westminster elections and a working vet, this speaker had been asked by Maggie to address the fringe meeting to provide some expert veterinary opinion on the kinds of injuries and sorts of suffering that can be caused by snares. Indeed, the audience listened intently as the invited speaker outlined how the 7 cm width of the snare's wire loop did not properly allow for struggle, resulting in the oft-reported desperate twists and turns of trapped animals before they collapsed exhausted and slowly died. The speaker went on to share how he had professionally seen cases of snares caught around the stomachs of foxes, where the wire had gradually cut through the bowels, and cases where deer and other nontarget species had been caught around the leg and starved to death. "This is a big issue, a big moral issue," the vet declared, before telling delegates that "we have to upset the balance on this vote." The speaker finally urged the delegates to lobby local branches and their members of Parliament or party candidates hard.

Beforehand, the fringe meeting had been formally opened by a long-standing and quite high-profile MSP very well known for her support of animal protection. Maggie had worked closely with this politician over several years. The MSP explained to the room that she was a formal sponsor of the fringe meeting but then apologized to us for the fact that she could not herself turn around and look at the slideshow. Such images just caused her too much distress. However, the MSP also told us that although the images could be upsetting, she knew that it was crucial that they be shown since they could help "change minds." The MSP then launched into a story about the time she came across an illegal fox hunt in her rural constituency and observed a vixen being ripped apart by the hound pack. Like foxhunting, the MSP stressed, snaring ought to be a "conscience matter." The MSP therefore urged those present to "break ranks" on this issue and petition for a ban. After this stirring introduction, Maggie briefly rose to explain the wider legislative context for their meeting and to formally name-check the two animal groups present. As the representative of an outside interest group,

Maggie stated that she wanted to keep her contribution short since she realized fringe meetings were meant to be a discussion forum for SNP delegates and party members. But before sitting back down, Maggie announced the results of the group's polling, which she said showed that up to 79 percent of the SNP membership supported a ban on snares—proof, Maggie concluded (if proof was needed), that despite their government's policy, the party itself was full of "compassionate, conscientious people."

A Q&A session followed these brief speeches, with many contributions from the floor notably angry or tearful. Indeed, the looped slideshow still playing in the background provided implicit and sometimes very direct reference points for those emotions. After the meeting, Maggie told me that she noticed many audience members kept wincing as they watched the slideshow and that some looked visibly unsettled. As Maggie pointed out, a lot of the contributions were less like questions and more like impassioned responses to what they had seen and heard.

"What have animals done to us, humans?" one delegate rose to say, yellow lanyard and conference pass swinging as they spoke. "Nothing!" the delegate answered. "What have we done to them?" The delegate nodded toward the projected images on the screen. "Lots!" Another delegate took the slideshow as a prompt to muse on the fact that as a child they had once shot a sparrow with a slug gun. "I still have the death of that bird on my conscience," the delegate mournfully confessed before proceeding to inquire whether any of the speakers had seen a recent television documentary about snaring (I believe the delegate was referring to the BBC program mentioned earlier, in which Barry featured). A third delegate put a hand up to suggest a plan of action. "We should be trying to get our own people and visitors to Scotland out into the countryside," the delegate urged; "then, they would come across some poor animal snared, and what would that do for tourism!" The next audience member prefaced their comments by announcing that they were not a delegate or party member but instead a representative of Dogs Trust, another animal group represented at the SNP conference. The spokesperson reminded everyone that it was not just wild animals that got caught in snares. Indeed, their organization received many reports of household animals terribly injured or killed in these traps. The insight motivated another seated delegate to rise to their feet and forcefully affirm, "Regardless of how you dress it up, snaring is barbaric. Anyone who has seen animals in snares would know it is horrible." Finally, another delegate rose and, with evident struggle to control her voice, started by informing the audience that her knees were shaking and that she was trying not to cry or to look at the slideshow any longer. The delegate explained that this response was because her daughter was on a land management course at that very moment and had already got into trouble for refusing to lay snares. Before sitting back down, the delegate passionately appealed for the public to become better educated on the issue.

This was a collective display of heart or moral passion. Even Maggie was surprised by the extent and intensity of impassioned support expressed in the room for a snare ban. Indeed, the fringe meeting appeared to Massie as a very public reaffirmation of the value of showing images of animal suffering to such audiences. Before the fringe meeting wound up, the delegates present agreed to lobby the leadership and other party members on the matter. The sponsoring MSP also encouraged them to write to local and national Scottish newspapers about a ban and particularly to highlight the plight of dogs getting caught in snares. Like Maggie, the MSP recognized that this might be the most effective way of getting a public reaction to their campaign. There was then a very conscious sense of the need for these responses to lead directly into strategic deliberation, not just plans of action but also further evidence gathering and consultative submissions.

Yet, the meeting could not end without an invocation to also balance interests. After advising the fringe meeting that a "little rebellion" was healthy in a political party, the invited speaker with a veterinary background added a caveat to their powerful identification of snaring as a "big moral issue." If one was going to defend the minister's decision to reject a snare ban, the speaker reflected in more conciliatory fashion, then one might point to the fact that the government did want to stop the poisoning of birds of prey. Perhaps, the speaker mused, the pullback on a snare ban was part of a deal with "shooting interests" to secure their support for the protection of raptors. After all, the invited speaker knowingly speculated, this would be an example of entirely "understandable politics."

Conscience Vote

In a study of British pacifists, Tobias Kelly (2015) outlines the historical role of conscience as a marker of rights and duties within notions of British citizenship. Kelly points out, for instance, the emphasis placed on the legal protection of conscience "as proof of British commitment to freedom" (2015, 699) during the Second World War, most notably manifest in the setting up of tribunals to process the application of "conscientious objectors" and to assess the sincerity of their pleas. The tribunal procedure, Kelly observes, led its judges and claimants to conventionalize conscience in certain ways, including its signs, performance, and modes of recognition. A proper conscientious objector was expected to display a "detached conviction that was understood as moderate and tempered" (697). Neither "too fervent nor too calculated," that conscience was typically indexed through evidence of attachment to a mainstream Protestant denomination or through a long family history of objection. Although "British law was unique in the mid-twentieth century in not requiring a formal religious affiliation for the recognition of conscience" (696), in practice restrained religious convictions were far more commonly welcomed by the tribunals. It was very difficult to get a political conviction for conscientious objection, such as one born out of socialist principles, accepted by the judges (709). Nevertheless, the idea of protection resonated, affirming the wider modern assumption not just that conscience was "potentially separate from religious belief or thought"(696) but also that it was most closely connected with the expression of inner moral life or with the subject's personal and hence most essential set of moral commitments.

As conversation during the SNP fringe meeting suggested, another crucial element in that notion of protection was the principle of the "conscience vote" or "free vote" in the parliamentary process of lawmaking. On occasion, members of Parliament at Westminster and Holyrood were formally permitted to vote based on that personal conscience rather than based on an official line laid down by their political party. At Westminster, the historical model for the conscience vote had for many years been the issue of the restoration of the death penalty. Although that penalty was abolished in 1964, restoration was by convention debated once in every Parliament until 1997. Likewise, debates connected with abortion often received free votes in the House of Commons. At Holyrood, the most well-publicized example of a free vote during my time with the animal group arose around discussion of the issue of assisted suicide. This resulted in a private members bill in 2015; the SNP government did not support the bill, but along with all other political parties in the Scottish Parliament, it allowed its legislators to vote "according to their conscience." In terms of legislation linked to animal protection, there was also some precedent of free votes on certain issues, such as tail docking in dogs and perhaps most well known around foxhunting. The latter was banned in both parliaments in the early 2000s (although, especially in the English case, with plenty of loopholes!). In these and other examples, the free or conscience vote was sometimes agreed across main parties. At other times, it was granted only by individual political parties.

So, when the MSP at the fringe meeting declared that in her opinion, a snare ban, like the previous legislation banning hunting with dogs in Scotland, should be adjudged a "conscience matter," she was drawing on a long legislative tradition. Indeed, party members generally expected that their MSPs and MPs would occasionally confront a legislative issue that challenged personal conscience and required them to vote according to what they believed to be morally right rather than in conformity with the policy position whipped by the party. As well as formally granted free votes, conscience matters could be informally cited to explain occasions when members of Parliament voted against policy positions even when a free vote was not allowed. Political parties generally tolerated this behavior when the vote against the party line was adjudged to be a rare or exceptional circumstance and a history of commitment to a particular moral issue or cause was well documented. The MSP in question had an established reputation as an advocate of animal protection, and so if she did vote against the SNP government's legislative plans, this would surprise no one.

Just as Kelly points out in the historical analysis of the decision-making process at tribunals to assess conscientious objectors, the notion of a tempered or moderate personal conscience was essential to these parliamentary judgments. A member who consistently voted against their party line or invoked conscience matters too often lacked that necessary restraint or could even be accused of self-interested motivations. But, as Kelly also highlights, that detached conscience, to be convincing, had to simultaneously retain a sense of involuntary action. "Freedom of conscience," Kelly records, "was most readily recognised [by the tribunal] when it was thought not to have been chosen freely" (2015, 697). Indeed, applicants for conscientious objection regularly presented conscience as something that was not controllable, as a matter of duty rather than choice, for instance (705; & see Weiss 2012). In such statements they appeared rather more like moral patients than moral agents; as Kelly says, "to make a claim of conscience was to make a claim to be unfree in some way" (717). The same kind of emphasis played out in the appreciation of conscience matters in the Scottish Parliament. This was especially the case when individual MSPs voted against the party line—that is, when there was no free vote. If one invoked personal conscience as the reason, then the strong expectation existed that a member of Parliament ought to have struggled with the decision but ultimately been left with no option. Once again, religious belief was the paradigmatic example. In the case of nonreligious sets of moral convictions, such as those tied to animal protection, an equivalent sense of immovable obligation, which the MSP concerned felt unfree to change, needed to be conveyed.

In terms of the idiom of heart and head popularly invoked by lawmakers, conscience matters were undoubtedly heartfelt. The moral stance taken up by an MSP might be reasoned and well thought out, for the expectation remained that heart and head should remain in balance to avoid extreme positions, but without emotion that stance risked looking too rational or coldly deliberated to appear not freely chosen.

As the SNP fringe meeting discussions additionally highlighted, conscience matters could clearly also be interpreted as an explicitly species-specific affair. Despite the patency or unfreedom essential to the experience of personal conscience, its invocation was generally taken to be indicative of a human individual's moral agency. In this register, the dominant after-the-fact moral patient position typically got reassigned to the nonhuman object of welfare concern—that is, to the trapped fox or badger, and in the case of one audience member at the fringe meeting, to the sparrow they once shot with a slug gun. Or, in a move more reminiscent of the definition of moral patient offered by Regan (2004), it got reassigned to a nonhuman subject taken to be without the capacity for conscience and hence not morally accountable. Human beings, the opening speaker at the fringe meeting declared, were "the only animals that can actually blush with shame," a claim of exceptionalism designed precisely to reinforce our sense of responsibility to do something about the "barbaric" practice of snaring. But, of course, the statement simultaneously affirmed the (human) patiency involved in conscience matters. Blushing is, after all, usually assumed to be an involuntary action, one that reveals our moral feelings despite ourselves. Indeed, as the speaker at the fringe meeting provokingly suggested, that physical action might perhaps have even betrayed the shame felt by the government minister who stood up in Parliament to defend the laying of snares as a necessary tool of land management. Blushing aside, these accounts of personal conscience invariably return us to the quality of that patient position. Phrases such as "I still have the death of that bird on my conscience" implied a transitive dimension in which conscience acted on the moral subject in autonomous fashion—but just as importantly in which conscience did so precisely because conscience too had suffered the effects of an event laying heavily on it.

Sensitive to all these conventional registers and to the dynamic movement between them, Maggie attempted to manufacture a scenario that could let conscience speak. More specifically, Maggie sought to make snaring a trigger for the sensation or experience of that personal conscience. For example, the policy director used polling results to inform her audience that the wider SNP membership was convinced by the case for a snare ban, statistical results that Maggie claimed already demonstrated a compassionate and conscientious nature. But the policy director also tried to "change minds" by provoking an involuntary action. This was manifest not just in the winces or blushes of party members as they watched a slideshow of animal suffering but also in the outpourings that followed its viewing. This reaction was necessarily underdetermined. Indeed, while Maggie might have strategically planned for it, any response was always a genuine surprise, and her surprise itself was a further confirmation that conscience had indeed spoken.

Part of Maggie's role, then, was to encourage party members and especially their legislators to vote with their conscience when that conscience revealed itself through an expressed concern for the protection of animals. As well as cultivating personal conscience through showings of video film and photographs, this involved identifying nascent signs of possible sympathies in the past words and actions of politicians. In fact, Maggie paid special attention to any incoming crop of new parliamentarians, scrutinizing their biographies and listening for word-of-mouth information that might point to their disposition toward animal protection. Although there were few MSPs with a level of commitment equal to that displayed by the SNP politician who sponsored the fringe meeting, Maggie knew that a larger number of MSPs from a variety of parties would listen to the case for animal protection and take it into consideration. These politicians might rarely vote with their conscience against the party whip, but they would consider and sometimes did put down or speak to compromise legislative clauses that at least nodded toward the position of the animal group. So, the time Maggie spent with each member of the parliamentary committee examining the Wildlife and Natural Environment Bill did not in the end lead to any of them eventually supporting a snare ban if their party opposed it. But it did result in several of the committee members supporting a timeline for a formal review of new snaring regulations and advocating for a clause on vicarious liability. Even if the politician concerned insisted, as they usually did, that they were simply responding to the evidence put forward, Maggie believed that it was an awakening of personal conscience that ultimately often tipped the balance in favor of such compromises.

Yet, the commitment of the animal group to cultivating the conscience of legislators was also matched by tactical appeals that bypassed reference to individual conscience altogether or sometimes even called on the personal conscience of legislators to be put to one side. For instance, Maggie told me that where a political party had a manifesto commitment to a particular animal welfare issue, she would often draw the attention of its legislators to that obligation. In these circumstances, the policy director would urge members to adhere to their promise, both as a matter of party loyalty and in the name of the compact made with the electorate. During the passage of the Wildlife and Natural Environment Bill, Maggie had numerous conversations with the Scottish Labour Party spokesperson on animal protection. Maggie explained to me that this MSP had clearly not been impressed or moved by the animal group's pitch for a snare ban. Indeed, the member had communicated a sense of moral obligation to protect the livelihoods or economic interests of their rural constituents. But when the party spokesperson finally informed Maggie that they were not personally minded to either support or recommend a snare ban, Maggie decided to "play hardball." Although Scottish Labour had not explicitly listed support for a ban as a manifesto promise, Maggie stressed that a series of informal commitments had been made. In fact, Maggie threatened to publish an old letter addressed to the animal group from the now-current Scottish Labour leader that clearly stated that commitment. The tactic, unusually confrontational for Maggie and full of risks for the group's longer-term relationship with that political party, paid off. Not only did the party concerned finally whip in favor of a snare ban, but its animal protection spokesperson ended up voting for the clause too, despite the fact that the spokesperson still strongly disagreed with the policy.

This, then, was another instance of the policy director acting with heart and head. As Maggie knew well, the virtue of party loyalty was the dominant ground against which matters of personal conscience typically played out. It typically defined the struggle that members of Parliament usually went through in deciding to follow or not follow their conscience. In fact, the animal protection spokesperson for Scottish Labour once told me that as far as they were concerned, adherence to the agreed party line was itself the overriding issue of conscience. While Maggie's move indicated that on occasion there could be some steel behind the animal group's policy work, it also pointed to a continuing pragmatic attitude toward moral stances within the legislative field. The free or conscience vote, by its very rarity, suggested that moral grounds were hardly ever sufficient reason for an individual member of Parliament to vote against the party line, just as they were hardly ever an admissible form of stakeholder submission to government consultation exercises or to parliamentary committees. Maggie might persist in making appeals to the heart, trying to awaken the personal conscience of legislators, but the policy director's own conscience or moral agency in these affairs was necessarily muted.

Indeed, Maggie remained very conscious that the continuing status of the organization as an interest group and her role as a representative for that interest group often carried more weight in the minds of politicians and lawmakers. An interest-laden perspective also often carried more weight in Maggie's mind. The policy director suspected, for example, that the stance of the Scottish Labour spokesperson was really informed by self-interest—that is, the desire of the MSP not to alienate the voters in their rural constituency. Whether this was the case or not, Maggie consistently acted in the knowledge that other people were likely asking themselves such questions. For instance, how can we accommodate the interests of animal groups alongside the interests of different stakeholders, such as the interests of the shooting industry or farmers or conservation groups? And how can that accommodation serve our own interests as a government or political party?

The point is best illustrated by referencing an extraordinary circumstance in the legislative process that took place a few years after the Wildlife and Natural Environment Bill entered law. While a snare ban was not part of that final legislation, in 2017 the Scottish Parliament did unanimously pass a bill banning the use of wild animals in circuses. Cross-party unanimity on animal welfare issues was oddity enough, but the most remarkable aspect of this legislation was the fact that it was introduced by the government and subsequently approved solely on "moral and ethical grounds." Such a move surprised many animal groups as well as MSPs when first proposed. It seemed to run counter to the protocols for lawmaking that everyone, including Maggie, had come to understand. However, the explanation for this turn of events was typically interest-laden in tone. The bill could be passed on moral and ethical grounds, it was announced, precisely because in this instance there were no competing interests to balance. As a Scottish government survey (quoted by the relevant minister) testified, more than 95 percent of respondents were in favor of a ban. In the light of this fact and the even more pressing reality that traveling circuses with wild animals hardly ever visited Scotland anymore, a consensus seemed to emerge that it would be a waste of government time and resources to collect or commission scientific evidence on the welfare issues concerned. Indeed, some believed that the legislation was passed on moral and ethical grounds exactly to avoid having to specify the evidential basis for a ban, which did not exist because the research was lacking. In this rather surreal instance, a conscience matter developed as the pragmatic tool of lawmaking for those circumstances when, it appeared, no one claimed an opposing stake.

MORAL SUBJECTS

In this book, I have sought to sympathetically engage with the priorities and struggles of moderate animal activism. However, it would not be fair to close this account without giving some further consideration to the more negative perspectives that a moderate stance attracted, especially from within the animal movement itself or from certain other animal groups. This is important not only because it helps throw that stance into some relief but also because colleagues were very much aware of those criticisms, including the implied counterframing of their outlook and activities. I therefore want to begin my conclusions with a brief examination of the moderate animal activist as a figure of suspicion. This necessarily involves a look at the common charge of complicity or collusion. Most obviously focused on the organization's lobbying or policy work and hence targeted at the role occupied by Maggie, these charges entirely recalibrated an understanding of that expert practice. But, as we will see, the charges also played a vital role in reaffirming the value of Maggie's work and more broadly the status of the Edinburgh-based organization as a moderate animal group.

Of course, the accusation of complicity or collusion has a long tail within the animal movement. It goes back at least as far as Peter Singer's original comments on the "animal welfare establishment" first offered in *Animal Liberation*. At the time of publication (1975), criticism was largely focused on large and well-established animal protection organizations. Singer's famous attack described a history of groups that had over time "lost their radical commitment" as a direct result of engaging with government, scientists, and commercial agencies as well as an outcome of wealth accrued (2015, 218). "Again and again," Singer narrates, "the societies compromised their fundamental principles for the sake of trivial reforms. Better some progress now than nothing at all, they said" (2015, 218). Although Singer later acknowledged that some of these organizations had since adopted a "more forceful stand against cruelty," assumed by Singer to be due to the initiative and promptings of the new wave of animal liberation and animal rights groups (2015, 219), the sense of that criticism remains. Indeed, on the ground, at least in a UK or Scottish context, it continues to define much of the interaction between animal groups and much of the ongoing tension. (For a fuller sociological account of the factionalism between animal groups in the UK, this time grounded in a presumed opposition between radical and professionalized outlooks, which assumes the latter to be moderate and hence in some fashion co-opted or morally compromised, see Wrenn 2019.)[1]

The animal group that I worked with continued to periodically face these charges even though colleagues were sympathetic to aspects of Singer's general thesis and felt very much part of that broad new wave that Singer once described, and even though their organization was very far from being large or well resourced. In terms of the field of Scottish-based animal groups or animal groups working within Scottish policy contexts, that tension was further exacerbated by the fact that everyone knew each other. Scotland is a small place, policy directors like Maggie frequently observed, and even smaller for those exclusively concerned to lobby on animal protection issues. Groups or their representatives regularly met at the same forums, contributed submissions to the same Scottish government consultations, and mobilized in response to the same animal-related news stories. They had often in the past also worked together on individual campaigns or been part of the same broader campaign alliances.

By and large, suspicions of moderate animal activism were drawn from a perspective that refused to compromise or that saw compromise itself as morally suspect action. Take, for example, the charge of collusion. A typical accusation would center on claims that the animal group knowingly cooperated with the government or with other interest groups identified as linked to animal exploitation. During my time with the animal group, this kind of charge was best illustrated by a fallout over the issue of licensed seal culls. After a media statement made by Maggie that claimed the organization's lobbying work was in part responsible for a significant managed drop in the numbers of seals killed annually, a direct-action group specializing in marine animal welfare immediately accused Maggie and the organization of colluding with Marine Scotland, the government agency that published those figures. Indeed, a storm of angry messaging on Facebook ensued. The same direct-action group asserted that instead of working with the government, the organization should have joined the zero-kill campaign and refused to negotiate. "You can't head towards a ban; you have to be for or against it," the group warned, before dismissing Maggie and her colleagues as "pathetic." Other animal groups weighed in to accuse the organization of helping Marine Scotland successfully spin the licensed cull arrangement by enabling the government agency to claim that it had the support of animal protection and conservation interests. "It is time the government and [the animal group] heard the voice of the people," another group dramatically added before someone else suggested that until Maggie's organization changed tactics, it should not be "welcomed back into the AR [i.e., animal rights] fold."

As we can see, the charge of collusion tended to stress the agentive role of the policy director and her organization in such dealings. It also tended to read ordinary achievements of moderate animal activism such as the establishment of close working relationships with government as something more sinister or clandestine. The charge was quite regularly accompanied by whispers and even sometimes by very public claims of secret deals or behind-closed-doors agreements. And as the term "collusion" further suggests, it included a suspicion of acting in partnership with "the enemy." Indeed, during altercations of this kind, spokespersons for opposing animal groups would advise Maggie that her group needed to "choose a side."

By contrast, the charge of complicity tended to downplay the language of betrayal and instead focus on the perceived naivete of moderate animal groups. Such organizations, it was conceded, might be acting in good faith, but their actions nevertheless assisted the enemy in an activity that was morally wrong. Here, figures such as Maggie and organizations that positively participated in the legislative process were portrayed in more passive roles. They were often presented as akin to unwitting collaborators or as pawns in a wider game of power politics that they did not fully understand. As in Singer's narration, any legislative wins or reforms linked to their lobbying usually got dismissed as "trivial." Regarded as a form of tinkering that made no real difference at the scale of the global struggle against systems of animal abuse, their work might even inadvertently aid and abet the perpetuation of those systems.

So, while the charge of collusion inflated the negative moral agency of moderate animal groups, the charge of complicity rendered them powerless or assigned them a subservient position within a far wider struggle. In the latter scenario, it was the negative moral agency of other actors that tended to get inflated. Curiously, the chief targets of these accusations of nefariousness were often not government ministers or politicians or even classic embodiments of "the enemy" such as farming or shooting associations or scientific bodies but instead ordinary civil servants. Particularly, it was the staff of the Animal Health and Welfare Division that most regularly drew those animal groups' ire or sometimes even more active animosity. Those government officers responsible for the mundane work of consultation, the preparation of bills for legislative passage, and the drafting of responses to public or stakeholder inquiry were held to blame for perceptions of Scottish government inaction or, worse, for deliberate miscommunication and prevarication. Civil servants were also identified as those who took advantage of or played on the naivete of the moderate animal activist.

I can recall numerous flare-ups in which the animal group was drawn by association into such charges of complicity. There was the occasion, for instance, when a few other animal groups accused civil servants of misleading them or altering the figures provided in a parliamentary-sponsored public forum on animal welfare, to which Maggie had been elected the current rotating chair. "The fact a Chair and an animal welfare charity then struck this [fact] from official minutes," one circulating accusation declared, "should give everybody an idea of just how rotten things are through there." This extraordinary claim was only a prelude, however, to the real attack. "It's actually the Scottish Govt. civil servants that are the main problem," the same group continued. "The comically named Animal Health and Welfare Division are a shower of real sweethearts. They fight tooth and nail to block and erode any real improvements for animals." This was followed by a final, damning assessment of those civil servants: what those uncompromising animal groups apparently had to deal with was "a nasty, nasty group of petty bureaucrats with the power of life and death."

Although usually unflappable, Maggie did sometimes get exasperated by these charges. In fact, she once admitted to having a "bête noire," the spokesperson of another small animal group in Scotland who regularly called Maggie out in a rather grandiloquent tone that consistently got under her skin. "You can always tell these types," Maggie offered by way of dismissal, "because they send their angry messages out at three a.m. when they should be asleep!" Likewise, the policy director sometimes got upset on behalf of the civil servants who took the brunt of these accusations. Working closely with the Animal Health and Welfare team over the years, Maggie appreciated that they had in fact a range of personal views about various animal protection issues and that they were understaffed and overworked. The sympathy was extended because Maggie also knew how upset some had been about these public claims made against them and because Maggie was aware that individuals had to deal with such treatment in the context of their own lives or family challenges.

Indeed, this kind of assessment was the crux of the difference that defined the moderate animal activist. Maggie and her colleagues understood the popularly invoked distinction between the professional and personal as a real or meaningful separation, as they did the separation between a person's moral views on animal protection issues and the rest of their outlook or attitude. Despite identifying as animal people and claiming to be ruled by moral passions and their love of animals, colleagues consistently insisted that they were more than just animal activists and likewise that others were more than one thing too. Hence, they adjudged the capacity to maintain relations with friends or with family members who did not hold animal-friendly views and to live with or marry someone who ate meat as positive demonstrations. The same ethos led them to value toleration and reasonableness and to exhibit a general reluctance to lecture others. In fact, for Maggie, the ability to recognize the person behind the civil servant or for that matter behind the scientist or the farmer or the gamekeeper was essential. While it certainly fed in a strategic fashion into skills of personability and hence into what made Maggie an effective lobbyist and policy director, the disposition was clearly appreciated on its own terms. It was also deemed to be precisely the ability that other kinds of nonmoderate animal activist, such as Maggie's bête noire, lacked.

Turned around and seen from the perspective of those other animal groups, this looked a lot like inconsistency, contradiction, or weakness. For the refusal to compromise on animal protection issues was also a refusal to accept these separations. It was, as Maggie further admitted, a refusal to engage. Despite her best efforts, spokespersons for these groups often dismissed the invitation to agree to disagree and the accompanying implied suggestion that they could still get along personally. While Maggie sometimes failed this test herself, part of her frustrations derived from the uncomfortable sensation of being shunned as person. Her bête noire consistently failed to see Maggie as something more than the complicit or colluding policy director of a moderate animal group. In fact, in that individual's eyes, Maggie felt regularly dismissed as someone who mattered only because they were taken to be morally suspect or compromised.

More robustly, Maggie and her colleagues saw the accusatory stance as itself morally flawed and suspect. Indeed, the policy director often threw the charge of naivete back at these groups. For instance, Maggie derided their habit of offering simple moral outrage or blanket condemnations as tactically ill-informed, or worse as an irresponsible indulgence. By way of elaboration, Maggie liked to refer back to a recent news story about the controversial shooting of a "giant Red Stag" that took place on Exmoor, in the Southwest of England (see chapter 4). At the time this story broke, it provoked a public outcry and led some animal groups to call for a ban on all forms of hunting. However, in her own blog post about the incident, Maggie was very careful to instead focus on the issue of a breach of shooting convention, the fact that the hunter concerned had shot the stag during the breeding season and hence contravened their own voluntary code. This kind of detailed intervention, Maggie opined, might lack the glamour of moral indignation or of calls for an outright ban, but it would ultimately be far harder for the shooting lobby to resist. The point was not that bans should never be campaigned for—rather, that a ban should be the policy or lobbying position only when it was likely to be successful or to prove effective as a point of negotiation.

As we saw in chapter 8, this gradualist chipping away at the position of opponents remained central to the ethos and practice of the animal group and its policy director. But more broadly, Maggie and the organization often responded to charges of collusion or complicity by returning to the theme of expertise and technical competence. Often, they claimed, other groups just misread a situation because they straightforwardly lacked knowledge or relevant training about the stages and protocols of the legislative process. At times, such comments came close to a straightforward assertion of cultural capital; spokespersons for other groups simply did not have the education or the experience that came from years of paying close attention to that process. At other times, the comments were fueled by a sense of sheer laziness. Maggie expected that these other spokespersons should know, for instance, that moral outrage or condemnation on its own terms did not usually form an admissible basis for a submission to a government consultation or to the committee stage hearing of parliamentary scrutiny of a bill. As well as displaying ignorance about the instruments of lawmaking, Maggie countered, those animal groups tended to be too suspicious of what they did not understand, to immediately dismiss the value of procedure and protocol, and to be too ready to mistake attention to detail for obfuscation.

The point invariably drew Maggie back to the example of civil servants and the personal attacks made against them. For in this case, Maggie observed, such accusations demonstrated not just a failure to appreciate the assigned role of civil servants in lawmaking and government but also a failure to appreciate the moral value or purpose of that work. Maggie referred not just to the British civil service principle of independence and nonpartisanship but also to the wider ethos of duty or subservience to office. This included the ideal that while civil servants might be responsible for drafting legislation or responding on behalf of government, they were not speaking for themselves or expressing a personal view in doing so. In other words, these animal groups failed to see or appreciate the studied impersonality of the state bureaucrat as well as the notion that a self-denial or deliberate subordination of personal moral enthusiasms could be a "positive moral achievement" (Du Gay 2008, 17). These animal groups were far from alone in regarding those virtues as morally questionable; as Du Gay (2000, 2006, 2008) points out, such an ethos appeared to clash with the growing presumption in British public life and in new brands of managerialism that ethical action could precede only through the exercise of moral autonomy or individual conscience-driven judgments. Nevertheless, for Maggie and her animal group, it was precisely that blindness to other forms of moral achievement, including civil servants' achievement of a "subaltern status" (Du Gay 2008, 2), that stood out. By implication, the ability to fully appreciate the moral labor involved in that subservience to office was another thing that distinguished the moderate animal activist.

Inconsistency

Among the charges against moderate animal activists, that of moral contradiction or inconsistency was recurring. But that type of accusation came not just from those identified on the more "extreme" end of animal activism; it additionally came from those identified as belonging to the mainstream. Indeed, facing accusations from both directions was also a defining experience of the moderate animal activist, one that reaffirmed their sense of being in the middle. As Maggie conceded, the public charges of collusion or complicity made against them by other animal groups had the useful side effect of reinforcing the impression, at least in the minds of government and other stakeholders, that the organization was moderate and hence worth engaging with. It would be hard to make the same strategic case for mainstream interrogations. However, their accusations of contradiction or inconsistency did have the benefit of reconfirming what colleagues also stood for, precisely as animal activists operating outside or beyond that mainstream.

Reflecting on the lessons learned from conducting ethnographic research with a diverse range of animal activists in India, Naisargi Davé (2017, 2023) dwells for a while on the pressures faced by those who occupy "any ethics in an oppositional or oblique relationship to the way things are" (2017, 37). More specifically, Davé focuses on a "tyranny of consistency" (see chapter 4). This includes the common expectation or demand placed on such moral subjects to adhere to "principles of continuity and accumulation," which judge any action against the subject's past action or inaction, and to "principles of representation/identity/analogy," which insist on the comparability of *this* (for instance, the life of a "maggot-ridden" dog healed by a rescuer) and *that* (the lives of those maggots sacrificed or "not helped" in the process [2023, 60]). Likewise, it includes the necessity to give an account of their ethics and particularly to explain "apparent inconsistencies or contradictions" (2017, 37). As well as the persistent querying of the act of rescue (why save this critter and not that one?), Davé notes the everyday challenges to lifestyle especially faced by the activist in North America who commits to become vegetarian or vegan (Davé 2019). But those who would make such demands of the animal activist, Davé observes, are not usually called on to account for their own contradictions (2017, 37). Instead, they exercise the tyranny of consistency from a normative position of power or authority. Indeed, for Davé, the normative has that defining characteristic: it is precisely "that which is allowed to be and remain in contradiction without existential consequence" (2017, 37; see also 2023, 58).

Drawing on Peter Sloterdijk's notion of "cynical reason" to further explain the normative impulse to query the animal activist, Davé suggests that the action is motivated by a desire to exhaust the moral subject, "literally, to make people tired so that they give up" (2017, 38; see also Davé 2023, 59). That exhaustion "might take the form of drowning and disappearing, or it might take the form of folding quietly back into the world as it is" (2017, 38). In fact, as Davé acknowledges, the latter movement can come as a relief. For the tyranny of consistency is not just an external demand constantly placed on the moral subject; it is also an interrogation from within. This is not surprising, since the activist, like the "ethically otherwise" stance that they subscribe to, ultimately "still belongs to, and emerges from, the world as it is" (2017, 38). Animal activists typically demand consistency of themselves.

Such observations would make sense to the colleagues that I worked with. As they regularly reported, once strangers found out that they worked for an animal group, more or less friendly interrogations of their moral stance usually ensued. These would invariably circle around perceived points of contradiction or inconsistency. The same conversations were also repeatedly had with friends and family. Indeed, I confess that, especially in the early days of my research, colleagues had to put up with my own use of contradiction thinking or cynical reason. Why restrict your moral concern to sentient animals only? Is it OK to kill and eat the meat of an animal if you have cared for it personally and ensured its welfare? What about eating lab-grown meat (at the time, there were a series of breaking news stories about the possibilities of meat being cultivated in laboratories without the need for livestock)? How would you respond if it were discovered that it was possible to measure an equivalent of feelings in plants? (This latter question was often structured around a lively discussion about the life and welfare of the weeping fig that stood in a large pot in one corner of the boardroom.) And if you were sick, would you take medicines or undergo treatments, knowing that the pharmaceutical products used were more than likely tested on animals? Et cetera. It was clear that such lines of questioning were entirely familiar and hence unsurprising to colleagues. This was either because they were frequently interrogated in such fashion or because they had asked versions of such questions of themselves (see Reed 2024).

In fact, the majority who committed to a vegan lifestyle typically did so out of a concern for consistency in their own lives. Individuals could no longer justify to themselves a vegetarian diet grounded in a commitment to do no harm if it included the consumption of eggs and dairy, especially once they appreciated the cruelties involved in those industries. Likewise, as vegans, colleagues dutifully strove to iron out inconsistencies in their purchasing habits (see chapter 4). Most obviously, this involved moves to ensure that they not only ceased to eat food derived from animals but also ceased to use other animal products. For some, this was a continual process of adjustment, since animal-derived ingredients could be found and were newly identified in a vast range of household products, and trace elements of animal product in even more. There was a direct expectation, therefore, that at least in this regard animal people ought to be consistent kinds of moral subjects.

Indeed, the consistency principle was also exercised in the form of critique. We might recall Barry's analysis of contradiction or "circus lies" among perpetrators of cruelty, observed through undercover investigation (see chapter 6). But less exceptionally, we might recall the familiar criticisms of colleagues. Like those in the broader animal movement, many shared a strong awareness of the inconsistencies or contradictions in the lives of others. This included familiar kinds of querying of the mainstream. How can they lavish so much love and care toward pets or companion animals yet exhibit such indifference toward the sufferings of other sentient beings? Or how can they speak of equality or against injustice but through their everyday practice deny the rights of animals? Sometimes articulating these ideas as charges of hypocrisy, at other times using expressions of exasperation or even disgust, colleagues called on wider Scottish or British society to abide by a higher moral standard. They also on occasion deployed contradiction thinking to point out inconsistencies in behavior or in the attitude of family or friends, or to make counterarguments to those put to them across the kitchen table, by the proverbial water cooler, or down the pub.

But as self-proclaimed moderate activists, colleagues were also concerned to closely police that kind of cynical reason. As we have seen, they did not want to be the kind of animal activists who loudly berated others for inconsistencies and moral failings. Indeed, part of the new vision of the organization was grounded in a conviction that the normative or mainstream stance on the world could be productively engaged with—for instance, that it was open to transformation precisely because the mainstream already contained signs of an animal-friendly disposition. From this perspective, a love of pets exhibited less the contradictions in normative attitudes toward animals and far more natural sympathies elsewhere suppressed. Although that project had mixed results and many colleagues remained ambivalent about the idea of attaching themselves too closely to a "mainstream movement," few associated their own ethically otherwise position with an absolute alienation from the world as it is. Instead of lecturing, preaching, or demonizing the mainstream, they generally agreed that one needed to work through encouragement and by example, to seek dialogue. That could require viewing the mainstream habit of contradiction thinking as itself a manifestation of communication, an expression not so much of cynicism or a desire to exhaust the moral stance of the other but rather of mutual curiosity.

But colleagues were far less likely to view the cynical reason of those who opposed them within the animal movement in the same way. As we have already seen, in significant aspect, colleagues at the Edinburgh animal group defined themselves as subjects who were the target of other animal activists' aggressive forms of contradiction thinking, where it felt like the ambition was to dismantle or ridicule the moral stance of the moderate activist. Their general reluctance to fully embrace the language and philosophy of animal rights partly derived from a perception of moral absolutism and its accompanying refusal to brook compromise or tolerate apparent inconsistency. Colleagues did not want to be that kind of animal activist. In fact, in many ways they did feel closer to the mainstream than to those activists who adopted a dismissive stance toward others within the animal movement.

However, there was no real sense of threat or impression that cynical reason emanating from whatever quarter might exhaust the resolve of the moral subject. Perhaps this was because moral reasoning itself was only ever a partial resource for the development or occupation of an ethically otherwise stance. As previously discussed, a few colleagues did on occasion suffer burnout and as a result withdraw from activist commitments for a period (see chapter 7). But that condition was never put down to the tyrannical demands of contradiction thinking. It was instead attributed to the costs and sacrifices of being committed to a cause, of feeling the injustice and bearing witness to the horrors of animal suffering daily.

Of course, as Davé points out, broader demands of consistency might still be read into such moments of emotional collapse or withdrawal. The burnout once experienced by Elaine, which she put down to an intense period of involvement in the investigative side of work, was a matter not just of caring too much but also of an accompanying sense that any individual act of care was never enough. Likewise, the often "frazzled" state that Cassie reported from the labor of animal rescue was caught up with a powerful awareness that there were always other animals out there still suffering. Such thoughts certainly plagued Cassie's composure and at varying times unsettled other colleagues too, just as they also sometimes motivated their animal activism.

Indeed, Davé makes the case that the starkness of the alternatives implied by that caring logic continually threatens to create a sense of unresolvable dilemma or to destabilize the moral subject within animal activism. More specifically, Davé draws attention to another "tyrannical formula," the common assumption "if something, then everything, and if not, then nothing" (2017, 38). But conducting fieldwork with animal activists in India also helped throw that assumption into some relief; it educated Davé in ways to persevere or deal with that corrosive way of figuring moral concern. For what some activists offered Davé was a metaphorical "shrug of the shoulders." Instead of focusing on the animals left unaccounted for, a few of these activists managed to resist the impulse to constantly relate the individual act of rescue to anything else. In the process, they prompted Davé to consider a reorientation toward an ethically otherwise stance, grounded in what Davé terms a more "immanent ethics." It was a lesson in making that "something" livable or "habitable all by itself" (2017, 38). At least for the period of enactment, these activists taught Davé how to consider that specific act of care as both all that mattered and nothing more than what it was.

Davé admits that this rendering might "all sound abstract" (2017, 39). It would certainly be a candidate for the kind of discussion that Maggie might dismiss as pub talk. However, Davé does insist on a few concrete specificities. First and foremost, Davé advises, this immanent ethics should not be read as collapsible into a liberal ethos that values something as better than nothing. That kind of resolution "still sees the something normatively, that is, in a relationship of potential consistency with everything or nothing, as if everything and nothing are viable options at all" (2017, 38). It also embraces a calculative logic of relative costs and benefits that many animal activists and indeed many other proponents of the ethically otherwise might hold responsible for wider injustices. Indeed, one suspects that this is the real reason for Davé's distancing statement on that liberal ethos. Certainly, as we have seen, it is a logic that many in the animal movement connected with trivial reform or unacceptable compromise and that colleagues themselves regarded with suspicion when deployed by those they identified as promoting or defending cruel practices. For example, this logic was used in claims made by the shooting lobby that they were motivated by a concern for animal welfare and in the arguments put forward by conservationists, which drew attention, for instance, to the net benefits of culling for the specific habitat that they were concerned to protect.

Yet I think for the animal group that I worked with, there was a way in which that liberal ethos of something being better than nothing did provide moments of genuine resolution or alleviation of ethical dilemma. This was the case for Maggie, who was most clearly responsible for having to make those kinds of calculations. There was even a way in which that ethos could be read to contain a form of immanent ethics. For in the minds of colleagues, the "wins" of the legislative process were *something*, livable achievements on behalf of animals that they could appreciate and just as importantly publicly promote as the outcome of the organization's labor. In her role as policy director, Maggie did inhabit those wins, at least in the sense of being a witness who was also partially responsible for their negotiated emergence. But unlike the individual and singular event of animal rescue that Davé mentions in discussion of immanent ethics, the care needed to cultivate a legislative win was a long, drawn-out process. As we have seen (see chapter 8), that process contained its frustrations and plenty of exposure to cynical reason. This included the contradiction thinking of other animal activists who viewed this win as nothing or sometimes as worse than nothing, to which Maggie typically responded with her own metaphorical shrug of the shoulders. A win was a win. While it clearly wasn't everything, neither was it nothing. Instead, it was something in between, habitable precisely because of its quality of being simultaneously better than and lesser than some other envisionable outcome.

In fact, as moderate activists, that was the quality that properly defined something as graspable. A reduced number of seals annually culled or of foxes trapped in snares really mattered to Maggie and the animal group. For them, that reduction was a moral accomplishment, even if they knew a ban on seal culls or on snaring would obviously be preferable. Likewise, a simple tightening of the code of practice around the use of traps in the countryside or a commitment built into legislation to review the code's implementation after five years could become achievements. Although not necessarily the desired outcome of lobbying, each was a win that colleagues could live with. For the somethingness of those wins was at once vitally tangible—in the sense of here was something rather than nothing—and abstract. The win might not be all that mattered, but it contained the promise of something more on the horizon, something worth working toward.

Fear of Wasps

It is probably a truism to state that observations about the consistency or inconsistency of the moral subject ultimately lie in the eyes of the beholder. What appears a glaring contradiction to me might not appear so to others and vice versa. And anyway, as Davé stresses, contradiction thinking is precisely that, a certain way of communicating with or addressing the moral subject that Davé invites us to resist (at least when occupying the subject position of the ethically otherwise).[2] In such instances, the pointing out of inconsistency appears to be an aggressive or cynical act, an attempt to catch someone out and in the process to challenge the integrity of their moral stance. But, of course, inconsistency in the human moral subject can also be positively appreciated. One has only to think of traditions for registering that subject's presence in literature and other forms of entertainment. It is often said, for example, that for both writers and readers of fiction, the moral status of a literary character typically only emerges on the page once additional, apparently incompatible layers or strands of the character are introduced.[3] In like manner, we often expect our moral subjects to contain contradiction or inconsistency or want to perceive it as an inevitability; in a sense, this was what defined the observations of Barry, the investigator's world-weary wisdom about circus lies (see chapter 6). In fact, the impression that the subject is more than one thing can be exactly what grants literary characters their broader status as properly moral subjects.[4]

As we have seen, there was a strong sense in which colleagues shared that value. They too wanted to regard others as well as themselves as more than one thing; indeed, that appreciation was a vital criterion for being a moderate activist. While not immune to contradiction thinking, colleagues were deeply suspicious of what they interpreted as its reductive impulse. A large part of their distrust of more "extreme" animal activists was in this regard aesthetic. Through the latter's rejection of compromise, those activists seemed to them to be demanding that moral subjects become entirely consistent. In the process of doing so, those accusing activists also appeared to turn themselves into that kind of one-dimensional character, ironically shorn of the strands or layers (of inconsistency) necessary for sympathetic treatment as moral subjects. Colleagues would admit that this assessment was not a million miles from the mainstream depreciation of animal activists as "the antis," a depiction that presented animal activists as hectoring or oppositional for the sake of being contrary. Yet moderate animal activists were still animal people; for them, their moderate stance was not a folding back into the world as it is. Their relationship with the mainstream might not be so antagonistic, but it remained awkward. Indeed, in many ways, awkwardness was another defining characteristic of the moderate activist. This included the awkwardness of their relationship to the twin poles of consistency and inconsistency.

At one level, what one got from Maggie, for instance, was consistency all the way down. As the policy director, Maggie knew how to be on message. Indeed, her knowledge of the animal protection brief was almost encyclopedic; I rarely saw Maggie unprepared or caught off guard. That mix of competence with politeness with a no-nonsense demeanor made Maggie a highly impressive operator in the policy field. But, as we have seen, Maggie was also someone who appreciated "personal relationship" and enjoyed the company of others even when their moral opinion appeared to directly clash with or contradict her own. For example, she could still have a pleasant dinner with a charming couple despite an awkward moment when it became clear that they were going to enthuse about shooting (see chapter 8). Likewise, Maggie continued to enjoy pastimes such as singing in a local choir and performing for an amateur dramatics group. There were occasions when Maggie invited colleagues to watch her perform. I remember with affection an evening of musical theater organized as a fundraiser for the animal group and loosely based around the poems of Burns, where Maggie surprised us all by striding onto stage dressed in a tricorne hat, long boots, and a sweeping cloak to narrate the tableaux of scenes. However, in general, Maggie stressed that these pastimes were completely unconnected to her animal activism, just as the latter was in turn unconnected to her role as wife, mother, and grandmother. Indeed, like many colleagues, Maggie insisted on keeping certain interests or roles apart, resisting any impulse to try to fully integrate them or to present herself as wholly defined by a commitment to animal protection.

So, from the perspective of the moderate animal activist, there appeared at least two kinds of moral subject in play or worth cultivating. One was the moral subject committed to the protection of animals and especially to the welfare of sentient animals. The other was the moral subject who emerged precisely as a result of exceeding that commitment or moral stance by displaying interests that apparently did not connect or that might even contradict.

Between the persons that Barry pretended to be for professional reasons and the moral subject that he professed to be, it was sometimes difficult to pin down who he really was. Yet, there remained certain attributes necessary to being an effective investigator that stood out. Interestingly, despite the nature of Barry's covert work, the quality of complicity or collusion was not one of them. Unlike Maggie and the field of lobbying and policy, no one either inside or outside the organization ever questioned Barry's integrity or accused him of inconsistency. Instead, colleagues and supporters pointed to the investigator's self-sacrifice and bravery. Indeed, of all those working at the animal group, it appeared that Barry integrated his commitment to the cause most completely. "Every aspect of my life privately and at work is about animal protection," Barry once told me (see chapter 6). "I am always thinking of new projects and the best ways to do things, the most effective way to live my life for animals."

But to me at least, there was always something a little intimidating about his integrity. Certainly, I found myself, as I imagine other colleagues sometimes did too, wondering how Barry sustained or managed that level of commitment and consistency. It was therefore with some relief that I finally received a confession of vulnerability, even if not in the direction I had anticipated. "The one thing I am really terrified of is wasps," Barry once revealed. "If I see a wasp in a room, I will run out and lock the door until someone has got rid of it." The admission, which Barry saw as a sign of weakness and as a blatant inconsistency, embarrassed him but at the same time made us laugh. It seemed so inexplicable that among all the possible fears the investigator might have to address, it should come down to this one. To Barry, being terrified of wasps seemed completely irrational. Yet to me there was also something rather glorious or comforting about it—that is, the image of the hardened investigator, infiltrator of numerous factory farms, laboratories, abattoirs, circuses, and shooting estates, fleeing the scene because of the simple presence of a buzzing wasp or two. I like to keep hold of that picture.

At least from my perspective, Maggie's insistence that the professional and personal should be kept apart never resulted in any glaring or stark sense of inconsistency. In fact, I can recall only one time when the policy director admitted feeling actively embarrassed by inconsistent behavior as it directly pertained to the values of the animal movement. This was due to a decision Maggie once took to bring in pest controllers, prompted by the discovery that there were rats living underneath her house. Maggie knew that this decision could be read as counter to the general ethos of animal protection and in contradiction with specific lobbying work devoted to campaigning against the use of lethal traps. Allowing these rats to be killed rather than, say, humanely caught and relocated could be judged as a denial of those animals' after-the-fact status as moral patients, even if Maggie could legitimately point out that humane trapping and relocation to the wild brought their own welfare dilemmas. But just as importantly, Maggie knew that her instinctive horror upon finding out that she had rats at home was counter to the animal love that defined animal people. Unsurprisingly, it was not a decision that Maggie widely shared or felt good about at the time. Indeed, Maggie feared that others might condemn her action. Yet as an obvious example of contradictory or inconsistent behavior, that decision could also be adjudged worthy of drawing out sympathy; it arguably shows Maggie in a new, more complicated light and hence perhaps as a properly moral subject.

But in this second sense of what makes a moral subject count, it was Cassie rather than Maggie that invariably sparked my curiosity (and sympathy). For Cassie's inconsistencies always seemed most pressingly contradictory in nature. This was only partially because Cassie continued to eat meat and cook it for her husband and family (see chapter 7) and only in part because she *loved* animals but *hated* maggots and parasites. In Cassie's mind, that distinction was simply a response to the harm and suffering that she witnessed the latter inflicting on other more sentient critters; nevertheless, the strength of emotion was odd given the animal movement's normative understanding that nonhuman species could not or should not be held morally accountable for their actions. Rather, what really struck and initially bemused me was the sheer array of enthusiasms that Cassie shared. I have briefly alluded to Cassie's long-term sense of commitment or connection with Native American or Indian American cultures (see chapter 1). From Cassie's perspective, that enthusiasm was entirely consistent with being an animal lover and patron of an animal group. Indeed, in that regard she claimed a mutual inspiration. We might choose to pick those claims apart—for instance, to highlight probable discrepancies between the ethos of animal protection and the ethos of relating with animals that she reported for Native American societies. Such an exploration of inconsistency might include an acknowledgment of the wider problematics of making such a comparison. However, we still have to accept that this was a connection that Cassie viscerally felt; for her, there was no obvious contradiction there. And even if I wanted to mark the comparison as misguided, I could not claim it was inconsistent in the sense of being unexpected or a surprise. As already discussed, such appeals to the exemplarity of indigenous practice and belief were a commonplace reference point in the animal movement.

Instead, what really jarred or took me by surprise was the fact that Cassie insisted on placing her passion for animals (and Indian American cultures) alongside an equally committed passion for the armed forces and especially the RAF. Ever since she was a teenage girl, Cassie explained, she had a dream of becoming a military pilot and seeing action. In fact, Cassie had been particularly fixated on the idea of flying Harrier Jump Jets. While a bad motorbike accident in her early twenties abruptly ended that ambition, Cassie could still be found regularly watching the jets land and take off at a nearby military air base. This side of her moral character initially bemused me. It seemed to bear no relationship to Cassie's animal activism or the caring role invested in rescue. More broadly, it seemed to directly clash with core values of that activism as they were typically expressed by colleagues. I am thinking not just of a commitment to the preservation or protection of life but also a commitment to a range of other issues that defined a progressive outlook, including an active suspicion toward participation in most recent UK-government-sponsored military conflicts or wars. This enthusiasm for a military career was thus not just a jarring inconsistency from my perspective; it would also be so from the perspective of many of those working at the animal group.

Highlighting inconsistency in this fashion can be made to tell several possible stories about moral subjects. One of them could be focused on how Cassie herself resolved this apparent contradiction—or, if Cassie didn't see any inconsistency on being asked, why not? As a matter of fact, Cassie was in turn bemused by my bemusement. She absolutely insisted that there was no contradiction. "I just admire people who have fought for what they wanted," Cassie explained. "I think you have to fight for things that are good in life; otherwise, they will get taken away from you." That it was about the combat or struggle appeared to me at once entirely unanticipated and immediately obvious. Of course, for Cassie, rescue was all about the fight, not just the fight to protect imperiled animals but also the struggle against evildoers, those who would sadistically harm those animals. As we have seen, it was in Cassie's mind a form of warfare (see chapter 7). There were real enemies out there whom one needed to stand up against, just as, Cassie added, Britain had once stood against Hitler (Cassie was unsure of the wisdom of more recent military campaigns though not of the presence of evildoers and evil cultures or of the need for citizens to be willing to join the armed forces). This response taught me something further about the undertones of moral imagination within the animal movement. It also reminded me once again of the strange alliances that were possible between animal people—the fact, for instance, that

Cassie could be good friends with Eilidh, the CEO of the animal group, who was in many ways an opposite moral character. Eilidh was cautious and actively resistant both to invocations of evil and to any reference to warfare as a language to describe the terms of animal activism.

However, while Cassie's explanations to some extent resolved my questions, the nature of the resolution left other inconsistencies apparently more blatant. Cassie's lifelong enthusiasm for Native American peoples and their histories, for example, suddenly felt harder to understand, especially when placed against a relatively unquestioning commitment to the armed forces and Britain's imperial record as an expansionist military power. Likewise, the evidently Christian inflections in Cassie's rendering of evil as a force in the world jarred even more strikingly against a clearly genuine sympathy for what Cassie took to be the cosmological underpinnings of human–animal relations in Native American cultures. For Cassie, the latter resonated strongly not just with the role sometimes assigned her as an animal whisperer but also with the special connection she had always felt toward wolves (see chapter 7). In other words, emphasizing one line of consistency seemed to open up another line of inconsistency.

As we have seen, this is itself another kind of story that one can tell about moral subjects. Contradiction thinking, for all its cynicism, appears remarkably ineffective as a measure concerned to iron out inconsistency or more aggressively to challenge the moral subject. Part of Cassie's bemusement at my response to her moral enthusiasms, which included pointing out apparent inconsistencies, lay in the fact that I seemed to expect her to struggle with these paradoxes. That expectation was met with a different kind of metaphorical shrug of the shoulders, as if to say, "This is just who I am; it's not a stance that I choose." Indeed, as we may recall, Cassie's bemusement really lay elsewhere, not with the nature of her convictions (or the logical relationships between them) but rather with the issue of their moral source. Where Cassie's passion for animal rescue or her dream of becoming a jump jet pilot or her love of wolves or of American Indian cultures came from was ultimately a much more pressing or compelling concern. As proper to someone who preferred to stress before-the-fact explanations of the moral patient position, she experienced these convictions or stances as conditions that befell her.

Yet it would be unfair to conclude the story there. For none of these colleagues wished the endpoint of our discussions to be a reflection on the inconsistency of (human) moral subjects. Certainly, in the case of Barry, the takeaway point turned out to be quite different. His admission of weakness, the investigator insisted, was ultimately a story about the self-recognition and overcoming of contradiction, not so much its ironing-out but its disciplining or management. "It's true that I am terrified of wasps," Barry reconfirmed, "but if I see one, I will do one of two things." Running out of the room was certainly the first of these options, but that was only viable, Barry explained, so long as there was someone else available to remove the wasp. "But if I'm by myself," Barry added, "then I will try to let it out." This latter action was conducted despite his fears and hence in that classic sense belonged to an identifiable moral agent. Indeed, the investigator partly shared the admission to discourse on what he took to be the scenario's wider moral lessons: "A lot of people just kill wasps. And, you know, that's something that a lot of people don't even think of. But I think it gives you an understanding of peoples' attitudes towards other [animals] that they don't understand. If you understand the life of a wasp and just how complex it is and how amazing their lives are, then maybe you would let the animal go without killing it. You know, it's all about respect." Barry's words also take us back to a core element of the animal group's new vision. Rather than only highlighting abuse and suffering, colleagues should set themselves the task of encouraging people to recognize and celebrate the sentient characteristics of animal others. In other words, they should invite the mainstream to make contact—that is, encourage people to allow themselves to be amazed. This is not simply to enable them to overcome barriers to respect through, say, deliberate moral reflection or choice. It is additionally to make respect itself a natural expression of the kind of moral patiency taken to emerge from a positive experience like understanding or amazement.

The Animal Question

Of course, within the animal movement, human subjects are not the only subjects of a life with a moral status. Much of this book has been concerned to explore that broader figuration of moral subjects as it plays out through the expert practice of moderate animal activism. I have done so in two ways: first, by returning to the appeal of that classic distinction between moral patient and moral agent, originally drawn from animal rights philosophy and particularly the treatise of Tom Regan, which continues in many ways to animate and shape both the organizational work and moral imagination of colleagues, and second, by spotlighting the equal appeal to before-the-fact explanations of such distinctions, where the patient status becomes redefined to stress the subject's moral condition as the recipient of actions, to whom something essential happens. In the latter case, I have partly found inspiration in the diverse and emergent literature on moral patiency. Indeed, the book has weaved across an eclectic range of literature that in various ways discusses the significance of the moral patient position, both for how we might reconsider moral agency and for what we might consider moral subjects to be.

We might recall, for instance, my early reference to the simple definition of moral subjects offered by AI and robotics ethicist Joanna Bryson (see chapter 1). According to Bryson (2018, 16), moral subjects can be taken "to be all moral agents or moral patients recognized by a society" (see chapter 1). Such a definition clearly extrapolates from the classic distinction developed by Regan. There is a shared emphasis, for example, on the issue of who or what deserves moral consideration. As an ethicist, Bryson's definition is ultimately concerned to enable more informed deliberation not just over the status of *our* moral subjects but additionally over their future status and any proposition for an enlargement in their scope or number. In fact, the comparatively recent focus on the moral patient within wider philosophical and ethical debate has been driven by concerns to radically redefine that moral community, moves that invariably lead to calls for a massive expansion of moral subjects and especially of those subjects whose welfare is deemed worthy of protection. Put another way, the case for that expansion has heavily leaned on the unfolding and developing categorization of moral patient.

But Bryson is not an ethicist directly concerned with the moral status of animals. Like a lot of the new and most sustained work on moral patiency, Bryson's focus falls instead on the question of the moral status of machines (see also Gunkel 2017), itself partly influenced by innovative developments in computer or information ethics, which takes a patient-oriented approach (see Floridi 1999, 2010). However, these arguments do usually draw back to broader comparisons between a range of (nonhuman) moral subjects. As Gunkel expresses it, "The question of moral patiency is, to put it rather schematically, whether and to what extent robots, machines, nonhuman animals, extra-terrestrials, and so on might constitute an *other* to which or to whom one would have appropriate moral duties and responsibilities" (2017, 93). The nature of reflections may take a different tone depending on which moral subjects our ethical attention primarily rests on. Yet, as Gunkel stresses, for better or worse, the original and paradigm-changing model for that consideration remains that found in animal rights philosophy (2017, 153). It is no coincidence, for instance, that Gunkel chose to name a book *The Machine Question*, a quite deliberate nod to the dominant and still-orienting shorthand notion of the "animal question."

Indeed, Gunkel's title is also a nod to a much longer heritage of philosophical discourse about animals (and machines). This includes reference to what animal rights philosophers often position as the notorious doctrine of the *bête-machine* or "animal-machine" put forward by Descartes. That doctrine famously denied the moral status of animals and did so precisely based on a positive comparison with "mindless automata" or artificial mechanisms (Gunkel 2017, 3). Gunkel's wider point is that the fates of animals and machines have been bound together since the beginning of modern philosophy and that their salvation as finally recognized moral subjects should perhaps be regarded as similarly entwined.

As Gunkel further observes, the thrust of much work on the animal question centers on a progressive ethics of inclusion and exclusion (2017, 16). While welcoming the patient-oriented example set by animal rights philosophers, which Gunkel accepts provides a useful starting point for considering the status of machines, Gunkel is also quick to highlight the exclusions contained in the moral arguments of figures such as Regan (and in the arguments of consequentialist philosophers such as Singer). First and foremost, Gunkel criticizes their machine phobia. A large part of the horror expressed by these philosophers toward Descartes's doctrine rests on the proposition that animals are machinelike. Instead of embracing the fact that "animal and machine share a common form of alterity," the case for animal liberation tends to insist on an absolute decoupling of that link, a claim that itself tends to reinforce the denial of the machine as any kind of moral subject (2017, 3–5). Ultimately suspicious of the broader ethics of inclusion and exclusion and of what Gunkel takes to be the essentially anthropocentric shape of its questions and answers, the author wants to experiment with the idea and practice of genuinely coparticipatory machine ethics. More specifically, Gunkel is suspicious of the consequences of importing the definitions of moral patient laid down by animal rights, such as that canonical question, "can they suffer" (2017, 153).

As we have seen, this latter suspicion is one that for very different reasons colleagues at the animal group broadly shared. While absolutely invested in the ethics of inclusion and exclusion, they expressed a series of reservations about the animal movement's overriding emphasis on animal suffering and particularly the antagonistic stance that often went with it. Embedded in the organization's new vision, this included a desire to open subjects up through positive encounters with animal others, or, to return to Barry's words above, to elicit a "respect for life" through being amazed rather than just distressed by animal sentient realities. In fact, the new vision drew on the ethics of inclusion and exclusion to make an unusual case for an inclusive attitude toward the mainstream (i.e., that normally defined in opposition to the ethically otherwise stance of animal activism) or even more radically to proffer the possibility of a mainstream animal movement. It is true that colleagues very much retained a conventional machine-phobic attitude in the ways in which they made a case for making contact as well as in the ways in which they continued to privilege human-inflicted animal suffering. To understand this, one has only to dwell a while on the way colleagues talked about "factory farming," where horror at the treatment of farmed animals easily slipped into horror at the industrial system or into horror at the too-close contact or coordination between animal body and machine—although it is worth noting that this phobia toned down considerably when colleagues switched to accounts of before-the-fact explanations of moral stances, as experienced in their own human animal bodies. Returning once again to Cassie's initial puzzlement over the sources of moral feeling (see chapter 1), we might remind ourselves that at one point she offered a machinic metaphor for both her own and other animal peoples' sustaining love of animals. "We say that we are hardwired," Cassie told me, "and that just means everything you are is for animals. Right from the start, that's all we do. We think of animals. We love all animals" (see chapter 7). In reaching for that metaphor, Cassie seemed to suggest that animal love was automatic or that animal people were somehow ethical automata.[5]

But the point of ending with reflections on the observations made by Gunkel is not to suggest a strong ethnographic connection to that machinic theme. Instead, I do so because Gunkel and that machinic perspective offer a useful sideways perspective on the ways in which the animal question is usually framed, both within the animal group and wider animal activism and within the debates contained in those classic philosophical texts on animal rights or interests, which colleagues either knew directly or had indirectly absorbed. Gunkel's observations and those of machine ethics in general provide me with a series of helpful provocations for thinking again about the deployment of moral patiency and that hegemonic distinction between moral agent and moral patient.

Chief among those provocations is that reworking or stretching of the category of moral subject. As well as including moral patients alongside moral agents in their definition, the introduction of machines and hence a consideration of moral subject as "something" instead of just "someone'"(Bryson 2018, 16) has been a useful stimulus to my own descriptions. (It is worth noting that Gunkel and Bryson have almost completely opposite responses to the machine question; however, for our purposes, the difference can remain mute.) This includes the various ways in which I have tried to describe and analyze the identified roles of moral subjects, both human and animal, within the ethical field of animal protection. Most conventionally, that has focused on an examination of the reported position of human subjects as either moral patients or moral agents and the reported position of animal subjects as the chief objects of (human) moral concern and duties or as the victim of (human) negligence and wrongdoing. Yet even within such a stark division, one that continues to dominate moral horizons, there remains considerable scope for dynamism (as we have seen). A lot of this book has been centered on plotting the expert work and narrated experiences of colleagues, a focus that I hope has not just complicated a picture of moral subjects as individuals seen in their aspect of moral agent but also complicated the picture of them as moral patients and as subjects of patiency. If the latter is "concerned not with determining the moral character of the agent or weighing the ethical significance of his/her/its actions, but with the victim, recipient, or receiver of such action" (Gunkel 2017, 154), then one might say that focus has in large part fallen on the moral character of human subjects as victims, recipients, or receivers in the widest sense—but with the important caveat that the status of the action taken to impact or do something to those (human) moral subjects is not always so closely defined. For instance, it is not necessarily initiated by a recognizable moral agent or a clearly defined singular moral source.

On the face of it, the status assigned to nonhuman animals appears far more restricted and usually confined to those marked as sentient. Within animal protection, these moral subjects are not normally assumed to be moral agents, though they are certainly taken to be agentive in other ways. However, even within the narrower definition of animals as moral patients, there is surprising movement or vigor to the position assigned them. As well as subjects deemed worthy of *our* moral concern and responsibility (say, in the formal sense laid out by Regan or in the campaign literature of the animal group), they are often taken to be moral subjects who provoke human others into action—for instance, through lively encounters with them. In this respect, their lack of status as moral agents does not necessarily prevent sentient and sometimes nonsentient animals being experienced as the stimulus for the actions of moral agents. Nor indeed does it prevent them being experienced as the stimulus for the extraordinary impressions that colleagues as animal people sometimes claimed to receive as moral patients.

For me, part of the attraction of rethinking the notion of moral subjects also lies in the ways it enables a reengagement with ordinary usage, both in an indirect or suggestive fashion and in quite literal manner. Because, of course, animal people and those in the wider animal movement do have a working understanding of the term "moral subject." This includes a common sense of themselves and other kinds of sentient animals as, to borrow Regan's term, subjects of a life in both its positive and negative expression and hence of a welfare worthy of preservation. But also embraces that ordinary sense of (usually but not exclusively human) moral agents and moral patients as biographical subjects. This book contains portraits of several such moral subjects taken to have a distinctive personality and a history of moral development or awareness worth telling. Some of them are further regarded as moral exemplars, archetypes of certain kinds of moral subjects within the animal movement and of certain kinds of expert roles. All these subjects are defined by kinds of relationship to moral choices and intensely felt moral passions as well as by kinds of moral struggle and dilemma. Finally, and just as importantly, they are subjects held to be in interaction with each other. In this respect, colleagues are joined by other kinds of characterful moral subjects, most obviously companion animals who also share the office but additionally those who share a home and who have often been the subjects of rescue.

In addition, one has a relationship to more generic, less characterful kinds of moral subjects—say, to sentient animals in general or to certain sentient species, such as those identified as especially harassed or exploited. Recall the special love of Cassie for wolves or the affection expressed by several colleagues for publicly vilified breeds of dogs, such as the Staffordshire bull terrier, or indeed the peculiar regard that Barry showed toward the most hunted or trapped animals in the British countryside. "I mean, foxes are my thing, you know, my chosen special animal," the investigator once shared. "I think they are the most persecuted animal in the country, and that's only because they have been demonized by those who wish to take pleasure in pursuing and killing them." Evidently distinguished as victims or classic kinds of moral patients within the ethical field of animal protection—Barry felt the need to add that, for him, foxes are "in fact quite innocent animals"—these generic kinds of subjects are nevertheless impactful. Likewise, we have explored a range of generic kinds of human moral subject, most notably that mass of apparently indifferent or unthinkingly cruel individuals collectively engaged with and colloquially known as the mainstream. Due to their attributed passivity, these individuals tend to be figured as unusual sorts of moral patient but with the caveat that they are always potentially transformable into animal-friendly moral agents. And of course, colleagues sometimes also speak of the presence of deliberately cruel or evil people (see chapter 7). This generic moral subject is a straightforward wrongdoer, held morally accountable for their actions, though the reported extremity and lust for cruelty exhibited by such moral subjects always render them potentially transformable into moral patients. That potential is linked to the commonly held suspicion that such moral subjects may in fact be victims of psychological disorder or cultural dysfunction, or it is linked to the perceived capacity of evil itself to be reappraised as a depersonalized force in the world.

But it should be noted that there is a series of slippages in the way in which I have redeployed the category of moral subjects here—most obviously, a slippage between ethnographic use and the use of that term in various literatures. And there are, of course, contradictions and limits in such loose usage. Most significantly, the whole move to reinterpret moral subjects as composed of both moral patients and moral agents at some point must run up against the broader question of the distinction's utility. As I highlighted right at the beginning of the book, the purchase of moral patiency as philosophical category or analytical concept is entirely dependent on the continuing purchase of the category and concept of moral agency. Indeed, as several observers have noted, the definition of moral patient actively leans into the definition of moral agent. It works precisely because it appears to turn perspectives around and "look at things from the other side" (Gunkel 2017, 93). However, "as long as moral patiency is characterised and conceptualised as nothing other than the converse and flip side of moral agency," Gunkel warns, "it will remain secondary and derivative" (2017, 107). For some, that derivative status is a problem because it ultimately reinforces the ascendancy of moral agency as a normative category, including its ascendant set of exclusions. Thinking about the machine question, for instance, Gunkel welcomes the act of inversion implied in the stress on moral patiency, but the author then advocates for a notional second step that might displace that opposition altogether (2017, 10). In the end, Gunkel calls on us to reach for or try to "articulate a thinking of ethics that operates beyond and in excess of the conceptual boundaries defined by the terms 'agent' and 'patient'" (2017, 8). What this might mean for a future understanding of moral subjects is left open to interpretation.

Speaking ethnographically, neither I nor the colleagues at the animal group that I worked with were always clear about where the boundary between the experience of moral agent and moral patient lay. That distinction could easily become blurred. As we have seen, this could even be the case when discussing such typical exemplifications of moral agency as choice making and the nature of personal commitment (see chapter 4).[6] Indeed, much of the description in the book has been concerned with the apparent instabilities or flux of these moral subjects. But rather than respond with a call to move beyond or eclipse that opposition between agent and patient, I have striven instead to reinvigorate the distinction by paying far closer attention to those moments when positions get invoked, altered, or transformed. For me, this is where the work of displacement or subversion lies—that is, in the specific sequencing of moral agent and patient positions, which includes blurring but also the shifts and reversals that mark the attribution and sensation of those positions both as forms of moral experience and as events in time.

As we have seen, such movement can take place between or across moral subjects as well as within them. In terms of the latter, one might refer again to the classic example of the kinds of reversal typically attributed to human conscience. In one instance, for example, a moral resolution said to be defined by conscience can appear as an exemplification of the individual agent's free and self-initiated action, yet in another instance that same resolution can appear to highlight the individual's essentially unfree status, for it is due to the dictates of conscience, over which moral subjects sometimes feel they have no control (see Tobias Kelly 2015). Here, conscience itself feels like a moral agent and hence in some respects a moral subject too, something that independently acts on or affects someone or something else and hence renders them into moral patients. But for colleagues at the animal group, that sort of blurring or reversal is further complicated by the shifts attributed to other kinds of detachable elements of the person. We have spent some time, for instance, examining the moral status attributed to personality, a slippery category that for colleagues simultaneously affirms and problematizes the commonalities between human and nonhuman sentient beings (see chapter 5). Here, behavior can be read as an expression of someone's individual personality, but equally personality can be read as a moral agent independent of that individual—that is, not through the moral effects personality has on that individual (as something, like a conscience, that turns the individual into a moral patient) but rather through the ways in which personality is sometimes taken to independently affect others.

It might be possible to account for such a sense of dissonance by appealing back to conventional types of distinction such as that between personality and character—for instance, the common assumption that one's personality reveals what is visible to the outside world, in contrast to character, which is harder to discern precisely because it is meant to show what you are like on the inside. However, this is neither how colleagues chose to interpret that dissonance nor exclusively how I understood the character or personality of moral subjects to be usually invoked. Indeed, archetypal kinds of moral subject within the animal movement, such as the professional figures of investigator, rescuer, and lobbyist, appeared to emerge via a zigzagging action between concepts of natural disposition, ideas of strategic self-development, and the notion that expert fields in some fashion cultivated that character or personality. More broadly, the archetype of animal people itself suggested a productive tension between agent and patient positions. On the one hand, animal people were understood to be people for animals—that is, moral subjects defined by a concern or duty to protect. On the other hand, they were understood to be animal people because of nonhuman animals—that is, moral subjects driven by the animal love that already, some would say somewhat mysteriously, defined them.

Notes

PROLOGUE

1. For an excellent account of the historical emergence of animal protection in the colonial context of British India as well as sets of ethnographic reflections on the current reverberations of that complex history for the movement in India today, see Davé 2023. Davé stresses the vital relationship between the cause and practice of animal protection and the "history and politics of empire" (and liberalism) (2023, 15), which makes it possible for certain constituencies in India to still view and dismiss the concerns of animal protection as "neocolonial meddling." But in addition, Davé explores other "affective histories," including the intersections between animal welfare and "Hindu anti-imperial nationalists" in the colonial period, which point forward to awkward links with what Davé terms present-day "Hindu fascist neo/imperialism" (2023, 31).

2. One could read the failure to notice the lack of diversity within the projected depiction of Anglo-American mainstream publics as a reflection of the fact that during my initial period of fieldwork, everyone presented cisgender identities and the staff body as a whole was acknowledged to be entirely white. But one might also point to broader contexts for such a failure. According to the 2011 government census, up to 90 percent of people in Scotland, when asked to select "what is your ethnic group," identified as either "White: Scottish" or "White: Other British." While in Edinburgh nearly 18 percent of the population identified as an "ethnic minority," it should be noted that over 9 percent of that figure came from people who ticked census boxes for either "White: Other," "White: Polish," or "White: Irish" (https://www.scotlandscensus.gov.uk/census-results/at-a-glance/ethnicity/). Of course, this contextualization explains only so much. The way in which the government census invites respondents to select race or ethnic categories of census identity shapes its own politics of race. And, as we will later explore, colleagues could always operate within wider horizons for a nation or imagined community, such as the United Kingdom, and thus in relationship to places, such as many major English cities, where government census results provide very different sets of figures. It is also very possible that the evident nondiversity in the cross section of Anglo-American mainstream publics depicted by the film speaks to much wider presumptions of whiteness within animal activism in Scotland or the rest of the UK.

1. ANIMAL PROTECTION AS A STORY IN THE ANTHROPOLOGY OF ETHICS

1. The term "animal protection" is in many ways regarded as a bridging category; it is suggestive to some of a more forceful approach toward animal advocacy than the more conservative identifier "animal welfare" but at the same time perceived as less strident and hence less controversial (as far as certain public attitudes go) than identifiers such as "animal rights" or "animal liberation." While they may disagree on much, nearly all animal groups in the UK can accept "animal protection" as a shared moniker.

2. I note that "animal people" is very far from being an agreed self-descriptor, even within other animal groups in Scotland or the rest of the UK, and I make no such claim to it being so. Indeed, I could envisage all kinds of reasons for rejecting that label. Speaking more widely, I am aware of other identifiers in common usage across the global animal

movement. Davé (2023, 14), for instance, tells us that "animalist" is an identifier "in use in India and elsewhere to refer to people who act politically alongside and on behalf of animals." Likewise, Sandoval-Cervantes (2023) reports that in Mexico animal activists self-identify as "animalistas." But I never heard the term "animalist" or variants of it being used in the office or by other animal groups in Scotland.

3. It is possible, as Ticktin (2015, 50–52) argues, to view the equivalence drawn between children and certain nonhuman animals as heavily reliant on dominant idioms of innocence, which, as Ticktin points out, typically ground and constrain an appreciation of the suffering subject, not just in animal activism but also in humanitarianism. Certainly, as we will see, the presumed innocence of animals remains central to the visibility of animal suffering within animal protection. However, in drawing attention to Regan's definition of moral patient and its wider invocation within traditions of animal activism, I want to suggest that our understanding of moral claims of innocence might be productively reconfigured; apart from anything else, in Regan's terminology the category of moral patient cannot be solely reduced to the status of suffering subject.

4. Davé (2014, 435) highlights that the complexities of those affective histories are partly shaped by the fact that the animal protection movement in India is still in important ways influenced by foreign animal activists and by animal groups from North America and Europe, and by the fact that certain animal activists from North America and Europe continue to move their activism to India (see Davé 2015). But Davé points out that those histories are also shaped by the ongoing need to negotiate and respond to the very different orientation of "homegrown animal welfare politics" centered on "cow protection" or *gauseva* (2014, 436; see also Davé 2023).

5. But as Davé (2023, 25) also points out, within animal activism in India, that surrender is always susceptible to reinstatements of sovereignty or of essential difference between human and animal, most obviously manifest through the compulsion to speak and act on behalf of animals. For Davé, the latter move is always compromised. Indeed, the liberatory ethics of indifference advocated by Davé consistently places value on sensorially defined states such as intimacy and on the potential of becoming grounded in sensorial registers such as the haptic (even if Davé is also cognizant of the capacity of touch to elsewhere exhibit "violent manifestations"—for instance, those informed by principles of caste [2023, 115]).

6. It is important to note that Mazzarella's work around the ethics of patiency and the specific paper attached to it is an ongoing project. Here I cite from the original 2020 version of that paper, but there is also an updated version available to readers (see Mazzarella 2021).

7. Mahmood reports, for instance, that "mosque participants do not consider fear of God to be natural, but something that must be learned" (2005, 144). That observation leads to a second reversal, this time in the postulated relationship between the interiority and exteriority of moral subjects. Instead of outward actions such as ritual prayer being seen as a sign of pious thought or an inner moral sense, there is scope, Mahmood suggests, to consider bodily acts of worship and other behaviors such as "social demeanour" as the generators or coordinators of "inward dispositions," including intentional thought and emotions (136). Mahmood wants us to envisage that with this moral-passional self the ethical formation of virtues can also be routed from exteriority to interiority (121).

8. Although Mittermaier does not use the neologism "patiency," the formal distinction between agent and patient is on occasions explicitly drawn (see Mittermaier 2010, 86).

9. "Everyone," Carrithers states at one point, "is possessed of moral agency-cum-patiency . . . a term which recognises that we both do and are done by" (2005, 434). But Carrithers's investigation into that "awareness of people as both acting and reacting, as both agents and patients in their social world" quickly detours into a broader conversation about the tensions between notions of personhood that allow for moral agency and the uncertainties of choice and those Maussian-inflected notions that stress the "dictates

of collective representations" (440). For Carrithers, the latter "allows for personhood only as patiency," a suggestion that reduces the dynamism of the moral patient position to unfreedom or pure passivity.

10. In Liu's schema, the three new dominant moral characters in contemporary China are the cadre or section chief (*chuzhang*), the boss (*laoban*), and the miss or hostess (*xiaojie*) (2002, 27). The dynamic relationship between these character types is for Liu in part determined by the way agent, patient, and instrument positions get reallocated. Liu observes, for instance, that in one plotting of that relationship, "the *laoban* should be considered an agent, defined as someone who makes a decision and takes responsibility for it; the *chuzhang* should be considered a patient, on whom the *laoban* carries out action; and the *xiaojie* is an instrument in the sense that the action takes effect only by means of her involvement" (48).

11. Later on, Strathern states that Gell's "'patient' overlaps with my [notion of the] person as a (personified) object of people's regard" (1999, 17). In the context of the ethical turn, one is reminded of the work of Webb Keane (2016) and especially Keane's efforts to reposition an anthropology of ethical life away from a sole focus on the first-person perspectives of moral subjects and toward a focus on the "second person" perspectives constituted via social interaction. For Keane, the stance of a moral agent is possible only because of its recognition by others; so, for instance, there is a kind of patiency or effectual action or "co-construction" built into what appears as the coherence of a subject's moral character (2016, 107).

12. As colleagues' general suspicion of Christianity might suggest, there was little awareness of the vital role certain Protestant groups and leaders, including missionaries, played in the early history of Anglo-American animal activism or any acknowledgment of the early influence on the movement of biblical concepts such as stewardship (see Davis 2016).

13. Of course, in the wider global animal movement, people do not always identify with progressive politics or even the politics of the left. Weiss (2016), for instance, traces contemporary shifts in political alignment or nonalignment in Israel. Weiss notes that while in the past "animal rights activism" was closely associated with human rights activism and especially the struggle against the treatment of Palestinians, in recent years there has been an active decoupling of that link (2016, 694). Animal rights activists are now far more likely to be single-issue activists, drawn from the right as much as the left, and the movement itself is "larger and more mainstream" (689). Indeed, each tradition of animal rights activism in Israel "critiques the other as perverting ethical priorities" (689), Weiss tells us. The older tradition insists on sustaining the link with human rights activism and articulates "their claims through the ethical regime of humanism and the commonality of suffering" (689), which provides a basis for equivalence between human and animal victims of human-inflicted forms of suffering. The newer tradition, by contrast, cultivates a position grounded in responsibilization and "the commonality of agency that foregrounds questions of guilt and innocence" (689). Here, humans and animals can be rendered nonequivalent kinds of subjects based on that comparative guilt or innocence. More specifically, Palestinians can be held to account or rendered unsympathetic subjects because of their reported actions, while animals remain innocent and thus sympathetic precisely because of what they have not done ("So far, a chicken has never tried to blow up a bus," Weiss's interlocutors explained on more than one occasion [2016, 700]). And for an extensive reflection on the present and historical intersections between animal welfare and contemporary "Hindu fascist neo/imperialism" in India, see Davé (2023).

14. As Laidlaw (2010a) points out, the broader interest of the animal liberation movement in religious traditions such as Jainism rarely involved skimming beyond the cosmological surface. It certainly did not tend to address disjunctions at the level of belief or practice. Noticing the consistent presence of campaigning animal rights groups at public

events organized by Jain diaspora in the UK and North America, Laidlaw was struck for instance by the lack of attention to obvious cosmological challenges. Most glaring, for Laidlaw, was the striking contrast that underlies the apparently shared principle of do no harm. As Laidlaw describes, practicing Jain renouncers are taught to perceive a world as "literally filled with sentient creatures" (2010a, 68) and to learn to know "the space around them as already inhabited'" (68); to pursue spiritual liberation, they must therefore "maintain a watchful awareness so that they never unnecessarily harm any living thing'" (63). However, this watchfulness and moral striving are conventionally premised on a very different set of assumptions from the Anglo-American concerns of animal protection. Not only is the Jain sphere of moral concern far wider—in that tradition, insects, plants, and elements of the material world are also adjudged to have senses and to be therefore capable of suffering (68)—but the orientation of its cosmological priorities is also radically other. In particular, the comparison disregards the Jain renouncer's knowledge that the soul of a living thing is recycled across species and that the extrahuman status of the soul's body is usually a sign of bad actions in a previous existence. "The point of cultivating experience of the world as teeming with life," Laidlaw points out, is precisely to arouse a sense of what Jains term "disgust" or "revulsion" with that world and hence motivate subjects to seek renunciation or ascetic withdrawal (69). As Laidlaw concludes, "A cosmology that invited intervention with the serious ambition of radically improving all this [i.e., the conditions of life of those nonhuman sentient creatures who suffer terribly], rather than changing oneself in order to escape it should on the face of it be more than superficially different" (77). Yet despite this disjunction, Laidlaw also admits that a growing number within the Jain diaspora in both the UK and North America do seem to claim that these religious principles can be brought into easy synthesis with the ethical vision of animal rights or the do-no-harm principle that informs the animal movement. Laidlaw goes on to explore several possible different explanations for how that convergence might happen. However, for our purposes the example is useful not just because it puts in perspective the claim of animal people to be influenced by such religious traditions but also because it serves to throw the moral and cosmological underpinnings of the animal movement itself into some further relief. But as Davé (2023, 37) points out, a closer consideration of the three "distinct but overlapping genealogies" behind animal activism in India might also throw the whole idea of the secular settlement of the global animal movement into question. As well as a heavy debt to a "liberal, Anglophilic Judeo-Christian" lineage and to a "Christian ethic of love" introduced by an Englishwoman, Crystal Rogers, often identified as an early founder of the movement in India (38), there is a "pagan Anglophobic dharmic Aryo-Hindutva" lineage (37) and a lineage born of "the theosophical tradition." As already reported, the problematic legacy of "vegetarian purism" within present-day forms of Hindu fascist neo/imperialism in India is part of one of those lineages (37), a fact not widely understood by animal people in the office.

15. There is some debate about the origins of the concept of a Rainbow Bridge, but it appears that it was first developed in the 1980s and 1990s by North American grief counselors and a range of animal people who sought to express their personal loss through artistic expression such as poetry and painting.

16. For a more critical assessment of professionalization and expertise in the animal movement in the UK and US, and a more suspicious rendering of the expert and moral outlook of moderate animal activism, see Wrenn (2019).

2. DOGS IN THE OFFICE

1. Of course, Regan and Singer remain the most well-known philosophers of that generation within the animal movement. They were certainly the only two figures that all colleagues at the animal group had heard of and whose work and arguments every-

one knew to varying extents, either through personal reading or secondhand, through indirect communications via fellow animal activists. However, some colleagues had read the works of different Oxford Group scholars, such as the theologian Andrew Linzey. As several commentators have pointed out, there were other important intellectual influences in that historical period too, including philosophical work on animals from contemporary feminist scholars that included critiques of gendered discourses within Singer and Regan (see Fraiman 2012). But these works were not mentioned by colleagues and did not appear on the shelves of the office library.

2. Wrenn (2019) offers a useful alternative rendering of the history of the animal movement in the UK and US, grounded in a much more skeptical attitude toward both expertise and the moderate stance of such organizations and their activists. In Wrenn's telling, the key drivers of change are professionalization and factionalism (2019, 48). It begins with a first wave of animal activism that starts in the Victorian period and runs through to the middle of the twentieth century. For Wrenn, this period is defined by the introduction of formal organizations, the struggle for financial stability, and the development of patronage and various forms of accommodation to the state and industry. Second-wave animal activism begins in late twentieth-century Britain and America and is marked by the expansion and growth of a more grassroots style of activism, which includes attempts to introduce antihierarchical and democratic forms of decision-making (67). From the 1960s onward, it is also marked by a focus on direct action tactics and by the rise of forms of vegan activism. Finally, third-wave animal activism really kicks in from the 1990s, which Wrenn marks as a "turning point in professionalized antispeciesism" and fundraising-driven organizational modeling (81). This is a period where Wrenn observes "a quiet shift from radical, abolitionist claims-making to the moderate language of compassion" (81) and a gradual move toward reformist language and aims or what Wrenn terms "reductionism" (94).

3. As many commentators point out, in these lines and other passages of poetry, Burns seemed to be deliberating revising the biblical emphasis on a God-given human dominion by asserting the notion of a "social union" between humans and other animals. Indeed, there appears to be a desire to expand "Rousseau's ideas expounded notably in the *Social Contract* (1762) to the whole sphere of flora and fauna" (Tholonait 2015).

4. For another attempt to present a radical relational ethics of interspecies being, drawing upon but also clearly differentiated from the legacy of Haraway and multispecies ethnography, one might consider the recent work of Naisargi Davé (2023). Instead of reconceptualizing the category of responsibility, Davé chooses to try to reclaim the category of indifference, partly prompted by a desire to diverge from a normative emphasis on the value of curiosity within anthropology and, Davé also stresses, within the work of Haraway (2023, 2). In Davé's words, "My understanding of indifference is Relational: of mutually existing in difference rather than being different beings seeking to grasp, gaze, admire and master the difference of others. . . . In that, *Indifference* [i.e., Davé book] argues for an interspecies relational ethic premised on mutual regard rather than curiosity, love, or animus" (2023, 6–7).

3. ENGAGING THE MAINSTREAM

1. For a genealogy of kindness as a key virtue or "gospel" of early animal advocacy, especially in the US, see Davis 2016.

2. Of course, the success of Oxfam's "be humankind" campaign rested on the promise of hands-on poverty alleviation, the most famous example of the rescue theme being the charity's much earlier "Give a pound to save a child" appeal. But what is also of note about the "be humankind" campaign is the emphasis it placed on the theme of public apathy; here kindness was invoked as an encouragement for people to fight against their own indifference to issues of global poverty.

4. THE ETHICAL CHOICE

1. The focus on largely secular explanations or justifications for the animal activist's vegetarianism and veganism also betrays the strong bias toward the typical orientation of Euro-American forms of progressive politics, where those choices are usually figured against the mainstream meat-eating majority. As Davé (2014, 436) points out, a look at animal activism in India necessarily involves some reconfiguration, especially since there those choices (i.e., vegetarianism and veganism) can equally be figured as not only mainstream but also more attached to religious and caste chauvinisms associated with Hindu nationalism. In that context, a progressive politics can sometimes be expressed and enacted through meat eating (see also Davé 2019, 72). Obviously, those concerns could also be pertinent in a UK context, where Hindu nationalism has appeal among at least some "Asian British" communities (I use the UK census designation here in awareness that this is not necessarily the identifier chosen by such individuals or their communities). However, such debates were not generally on the horizons of colleagues at the animal group. If mentioned at all, the links between vegetarianism/veganism and Hinduism were positively made (see chapter 1). Any more negative judgments on the associations between religion and meat eating were generally restricted to commentary on the historical influence of Christianity or occasionally to specific religious minority practices of slaughter in the UK, usually linked with Jewish or Muslim traditions.

6. DISCIPLINES OF INVESTIGATION

1. I do not know the extent to which such rougher measures to soothe lab animals remain normative. Certainly today, animal scientists and laboratory technicians also speak the language of animal welfare. As well as abiding by numerous welfare protocols, scientists and technicians have their own strategies for navigating animal activism and activists' portrayals of lab work and working relations with lab animals (see especially Nelson 2018; Sharp 2018).

7. EVIL PEOPLE (AND THE BONDS OF RESCUE)

1. The much-cited Five Freedoms framework emerged out of a 1965 British parliamentary inquiry into the welfare of animals in intensive livestock systems. Later extended to incorporate both the physical and mental needs of animals, it was fully codified in 1993 and soon became a generally accepted baseline standard and lobbying platform in animal welfare circles internationally. The Five Freedoms are freedom from hunger and thirst; freedom from discomfort; freedom from pain, injury, and disease; freedom to express normal behavior; and freedom from fear and distress (FAWC 1993). More recently, this framework has been somewhat superseded by standards stressing protections for more positive experiences of animal life, but the Five Freedoms model is still regularly referenced.

2. In ethnography of various animal sanctuaries in the US, Abrell also picks up on claims that certain rescuers were animal whisperers or had particular skills in interspecies care-based relationships (21016a, 164). Abrell cites the example of Susie Coston, a director and founder of Farm Sanctuary, one of the most well-known rescue centers, not just in North America but also internationally. In fact, at least one of the Edinburgh animal group's staff had taken sabbatical leave to visit and intern at the Farm Sanctuary's base in upstate New York.

3. In an exploration of recent shifts within the doctrine and practice of humanitarianism in the US, Ticktin (2015) observes an increased willingness to include certain forms of animal suffering alongside archetypal forms of human suffering within the humani-

tarian's orbit of moral concern and intervention. At the same time, Ticktin claims, one can observe an increased willingness to condemn those held responsible for those forms of animal suffering. In fact, Ticktin states that these medical and legal technologies of humanitarian care have increasingly been "used to expel [human] subjects from the category of humanity, by categorising them as 'inhumane'" (2015, 62). The irony here, as far as Ticktin is concerned, is that these cruel or inhumane subjects often end up being rendered as animalistic; here, animal suffering "functions as an indicator of human criminality, of people who become animal" (64). But at least as far as Cassie and her rescue group are concerned, the cruel or evil subject does not thereby become animal—quite the opposite. Although sometimes described as soulless, the cruel or evil subject's animality is not typically foregrounded.

4. For a fuller excavation of animal love and its politics within and beyond animal activism in India, see Davé (2022), especially Davé's arguments about love as a politics of distinction ("because when we love it is the one or ones who are special to us that we save" [657]) and its relationship to invocations of love as transcendence as well as to the idea or accusation (among certain publics in India) that the animal activist's love is a "failed affect" since it is too "categorical" in nature (i.e., addressed to all animals) and hence lacks distinction (662). In this chapter, I have been largely concerned with the rescuer's expression of love for all animals and its interaction with an expressed hatred for certain kinds of human subject, sometimes rendered as cruel or evil. But as we have seen, at least in the case of Cassie, that opposition can sometimes be underwritten by more discerning or differentiating forms of love, where it is recognized that certain rescued animals are special and that that loving relationship has the capacity to sometimes eclipse not just the rescuer's categorical love for all animals but also their differentiating hatred for evil people.

5. Sandoval-Cervantes (2023) draws out this necessary or dependent relationship of antagonistic equivalence between *animalista* or rescuer and perpetrator in the further context of animal activism in Mexico. Indeed, Sandoval-Cervantes claims that the "*animalista* must rely on a triad consisting of animal (the physical existence and representation of an individual injured animal), activist (the rescuer or someone involved in the animal's rehabilitation), and the abuser (the human that injured the animal) to consolidate" the kinds of "intimate relationship" between activist and individual nonhuman animal valorized and foregrounded by *animalista* (2023, 5). Additionally, that triadic relationship is central to petitioning and giving evidence in legal cases of animal abuse in Mexico. Sandoval-Cervantes states that "even when an animal has a name, a face, and a voice, for their stories to be convincing when told, they also need their abusers to have 'names and faces'" (25).

6. For a fuller account of the struggles attached to getting public recognition for the loss of companion animals and acknowledgment from near and distant (human) others for the status of those deceased companions as "grievable" lives, I would recommend the exploration by Desmond (2016) of contemporary mourning practices in the United States. As well as examinations of the rise of pet cemeteries and the controversy created by the emergence of pet obituaries and other forms of public expression of grief, Desmond explores the common charge that such emotions are "excessive" (2016, 100) and, in the case of pet obituaries, that they are also distasteful because they suggest a parallel to the loss of human companions (132–33). For Desmond, these debates are in part about new and future contestations over the definition of "cross-species kinship and familiar configurations" (102), since the justification for such mourning practices is usually linked to claims about the companion animal's status as family member. However, the depth of Cassie's grief in losing Smudge draws less on idioms of family and more on idioms of friendship; it is the quality of their relationship as "soulmates" that makes Smudge's passing, especially in such violent circumstances, so hard to bear.

8. BEING MODERATE IN A WORLD OF INTERESTS

1. Regan also occasionally deploys the idiom of heart and head, a point reemphasized in the 2004 foreword of *The Case for Animal Rights*. In this instance, the reference is introduced to defend the author against the charge of certain feminist critics that Regan's work privileges reason over moral sentiment or emotion (2004, xliii). Regan quotes these lines from an earlier essay: "There are times, and these are not infrequent, when tears come to my eyes when I see, or read, or hear of the wretched plight of animals in the hands of humans. Their pain, their suffering, their loneliness, their innocence, their death. Anger. Rage. Pity. Sorrow. Disgust. . . . It *is* our hearts, not just our heads, that call for an end to it all" (1985, 2).

2. For an ethnographic treatment of the policy and electioneering field from the perspective of Scottish Conservatism, the traditional party of British unionism, see the work of Smith (2014). Capturing what Smith terms the "banal activism" of Scottish Tories during a historical period of electoral marginalization in Scotland, this book also provides a different treatment of moderateness or moderation as political virtue (see also Smith and Holmwood 2013).

3. As Ticktin (2015, 60) points out for a US context, "the expansion of humanitarism to nonhumans has enabled animal suffering to become visible in new medical frames," such as the evidential forms of veterinary forensic science—and, of course, to become newly visible in legal frames too, including the framework of lawmaking. Although closely controlled, the expansion of concern for animal suffering within the legislative process at the Scottish Parliament, partly informed by principles of humanitarianism, has likewise relied heavily on evidence from veterinary medicine and on legal definitions of responsibility.

9. MORAL SUBJECTS

1. There are numerous other possible ways of figuring factionalism within the animal protection movement in the UK. Therese Kelly (2022), for instance, discusses "radical animal rights" activism and "vegan activism" as a pertinent distinction for interlocutors in a recent study of the movement in Bristol. In this figuration, vegan activism centers on a politics of giving witness to forms of human-inflicted animal suffering and on personal ethical self-transformation through not eating meat or dairy products. By contrast, radical animal activists focus on direct action and a prefigurative politics grounded in wider anticapitalist and anarchist protest. For them, veganism is a common practice, although it is "not seen as a moral imperative, but [instead] as a tactic to use within the wider fight for both animal and human liberation" (2022, 6). Here, the charge of complicity often centers on attacks on the single-issue focus of vegan activism and its perceived failure to consider the structural interconnections between the oppression of humans and that of other animals.

2. To be fair, Davé (2023) has more recently recognized the problem of critiquing contradiction thinking in only selective directions or of doing so while still deploying it elsewhere: "I keep running into this problem. I want to show the fallacies and impoverishment of contradiction thinking, yet I find myself doing so by calling out the contradictions of others. That itself, I realize, is a contradiction but—I hope to convince—a potentially *vital* one. The problem returns me to a question I left unaddressed earlier. To put it simply: Is there 'good' contradiction but also 'bad' contradiction? I say yes, and that the difference is a vital one" (2023, 66). However, the explanation offered for an exemption clause in the critique of the "tyrannical" practice of contradiction thinking does raise a number of questions about who has the right to insist on what constitutes good or bad contradiction.

3. I am reminded of this expectation once again when reading the novelist and short story writer George Saunders's book about the experience of teaching creative writing at Syracuse. Among the exercises Saunders regularly ran for these trainee writers was one that broke a short story by Chekhov into page-by-page installments. After reading each page, student-writers were invited to stop and reflect on what had been learned so far about the moral character of the protagonist. The exercise was intended to remind them that from the perspective of their prospective readers, care for literary character only emerges through the increasing particularity of that person. "Specificity makes character" is the lesson (Saunders 2021, 87), and at the heart of that particularization is a growing awareness of the character's behavior or thinking as inconsistent or contradictory.

4. Interestingly, Davé (2023), who draws quite extensively on observations from novels in *Indifference* (Davé's book), does not really pick up on the normative value placed on the status of literary character as an inconsistent or contradictory moral subject. This might in part be because Davé's immanent ethics, which strives to "take away the concept of the ethical actor, accumulating or depreciating over time," and instead to "imagine it never existing" (2023, 61), is really reaching for something closer to an existential treatment of moral character existing in the moment of its expression (for instance, in the mutual regard or responsiveness of rescue).

5. It is worth pointing out that Davé (2023, 19) highlights at least one other example of the machinic metaphor for ethical commitment being invoked, this time by a high-profile and somewhat controversial politician and leading animal activist in India, Maneka Gandhi. In response to the anthropologist's inquiry into what motivates such extraordinary levels of care and activism, Davé states that Gandhi insists that "*she* has nothing to do with it." For, Gandhi says, she is "a machine that is designed to do this, exactly this, only this," and that "it is a machine so sensitive that its skin literally prickles with another's pain. But there is no inside to the machine. There is just skin" (19). The claim to moral patiency could not be more evident. Indeed, Davé reports that Maneka Gandhi's core principle for animal activism was "surrender" (20).

6. As Tom Regan (2004) also recognizes, oscillation or alternation must be a necessary part of moral experience—at least, that is, for most human moral subjects. Regan expects that human subjects will occupy both moral patient and moral agent positions across their lifespans. Other kinds of sentient animal may not be assumed to do so, but even here, the moral subject retains a few instabilities. Note the classic worrying within animal rights philosophy, and within the animal movement more broadly, over which animals should be included as moral patients. And note the occasional willingness among colleagues at the animal group to grant certain critters, most commonly companion animals, a virtuous character and hence secondary status as moral agents.

References

Abrell, Elan. 2016a. "Saving Animals: Everyday Practices of Care and Rescue in the US Animal Sanctuary Movement." PhD thesis, City University of New York.

Abrell, Elan. 2016b. "Lively Sanctuaries: A Shabbat of Animal Sacer." In *Animals, Biopolitics, Law: Lively Legalities*, edited by Irus Braverman, 135–54. New York: Routledge.

Alexandrakis, Othon, ed. 2016. *Impulse to Act: A New Anthropology of Resistance and Social Justice*. Bloomington: Indiana University Press.

Alger, Janet M., and Steven F. Alger. 1999. "Cat Culture, Human Culture: An Ethnographic Study of a Cat Shelter." *Society and Animals* 7 (3): 199–218.

Balcombe, Jonathan. 2006. *Pleasurable Kingdom: Animals and the Nature of Feeling Good*. London: Palgrave MacMillan.

Balcombe, Jonathan. 2011. *The Exultant Ark: A Pictorial Tour of Animal Pleasure*. Berkeley: University of California Press.

Bell, Matthew. 2010. "Is Mary Bale the Most Evil Woman in Britain?" *The Independent*, October 23, 2010.

Benussi, Matteo. 2022. "Emancipating Ethics: An Autonomist Reading of Islamic Forms of Life in Russia." *Journal of the Royal Anthropological Institute* 28 (1): 30–51.

Boddice, Rob. 2009. *A History of Attitudes and Behaviours toward Animals in Eighteenth and Nineteenth Century Britain: Anthropocentrism and the Emergence of Animals*. Lewiston, NY: Edwin Mellen.

Boyer, Dominic. 2008. "Thinking through the Anthropology of Experts." *Anthropology in Action* 15 (2): 38–46.

Bryson, Joanna. 2018. "Patiency Is Not a Virtue: The Design of Intelligent Systems and Systems of Ethics." *Ethics and Informational Technology* 20 (1): 15–26.

Candea, Matei. 2010. "'I Fell in Love with Carlos the Meerkat': Engagement and Detachment in Human-Animal Relations." *American Ethnologist* 37 (2): 241–58.

Candea, Matei. 2013. "Habituating Meerkats and Redescribing Animal Behaviour Science." *Theory, Culture and Society* 30 (7–8): 105–28.

Candea, Matei, Paolo Heywood, Adam Reed, and Thomas Yarrow. 2024. "Ethnographies of Interest: Between Enthusiasm and Instrumentalism." Unpublished manuscript, last saved December 2023 (Microsoft Word file).

Cannell, Fenella. 2011. "English Ancestors: The Moral Possibilities of Popular Genealogy." *Journal of the Royal Anthropological Institute* 17 (3): 462–80.

Carrithers, Michael. 2005. "Anthropology as a Moral Science of Possibilities." *Current Anthropology* 46 (3): 433–56.

Carrithers, Michael, Steven Emery, and Louise J. Bracken. 2011. "Can a Species Be a Person? A Trope and Its Moral Entanglements in the Anthropocene Era." *Current Anthropology* 52 (5): 661–85.

Carsten, Janet. 2007. "Connections and Disconnections of Memory and Kinship in Narratives of Adoption Reunions in Scotland." In *Ghosts of Memory: Essays on Remembrance and Relatedness*, edited by Janet Carsten, 83–103. Oxford: Blackwell.

Cassidy, Rebecca. 2002. *Sport of Kings: Kinship, Class and Thoroughbred Breeding in Newmarket*. Cambridge: Cambridge University Press.

Chua, Liana. 2018. "Too Cute to Cuddle? 'Witnessing Publics' and Interspecies Relations on the Social Media-scape of Orangutan Conservation." *Anthropological Quarterly* 91 (3): 873–903.

Collingwood, R. G. 1942. *The New Leviathan, or Man, Society, Civilization, and Barbarism*. Oxford: Clarendon.

Collingwood, R. G. 1944. *An Autobiography*. London: Penguin.

Csordas, Thomas. 2013. "Morality as a Cultural System." *Current Anthropology* 54 (5): 523–46.

Das, Veena. 2010. "Engaging the Life of the Other: Love and Everyday Life." In *Ordinary Ethics: Anthropology, Language and Action*, edited by Michael Lambek, 376–99. New York: Fordham University Press.

Davé, Naisargi. 2014. "Witness: Humans, Animals, and the Politics of Becoming." *Cultural Anthropology* 29 (3): 433–56.

Davé, Naisargi. 2017. "Something, Everything, Nothing; or, Cows, Dogs, and Maggots." *Social Text* 130 (1): 37–57.

Davé, Naisargi. 2019. "Tyranny of Consistency." In *Messy Eating: Conversations on Animals as Food*, edited by Samantha King, R. Scott Carey, Isabel MacQuarrie, Victoria Niva Millious, and Elaine M. Power, 68–83. New York: Fordham University Press.

Davé, Naisargi. 2022. "Love and Other Injustices: On Humans, Animals and an Ethics of Indifference." *Comparative Studies of South Asia, Africa and the Middle East* 42 (3): 656–67.

Davé, Naisargi . 2023. *Indifference: On the Praxis of Interspecies Being*. Chapel Hill, NC: Duke University Press.

Davis, Janet M. 2016. *The Gospel of Kindness: Animal Welfare and the Making of Modern America*. Oxford: Oxford University Press.

della Porta, Donatella, and Sidney Tarrow, eds. 2005. *Transnational Protest and Global Activism*. Lanham, MD: Rowman and Littlefield.

Desmond, Jane C. 2016. *Displaying Death and Animating Life: Human-Animal Relations in Art, Science and Everyday Life*. Chicago: University of Chicago Press.

Dretske, Fred. 1988. *Explaining Behavior: Reasons in a World of Causes*. Cambridge, MA: MIT Press.

Du Gay, Paul. 2000. *In Praise of Bureaucracy*. London: Sage.

Du Gay, Paul, ed. 2005. *The Values of Bureaucracy*. Oxford: Oxford University Press.

Du Gay, Paul. 2006. "Re-instating an Ethic of Office? Office, Ethos and Persona in Public Management." Working Paper no. 13, ESRC Centre for Research on Socio-Cultural Change (CRESC): 1–29.

Du Gay, Paul. 2007. *Organising Identity: Persons and Organizations after Theory*. London: Sage.

Du Gay, Paul. 2008. "Max Weber and the Moral Economy of Office." *Journal of Cultural Economy* 1:129–44.

Duranti, Alessandro, ed. 2004. *A Companion to Linguistic Anthropology*. Malden, MA: Blackwell.

Ellicott, Claire. 2010. "Stroke of Pure Malice: Grey-Haired Lady Stops to Pet a Cat . . . Then Casually Dumps It in a Bin." *The Daily Mail*, August 24, 2010.

Escobar, Arturo. 2004. "Other Worlds Are (Already) Possible: Self-Organization, Complexity and Post-Capitalist Culture." In *The World Social Forum: Challenging Empires*, edited by Jai Sen and Peter Waterman, 393–404. New Delhi: Viveka Foundation.

Faubion, James. 2011. *An Anthropology of Ethics*. Cambridge: Cambridge University Press.

Faubion, James. 2013. "The Subject That Is Not One: On the Ethics of Mysticism." *Anthropological Theory* 13 (4): 287–307.

FAWC (Farm Animal Welfare Council). 1993. *Second Report on Priorities for Research and Development in Farm Animal Welfare.* London: DEFRA.

Fernando, Mayanthi. 2014. *The Republic Unsettled: Muslim French and the Contradictions of Secularism.* Chapel Hill, NC: Duke University Press.

Floridi, Luciano. 1999. "Information Ethics: On the Philosophical Foundation of Computer Ethics." *Ethics and Information Technology* 1 (1): 37–56.

Floridi, Luciano. 2010. "Information Ethics." In *Cambridge Handbook of Information and Computer Ethics,* edited by Luciano Floridi, 77–100. Cambridge: Cambridge University Press.

Fraiman, Susan. 2012. "Pussy Panic versus Liking Animals: Tracking Gender in Animal Studies." *Critical Inquiry* 39 (1): 89–115.

Franklin, Sarah. 2007. *Dolly Mixtures: The Remaking of Genealogy.* Chapel Hill, NC: Duke University Press.

Friese, Carrie. 2013. "Realizing Potential in Translational Medicine: The Uncanny Emergence of Care as Science." *Current Anthropology* 54 (S7): S129–38.

Friese, Carrie. 2019. "Intimate Entanglements in the Animal House: Caring for and about Mice." *Sociological Review* 67 (2): 287–98.

Friese, Carrie, and Joanna Latimer. 2019. "Entanglements in Health and Well-Being: Working with Model Organisms in Biomedicine and Bioscience." *Medical Anthropology Quarterly* 33 (1): 120–37.

Gell, Alfred. 1998. *Art and Agency: An Anthropological Theory.* Oxford: Clarendon Press.

Ginsborg, David S. 2020. "We Don't Know What We're Saying": Politics and Subjects among the Ultras of Centro Storico Lebowski Football Club in Florence, Italy." PhD thesis, University of Cambridge.

Graeber, David. 2002. "The New Anarchists." *New Left Review* 13:61–73.

Graeber, David. 2009. *Direct Action: An Ethnography.* Oakland, CA: AK Press.

Gray, Kurt, and Daniel M. Wegner. 2009. "Moral Typecasting: Divergent Perceptions of Moral Agents and Moral Patients." *Journal of Personality and Social Psychology* 96 (3): 505–20.

Gruen, Lori. 2011. *Ethics and Animals: An Introduction.* Cambridge: Cambridge University Press.

Guenther, Katja M. 2020. *The Lives and Deaths of Shelter Animals.* Stanford: Stanford University Press.

Gunkel, David. J. 2017. *The Machine Question: Critical Perspectives on AI, Robots, and Ethics.* Boston: MIT Press.

Haraway, Donna J. 2003. *The Companion Species Manifesto: Dogs, People, and Significant Otherness.* Chicago: Prickly Paradigm.

Haraway, Donna J. 2008. *When Species Meet.* Minneapolis: University of Minnesota Press.

Hearne, Vicki. 1986. *Adam's Task: Calling Animals by Name.* New York: Knopf.

Hirschman, Albert. 1977. *The Passions and the Interests: Arguments for Capitalism before Its Triumph.* Princeton, NJ: Princeton University Press.

Holbraad, Martin. 2018. "Steps Away from Moralism." In *Moral Anthropology: A Critique,* edited by Bruce Kapferer and Marina Gold, 27–48. London: Berghahn.

Holmberg, Tora. 2008. "A Feeling for the Animal: On Becoming an Experimentalist." *Society and Animals* 16 (4): 316–35.

Hopgood, Stephen. 2006. *Keepers of the Flame: Understanding Amnesty International.* Ithaca, NY: Cornell University Press.

Howell, Phillip. 2015. *At Home and Astray: The Domestic Dog in Victorian Britain*. Charlottesville: University of Virginia Press.

Hurn, Samantha. 2017. "Animals as Producers, Consumers and Consumed: The Complexities of Trans-species Sustenance in a Multi-faith Community." *Ethnos: Journal of Anthropology* 82 (2): 213–31.

Hyland, M. J. 2011. "The Trial of Mary Bale." *Financial Times*, March 25, 2011.

Jones McVey, Rosie. 2023. *Human-Horse Relations and the Ethics of Knowing*. London: Routledge.

Juris, Jeffey S. 2008: *Networking Futures: The Movements against Corporate Globalization*. Durham, NC: Duke University Press.

Karlsson, Mikael M. 2002. "Agency and Patiency: Back to Nature?" *Philosophical Explorations* 5 (1): 59–81.

Kean, Hilda. 1998. *Animal Rights: Political and Social Change in Britain since 1800*. London: Reaktion Books.

Keane, Webb. 2016. *Ethical Lives: Its Natural and Social Histories*. Princeton, NJ: Princeton University Press.

Kelly, Therese M. 2022. "What's Wrong with Veganism? An Exploration of the Differing Ethical Projects of Mainstream Vegan and Radical Animal Rights Activists in Bristol, UK." PhD thesis, University of Manchester.

Kelly, Tobias. 2015. "Citizenship, Cowardice, and Freedom of Conscience: British Pacifists in the Second World War." *Comparative Studies in Society and History* 57 (3): 694–722.

King, Samantha, R. Scott Carey, Isabel MacQuarrie, Victoria Niva Millious, and Elaine M. Power, eds. 2019. *Messy Eating: Conversations on Animals as Food*. New York: Fordham University Press.

Kirk, Robert G. 2016a. "Care in the Cage: Materializing Moral Economies of Animal Care in the Biomedical Sciences, c. 1945–." In *Animal Housing and Human–Animal Relations: Politics, Practices and Infrastructures*, edited by Kristian Bjørkdahl and Tone Druglitrø, 167–184. Oxon: Routledge.

Kirk, Robert G. 2016b. "The Birth of the Laboratory Animal: Biopolitics, Animal Experimentation, and Animal Wellbeing." In *Foucault and Animals*, edited by Matthew Chrulew and Dinesh Joseph Wadiwel, 191–221. Leiden: Brill.

Kirksey, S. Eben, and Stefan Helmreich. 2010. "The Emergence of Multispecies Ethnography." *Cultural Anthropology* 25 (4): 545–76.

Knight, John, ed. 2005. *Animals in Person: Cultural Perspectives on Human-Animal Intimacies*. London: Routledge.

Krøijer, Stine. 2015a. *Figurations of the Future: Forms and Temporalities of Left Radical Politics in Northern Europe*. New York: Berghahn.

Krøijer, Stine. 2015b. "Revolution Is the Way You Eat: Exemplification among Left Radical Activists in Denmark and in Anthropology." *Journal of the Royal Anthropological Institute*, n.s., 78–95.

Laidlaw, James. 2002. "For an Anthropology of Ethics and Freedom." *Journal of the Royal Anthropological Institute* 8 (2): 311–32.

Laidlaw, James. 2010a. "Ethical Traditions in Question: Diaspora Jainism and the Environmental and Animal Liberation Movements." In *Ethical Life in South Asia*, edited by Anand Pandian and Daud Ali, 61–80. Bloomington: Indiana University Press.

Laidlaw, James. 2010b. "Agency and Responsibility: Perhaps You Can Have Too Much of a Good Thing." In *Ordinary Ethics: Anthropology, Language and Action*, edited by Michael Lambek, 143–63. New York: Fordham University Press.

Laidlaw, James. 2014. *The Subject of Virtue: An Anthropology of Ethics and Freedom*. Cambridge: Cambridge University Press.

Lambek, Michael. 2010a. "Introduction." In *Ordinary Ethics: Anthropology, Language and Action*, edited by Michael Lambek, 1–36. New York: Fordham University Press.

Lambek, Michael. 2010b. "Towards an Ethics of the Act." In *Ordinary Ethics: Anthropology, Language and Action*, edited by Michael Lambek, 39–63. New York: Fordham University Press.

Lansbury, Coral. 1985. *The Old Brown Dog: Women, Workers, and Vivisection in Edwardian England*. Madison: University of Wisconsin Press.

Laszczkowski, Mateusz. 2019. "Rethinking Resistance through and as Affect." *Anthropological Theory* 19 (4): 489–509.

Law, John. 2010. "Care and Killing Tensions in Veterinary Practice." In *Care in Practice: On Tinkering in Clinics, Homes and Farms*, edited by Annemarie Mol, Ingunn Moser, and Jeanette Pols, 57–72. Bielefeld: Transcript.

Lévi-Strauss, Claude. 1966. *The Savage Mind*. London: Weidenfeld & Nicholson.

Lienhardt, Godfrey. 1961. *Divinity and Experience: The Religion of the Dinka*. Oxford: Oxford University Press.

Liu, Xin. 2002. *The Otherness of Self: A Genealogy of the Self in Contemporary China*. Ann Arbor: University of Michigan Press.

Lynch, Michel E. 1988. "Sacrifice and the Transformation of the Animal Body into a Scientific Object." *Social Studies of Science* 18:265–89.

Maeckelbergh, Marianne. 2009. *The Will of the Many: How the Alterglobalisation Movement Is Changing the Face of Democracy*. London: Pluto.

Maeckelbergh, Marianne. 2011. "Doing Is Believing: Prefiguration as Strategic Practice in the Alterglobalization Movement." *Social Movement Studies* 10 (1): 1–20.

Mahmood, Saba. 2005. *Politics of Piety: The Islamic Revival and the Feminist Subject*. Princeton, NJ: Princeton University Press.

Marvin, Gary. 2005. "Disciplined Affections: The Making of an English Pack of Foxhounds." In *Animals in Person: Cultural Perspectives on Human-Animal Intimacies*, edited by John Knight, 61–78. London: Routledge.

Mattingly, Cheryl. 2012. "Two Virtue Ethics and the Anthropology of Morality." *Anthropological Theory* 12 (2): 161–84.

Mattingly, Cheryl. 2014. *Moral Laboratories: Family Peril and the Struggle for a Good Life*. Berkeley: University of California Press.

Mazzarella, William. 2020. "On Patiency, or, Don't Just Do Something, Stand There!" Unpublished manuscript, last saved October 2020 (PDF file).

Mazzarella, William. 2021. "On Patiency, or, Don't Just Do Something, Stand There!" Unpublished manuscript, last saved August 2021 (PDF file).

Mittermaier, Amira. 2010. *Dreams That Matter: Egyptian Landscapes of the Imagination*. Berkeley: University of California Press.

Mittermaier, Amira. 2012. "Dreams from Elsewhere: Muslim Subjectivities beyond the Trope of Self-Cultivation." *Journal of the Royal Anthropological Institute* 18 (2): 247–65.

Molland, Noel. 2004. "Thirty Years of Direct Action." In *Terrorists or Freedom Fighters? Reflections on the Liberation of Animals*, edited by Steven Best and Anthony J. Nocella II, 67–74. Brooklyn: Lantern.

Nelson, Nicole C. 2018. *Model Behaviour: Animal Experiments, Complexity, and the Genetics of Psychiatric Disorders*. Chicago: University of Chicago Press.

Noske, Barbara. 1997. *Beyond Boundaries: Humans and Animals*. Montreal: Black Rose Books.

Ogden, Laura A., Billy Hall, and Kimiko Tanita. 2013. "Animals, Plants, People, and Things: A Review of Multispecies Ethnography." *Environment and Society: Advances in Research* 4:5–24.

Parker, Andrew. 2010. "It's a Fur Cop: We Find Woman Who Dumped Family's Pet Cat in a Wheelie Bin." *The Sun*, August 25, 2010.

Parkin, David, ed. 1985. *The Anthropology of Evil*. Oxford: Blackwell.

Parreñas, Juno Salazar. 2018. *Decolonizing Extinction: The Work of Care in Orangutan Rehabilitation*. Chapel Hill, NC: Duke University Press.

Phelps, Norm. 2007. *The Longest Struggle: Animal Advocacy from Pythagoras to PETA*. New York: Lantern Books.

Polletta, Francesca. 2002. *Freedom Is an Endless Meeting*. Chicago: Chicago University Press.

Razsa, Maple. 2015. *Bastards of Utopia: Living Radical Politics after Socialism*. Bloomington: Indiana University Press.

Reed, Adam. 2016a. "Crow Kill." In *Animals, Biopolitics, Law: Lively Legalities*, edited by I. Braverman, 99–116. London: Routledge.

Reed, Adam. 2016b. "Commons Feeling in Animal Welfare and Online Libertarian Activism." In *Releasing the Commons: Rethinking the Futures of the Commons*, edited by Ash Amin and Phillip Howell, 49–65. London: Routledge.

Reed, Adam. 2017a. "An Office of Ethics: Meetings, Roles and Moral Enthusiasm." *Journal of the Royal Anthropological Institute* 23 (S1): 163–79.

Reed, Adam. 2017b. "Snared: Ethics and Nature in Animal Protection." *Ethnos* 82 (1): 68–85.

Reed, Adam. 2024. "Life on the Lifeboat: Stylizing Ethical Dilemmas in the Philosophy and Activism of Animal Protection." *Social Analysis: The International Journal of Anthropology* 67 (3): 90–101.

Regan, Tom. 1985. "The Case for Animal Rights." In *Defense of Animals*, edited by Peter Singer, 13–26. Oxford: Basil Blackwell.

Regan, Tom. 2004. *The Case for Animal Rights*. Berkeley: University of California Press. First published 1983.

Ryder, Richard. 2000. *Animal Revolution: Changing Attitudes towards Speciesism*. Oxford: Berg.

Samanani, Farhan. 2021. "Power in the Minor Mode: Rethinking Anthropological Accounts of Power alongside London's Community Organisers." *Critique of Anthropology* 41 (3): 284–302.

Sandoval-Cervantes, Iván. 2023. "Gaining Voice through Injury: Voice and Corporeality in Animal Rights Activism in Ciudad Juárez, Mexico." *Cultural Anthropology* 38 (4): 541–566.

Saunders, George. 2021. *A Swim in a Pond in the Rain (in which Four Russians give a Master Class on Writing, Reading and Life)*. New York: Bloomsbury.

Sharp, Lesley. 2018. *Animal Ethos: The Morality of Human-Animal Encounters in Experimental Lab Science*. Berkeley: University of California Press.

Singer, Peter. 2015. *Animal Liberation*. London: Penguin. First published 1975.

Smith, Alexander. 2014. *Devolution and the Scottish Conservatives: Banal Activism, Electioneering and the Politics of Irrelevance*. Manchester: Manchester University Press.

Smith, Alexander Thomas T., and John Holmwood. 2013. *Sociologies of Moderation: Problems of Democracy, Expertise and the Media*. Oxford: Wiley Blackwell.

Smith, Mick. 2011. "Dis(appearance): Earth, Ethics and apparently (In)significant Others." *Australian Humanities Review* 50:23–44.

Song, Hoon. 2010. *Pigeon Trouble: Bestiary Biopolitics in a Deindustrialised America*. Philadelphia: University of Pennsylvania Press.

Spencer, Colin. 1993. *The Heretic's Feast: History of Vegetarianism*. Hanover, NH: University Press of New England.

Spencer, Colin. 2016. *Vegetarianism: A History*. London: Grub Street.

Strathern, Marilyn. 1992. *After Nature: English Kinship in the Late Twentieth Century*. Cambridge: Cambridge University Press.

Strathern, Marilyn. 1999. *Property, Substance and Effect*. Cambridge: Cambridge University Press.

Svendsen, Mette N. 2021. *Near Human: Border Zones of Species, Life and Belonging*. New Brunswick, NJ: Rutgers University Press.

Taylor, Charles. 1989. *Sources of Self: The Making of the Modern Identity*. Cambridge: Cambridge University Press.

Tholonait, Yann. 2015. "Robert Burns: Nature's Bard and Nature's Powers." In *Environmental and Ecological Readings: Nature, Human, and Posthuman Dimensions in Scottish Literature and Arts*, edited by Philippe Laplace, 75–92. Besancon: Presses Universitaires de Franche-Comte.

Ticktin, Miriam. 2015. "Non-human Suffering: A Humanitarian Project." In *The Clinic and the Courtroom*, edited by Tobias Kelly, Ian Harper, and Akshay Khanna, 49–71. Cambridge: Cambridge University Press.

Walters, Kerry S., and Lisa Portmess. 1999. *Ethical Vegetarianism: From Pythagoras to Peter Singer*. New York: State University of New York Press.

Webster, Joseph. 2020. *The Religion of Orange Politics: Protestantism and Fraternity in Contemporary Scotland*. Manchester: Manchester University Press.

Weiss, Erica. 2012. "Principle or Pathology? Adjudicating the Right to a Conscience in the Israeli Military." *American Anthropologist* 114 (1): 81–94.

Weiss, Erica. 2016. "'There Are No Chickens in Suicide Vests': The Decoupling of Human Rights and Animal Rights in Israel." *Journal of the Royal Anthropological Institute* 22 (3): 688–706.

Williams, Bernard. 1993. *Shame and Necessity*. Berkeley: University of California Press.

Wrenn, Corey Lee. 2019. *Piecemeal Protest: Animal Rights in the Age of Nonprofits*. Ann Arbor: University of Michigan Press.

Zigon, Jarrett. 2008. *Morality: An Anthropological Perspective*. Oxford: Berg.

Index

Page numbers followed by letter *f* refer to figures.

Abrell, Elan, 76–77, 201, 282n2
affect theory, and moral patiency, 23–24
AI (artificial intelligence): moral status of, 18–19. *See also* robots/machines
Amnesty International, 58
animal(s), nonhuman: affective power of, 21, 22; alien status of, sci-fi hoax film on, 3, 5; and children, commonalities of, 278n3; exploitation of, Christianity and, 34, 39, 279n12; and machines, commonalities of, 270–71; as moral agents, resistance to idea of, 19; as moral patients, 16, 20–21, 273. *See also* companion animals; encounter(s), with nonhuman animals; *specific animals*
Animal Aid, 12
Animal Angels, 10
animal cruelty/abuse: vs. animal love, as innate passions, 211–12; in circuses, 156–57, 182; confrontation with, animal rescuer and, 154, 204, 205–6, 213; contradictions in lives of perpetrators of ("circus lies"), 181–85, 260, 263–69; cultures of, 189, 206–9; and early history of animal protection, 55; evil dimensions of, 190, 198, 205–12; explanations for, 195–96, 274; in laboratories, 176–78, 182; moral patiency and, 274; questioning of, 12; and state of humanity, concern about, 188–89, 212–13; tabloid report on (Cat Bin Lady story), 191–98, 192*f*; undercover investigator as witness to, 45, 157–59, 171–72; undercover investigator's perspective on, 179, 181–86, 188–89, 190
Animal Health and Welfare Act of 2006 (Scotland), 229
Animal Health and Welfare Division of Scottish government, 219, 221; radical animal activists' animosity toward, 254–55; and Wildlife and Natural Environment Bill, 235
animalist, use of term, ix, 278n2
animal liberation, 56, 253, 277n1

Animal Liberation (Singer), 9, 15, 112, 252–53
Animal Liberation Front, 56
animal love: vs. animal cruelty, as innate passions, 211–12; animal people's reflections on, 199–200, 272; dormant, belief in, 86; and fellow feeling, 106; "gloved," 75; and moral development, 8; moral patiency and, 112; multispecies ethnography on, 74; reciprocated, 203–4; and rescue work, 210–11, 213–14, 283n4; undercover investigator's inability to express, 170–71
animal-machine *(bête-machine)*, doctrine of, 270, 271
animal (protection) movement: "anti" tendency within, 88–91, 110, 264; aspirations for humanity and, 55, 187, 188–89; classic targets of critique within, 5; colonizing project and, 4, 277n1; consciousness-raising as model for, 124, 125–27; vs. conservation, 88, 90, 140–41, 147; critical encounters with animal suffering and, 23, 55, 187–88, 271; diversity within, 7; ethical work associated with, 3, 34; experiences of moral patiency and, 25–26; factionalism within, 253–58, 261, 263, 264, 284n1; founding manifesto of, 9; history of, 55–58, 65, 112, 281n2; humanitarianism and, 57, 279n13, 282n3, 284n3; legislative wins in Scotland, xv, 263; literature on, ix–x, 15–17, 55, 71–72, 74–76; moral subjects within, 273; new wave of, role of undercover investigator in, 158; normative impulse to query, 258–59; passion guiding, 33; philosophical questions framing debate in, 14; political alignment of, 35, 279n13; professionalization of, 7, 42, 58, 281n2; and responsibility, fluctuating claims around, 39; sense of unresolvable dilemma in, 261–62; "spiritual" dimension of, 36; transnational network of, 4, 65; use of term, 7, 8, 277n1

animal people: and category of evil,
approaches to, 46, 190–91, 196, 198,
209–10; childhood experiences and moral
development of, 8, 10, 11, 12–13, 17–18,
89, 148, 163, 229; concern for consistency
among, 259–60; moderate animal
activists as, 264; moral agency of, 21,
41, 46, 276; moral drive of, puzzlement
over, 12–13, 21–22; moral patiency of,
21, 25–26, 27, 29, 46, 170, 273, 276;
self-presentations/moral biographies of,
8–12, 17–18, 89–90, 114–19, 128–38,
156–57, 163–64, 199–200, 229–31;
sense of exceptionalism among, 10–11,
12–13; suspicion of organized religion/
Christianity, 34–35, 39, 88, 132, 215; use
of term, 8, 277n2
animal rescue/animal rescuer(s), xvii, 9–10,
191, 199–202; as animal whisperers, 202,
203, 282n2; as compulsion/instinct, 8,
201; and confrontation of evil, 154, 204,
205–6, 213; emotive stories of, PETA's use
of, 54; ethical lives of animal people and,
46, 191; as frontline soldiers of animal
protection, 210; and impulse to retaliate,
211–12; and interspecies communication,
203–4, 282n2; persistent querying of, 258;
personality of, 154; reframing as kind
act, 102, 104; and self-sacrifice, 213–14;
transnational network for, xvii, 9–10, 191,
200, 203, 205–7
animal rights: in context of committed
relationships, 72–73; human rights
activism and, 279n13; ideas of, in history
of animal protection, 56, 253; language of,
Edinburgh group's reluctance to embrace,
261; use of term, 277n1
animal sanctuary movement: contradictions
within, 76–77; principles of, 201
animal suffering: affective powers of,
17–18, 21, 23–24, 27, 240–45; animal
movement's overriding emphasis on,
reservations about, 271; and choice to
become vegetarian/vegan, 130–31, 142;
climate change and, 187; conservation
and, 141; encounters with, and moral
development, 11; focus on, vs. admiration
for animal sentient powers, 92, 99, 102;
human causes of, as focus of animal
protection, 187–88; images of, as tool
of public messaging, 54, 98; reducing,
early anticruelty organizations and, 55;
systemic causes of, reflections on, 9

animal testing: campaigns against, Edinburgh
group and, 80–81, 101, 109, 231;
medical treatments and, ethical choice
regarding, 123; as touchstone issue
for moral development, 9; undercover
investigations of, 164–65. See also
laboratories
animal welfare: Five Freedoms and, 187,
282n1; use of term, 277n1
Animal Welfare Inter-Group of European
Parliament, 64
"antis," animal organizations depicted as,
88–91, 110, 264; vs. Edinburgh group's
new inclusive vision, 91, 119
Aristotle, 218
Art and Agency (Gell), 30–32

Balcombe, Jonathan, 71, 93, 99
Bale, Mary, 191–98, 192*f*
Barry (undercover investigator; pseudonym),
xvii, 156–86; on animal cruelty, 179, 181–
86, 188–89, 190; in BBC documentary
film, 220, 232; career of, 163–65; chosen
special animal of, 274; on "circus lies,"
181–85, 260, 263–69; contradictions in
life of, 185–86, 265, 268–69; dedication to
animals, 160, 161, 169, 191, 265; expertise
of, 45, 156, 158–59, 160, 161–62, 165–69;
investigatory trips to shooting estates, 43,
165–69, 241; moral authority of, 158–59,
179; moral biography of, 156–57, 163–64;
office of, 49; personalities adopted by,
154, 157, 159, 180, 181, 265; practices
of habituation, 172–79; relationship to
coworkers in undercover operations,
174–79; on respect for life, 269, 271; self-
sacrifice of, 159, 160–61, 169–72, 175–76,
186, 265; social background of, 161, 164;
work with policy director, xvi–xvii, 49,
159–60, 241, 243
bête-machine (animal-machine), doctrine of,
270, 271
Blair, Tony, 232
Born Free Foundation, 164
Bracken, Louise J., 146–49
Britain: Brexit referendum in, idiom of
heart and head in, 224; as "country
of animal lovers," cliché of, 188, 189;
early anticruelty organizations in, 55;
indifference to animal cruelty in, 207
British Union against Vivisection
(BUAV), 55
Bryson, Joanna, 18–20, 270, 272

Buddhism: as animal-friendly form of "spirituality," 36, 57; and vegetarianism/veganism, 142–43

bullfighting: activism against, 64; arguments made in defense of, 35

Burns, Robert (Robbie): "To a Mouse," 59, 60; "The Wounded Hare," 60–61

Burns night celebrations, 60–62, 67

Candea, Matei, 74, 75, 173–74, 175

Captive Animals' Protection Society (CAPS), 53

Carrithers, Michael, 28, 146–49, 278n9

The Case for Animal Rights (Regan), 15–17, 25–26, 284n1

Cassidy, Rebecca, 75

Cassie (patron of Edinburgh group; pseudonym), xvii; on American Indian peoples, 35, 266, 268; on animal love, 199–200, 272; on animal people, 12–13, 21–22, 27; and animal rescue, xvii, 199–204; burnout experienced by, 261; friendship with Eilidh, 201–2, 203, 205, 210, 215, 268; home of, 198–99; inconsistencies of, 266–68; and interspecies communication, 201, 203; love for wolves, 208, 268, 274; moral biography of, 9–10, 11; as published writer, 201; relationship with Smudge (cat), xvii, 203–4, 214–16, 283n6; self-sacrifice in rescue work of, 213–14; understanding of evil, 205–12; vegetarianism of, 210

cat(s): interspecies relationships with, xvii, 138–39, 145, 203–4, 214–16, 283n6; in Majorca, 140; in sci-fi hoax film, 3

Cat Bin Lady story, 191–98, 192*f*

childhood experiences: and capacity for empathy, 241; and moral development of animal people, 8, 10, 11, 12–13, 17–18, 89, 148, 163, 229

children: animal cruelty perpetrated by, 207, 209; education programs for, Edinburgh group and, 85, 209; feelings of, and origins of moral development, 8; as moral patients, 16, 18; and nonhuman animals, equivalence drawn between, 278n3

China, new moral characters in, 279n10

Christianity: and animal exploitation, historical tradition of, 34, 39, 60, 279n12; animal people's suspicion of, 34–35, 39, 88, 132, 215

circus(es): ban on wild animals in, xv, 250–51; undercover operation in, 156–57, 162, 175, 182

climate change, and animal suffering, 187

cloning, 75

Collingwood, R. G., 28, 29, 30, 99

collusion/complicity, moderate animal activists accused of, 252–56, 258, 284n1

colonialism, and animal protection, 4, 277n1

communications officer: role of, 81, 114, 196–97. *See also* Euan

companion animals: as aliens, sci-fi hoax film on, 3, 148; and ethics of relating, 70–73; loss of, grief associated with, 214–16, 283n6; as moral subjects, 273, 285n6; personality traits of, 152; and personhood, 148; relationships with, 48–49, 68–69, 203–4, 214–15; rescue, xvii, 191, 199–204

Companion Species Manifesto (Haraway), 72

Compassion in World Farming, 53

compromise/negotiation: capacity for, policy director and, 155; and charges of collusion, 253–54; Edinburgh group and, 88, 123; moderate animal activism and, 218; radical activists' rejection of, 264

conscience: legal protection of, 245–46; moral patiency and, 247–48, 275

conscience vote, 246–51

conscientious objectors, 245–46, 247

consciousness-raising, 123–27; feminist, 125, 126; as model for animal activism, 124, 125–27

conservation, critique of, 88, 90, 140–41, 147

consistency, tyranny of: Davé on, 123–24, 258–59, 261; radical animal activism and, 258, 261; vegetarian/vegan lifestyle and, 259–60

contradiction/inconsistency: Edinburgh animal group and charges of, 258–59, 261; literary subjects and, 263–64, 285n4; moderate activism and appreciation of, 264; moral subjects and, 263–69, 285n6

contradiction thinking: Davé on, 263, 284n2; ineffectiveness of, 268; reductive impulse of, 264

cosmetics: ethical choices regarding, 121; testing on animals, campaign against, 80–81, 101, 109

Coston, Susie, 282n2

Craig (science and research manager; pseudonym): on animal-friendly future, 143–44; on animal personalities, 150; and "Animals A-Z"/Amazing Animal Facts, 94–95, 99, 150, 151; and ethos of positivity, 96, 97; favorite author of, 71; on health vs. animal welfare concerns, 123; and mainstream, project to engage, 85, 89, 91, 93–94; moral biography of, 89–90; on personality as sign of sentience, 149; on pressure to be professionally angry, 90–91; relationship with Paddy (dog), 48, 96, 106, 148; support for author's project, xvi, 42, 43

Cross-Party Group for Animal Welfare, Scottish Parliament, 43

Csordas, Thomas, 194–95, 204–5

cultures of cruelty, 189, 206–9

dairy farming, critiques of, 5, 142

Davé, Naisargi: on affective histories, 23–24; on contradiction thinking, 263, 284n2; on ethos of surrender, 24, 26–27; on genealogies of animal activism in India, 277n1, 278n4, 280n14, 282n1; on immanent ethics, 261–62, 285n4; *Indifference,* ix–x, 24, 281n4, 285n4; inspiration drawn from, xi, xii; on interspecies being, x, 281n4; on love as politics of distinction, 283n4; narrative account of, xi; notion of moral patiency in work of, 23–24, 26–27; on term "animalist," 278n2; on "tyranny of consistency," 123–24, 258–59, 261

David (social media strategist; pseudonym), 48; critical encounter with animal suffering, 17–18; and ethos of positivity, 196; at Girls' Day Out event, 83, 109; moral biography of, 12, 17–18; role of, 134; support for author's project, xvi; on utopian visions, 144; vegetarianism of, 136–37, 142

Descartes, René, 270, 271

Desmond, Jane C., 283n6

desubjectivation, 180

detachment, cultivated, 74–76

Divinity and Experience (Lienhardt), 28–29

dog(s): in Edinburgh group's offices, 48–49, 68–69, 73, 77; entangled history as companion species, 70; Haraway's, 70, 72; in sci-fi hoax film, 3

Dog's Bazaar, 65–68, 66f

Dogs Trust (animal group), 244

do-no-harm principle: alternative communities based on, utopian visions of, 142–45; vs. conservation, 140–41; historical context for, 141; Jainism and, 280n14; vs. medicines, 123; vegetarianism/veganism and, 113

Dretske, Fred, 20

Du Gay, Paul, 234, 257

Durkheim, Émile, 24, 129

Eastern Europe, animal rescue from, xvii, 9–10, 191, 200, 203, 205–6, 214

Edinburgh animal group: annual reports of, 40f, 59–60, 64–68, 66f, 174f, 190f; antisnaring campaign of, xv, 52, 64f, 165, 236, 240–45, 248–50; Burns night celebrations of, 60–62, 67; celebrity endorsements of, 100–101; charges of collusion against, 253–54, 258; charges of inconsistency against, 258–59, 261; coalition partners of, 4, 52–53; Dog's Bazaar of, 65–68, 66f; dogs in offices of, 48–49, 68–69, 73, 77; and ethos of positivity, 44, 91–104, 95f, 124, 189, 196, 198, 240; focus of, x, 39, 41–42; as generalist organization, 7, 49–52, 57, 58, 63; at Girls' Day Out event, 79–83, 86, 101, 108–10; gradualist approach of, 53, 73, 85, 256–57; heart-and-head stance of, 225; history of, 3–4, 55–58, 63; kind acts promoted by, 44, 100–104; launch of rebranded organization, 1, 83–86, 94, 119–20; legislative campaigns of, 39, 159–60, 228–29, 233, 240–50, 262–63; library of, 15, 48, 71–72; and mainstream, project to engage, 4–5, 41, 44, 54, 78, 79–87, 89, 91, 93–94, 108–10, 113, 119–20, 189–90, 209–10, 260–61, 269; and making contact, narrative of, 1–6, 60, 71, 72, 77, 85–86, 107, 148, 269, 271; moderate stance of, x, 8, 41, 46, 53–54, 122–23, 217, 219–20, 252; moral sources for, 34–37; National Anti-Vivisection Society (NAVS) and, 55, 56, 63; national identity and, 3, 58–65; new inclusive vision of, 41, 83–87, 91, 104, 119, 121–22, 137, 198, 260, 269, 271; offices of, 48–49, 56; oppositional work of, 89, 90–91; organizational transformations in, xv, 41–42, 57; PETA compared to, 53–54, 57; philosophical influences on, 280n1; policy handbook of, 50–52, 54; pragmatic view of animal protection, 14, 53, 73, 88, 122, 144, 231;

professionalization of, x, 57–58; questions animating meetings of, 13–14; regulative successes of, xv, 250–51; renaming of, 57, 65, 100; and rescue work, 191; resistance to preaching, x, xii, 45, 91, 104, 137, 190, 217, 260; sci-fi hoax film released by, 1–6, 27; staff of, 48–49, 57–58, 122; supporter base of, 3, 57, 67; support for author's project, xv, xvi–xvii, 42; tabloid report on animal cruelty (Cat Bin Lady story) and, 193–98; tolerance of, 35–36; website of, 81, 83, 94, 98, 103, 134. *See also names (pseudonyms) of specific staff members*

Egypt: animal cruelty in, 207; Sufi community in, 28; women's mosque movement in, 26, 278n7

Eilidh (CEO of Edinburgh group; pseudonym): on animal cruelty, 196; on animal sentience, 92; on "antis," animal groups perceived as, 88–89; Cat Bin Lady story and, 193, 197–98; child protection background of, 86; and ethos of positivity, 91, 94, 96, 97, 98, 100, 104, 124, 189, 240; friendship with Cassie, 201–2, 203, 205, 210, 215, 268; inability to engage with realities of evil, 210; and launch event at Edinburgh Castle, 83–86, 94, 119–20; and mainstream, project to engage, 41, 81, 84, 85, 113, 119, 189–90, 209; moral biography of, 8–9, 164; and new inclusive vision, 119, 121–22; office of, 48; personality of, 151; spiritual beliefs of, 36; support for author's project, xvi, 42, 43

Elaine (personal assistant to CEO; pseudonym), xvi; and animal rescue, 202; burnout experienced by, 261; and interspecies communication, 203; moral biography of, 10–11; office of, 48; personality of, 152, 181; role of, 48–49; spiritual beliefs of, 36

Emery, Steven, 146–49

encounter(s), with nonhuman animals: Edinburgh group's emphasis on, 1–6, 60, 71, 72, 77, 85–86, 107, 148, 269, 271; and moral awareness, 5–6, 9–10, 11, 12–13, 17–18; moral patiency assumed in, 86; multispecies ethnography on, 70–76; mutuality in, 139; responsibility in, 140; sci-fi hoax film urging, 1–6, 148, 188; and sense of exceptionalism, 10–11

ethical choice: consciousness-raising and, 123–27; hidden motivations behind, 133–35; inherited status of, 45, 135, 136, 138; and

medicines, consumption of, 123; moral agency and, 111–12; moral patiency and, 112, 124; as self-willed vs. predetermined, 128; and vegetarianism/veganism, 44, 112–13, 117–18, 120, 122, 123, 127, 130–32, 135–38, 139, 141–42

ethical turn in anthropology, 33–34, 128–29; category of evil as blind spot in, 204–5; debates within, 135; and descriptions of self-interpretation and self-control, 179–80; Foucault and, 128–29, 135, 180; and politics of prefiguration, 218

ethics: of animal protection, encounters with nonhuman animals and, 5–6, 9–10, 11, 12–13, 17–18; immanent, 261–63, 285n4; of relating, 70–73; virtue, 34. *See also under* ethical; moral

Euan (communications officer; pseudonym), xvi; on Cat Bin Lady story, 197, 198; expertise of, 114; and Girls' Day Out event, 79, 81, 109; and metaphor of addiction, 113, 124; moral biography of, 114–19, 153; office of, 48; personality of, 151, 153; relations with journalists, 81, 114, 196–97; vegetarianism of, 114–19, 122, 127–28, 153

European Parliament: Animal Welfare Inter-Group in, 64; Members of (MEPs), lobbying of, 10, 63

European Union: campaign to ban cosmetics testing in, 80; and transnational animal protection movement, 4, 65

evil: animal cruelty and sense of, 190, 198, 205–12; animal people's approaches to, 46, 190–91, 196, 198, 209–10; as blind spot of anthropology's ethical turn, 204–5; category of, skepticism regarding, 194–95; confrontation with, animal rescuer and, 154, 204, 205–6, 213; cultures of, 189, 206–9; inability to engage with realities of, 209–10; perpetrators of, 208–9, 211, 212

expert(s)/expertise: in animal movement, 33, 42; animal-related, studies of, 75–77; calls to "humanize," 42; of communications officer, 114; Harraway's approach to, 73; of IT officer, 129, 134; and personality, 147, 154–55; of policy director/lobbyist, 46, 219, 228, 229, 231, 236–39, 240–41, 248–50, 264; reliance on, in legislative process, 222; of science and research manager, 90–91; of undercover investigator, 45, 156, 158–59, 160, 161–62, 165–69

Explaining Behaviour (Karlsson), 20
The Exultant Ark (Balcombe), 93, 99

family: and ethical choices, 45, 135, 136, 138; vegetarianism/veganism and relations with, 135–38
Farm Sanctuary, 282n2
Faubion, James, 180
feelings: and animal protection, 33, 106; control of, undercover investigator and, 45; and origins of moral development, 8, 11. *See also* animal love
fellow feeling, 105; within animal movement, 106; images and, 106, 107
feminist movement, consciousness-raising in, 125, 126
fieldwork, 3, 43
film(s): about animal cruelty, and turn to vegetarianism, 130–31; BBC documentary, on Scottish shooting estates, 220–21, 224–25, 232; PETA's strategic use of, 54, 98; sci-fi hoax, 1–6, 27, 71
Five Freedoms, 187, 282n1
Foucault, Michel, 128–29, 135, 180
foxhunting, in Britain, 189, 190*f*; ban on, 246
Franklin, Sarah, 75
Fraser (fundraising director; pseudonym), xvi, 48; ethical choices of, 122; new inclusive vision and, 119
fur trade, evil in, 208

Gandhi, Maneka, 285n5
Gelassenheit, concept of, 24
Gell, Alfred, 30–32
generalist animal protection organization: challenges of working for, 44, 52; Edinburgh group as, 7, 49–52, 57, 58, 63
Girls' Day Out event, 79–83, 86, 101
God/divine: as moral source, 34. *See also* religion
Goodall, Jane, 174
good(s)/goodness: ethical choice and, 128; as evolutionary necessity, 133–35; as "felt force," 34; as form of moral patiency, 46. *See also* evil
gradualist tactics: Edinburgh animal group and, 53, 73, 85, 256–57; National Anti-Vivisection Society (NAVS) and, 55, 56

grassroots animal group(s), 7
grief, after loss of companion animal, 214–16, 283n6
guilt, as authentic experience of moral subject, 29
Gunkel, David, 270–71; on moral patiency, 272–73, 274–75

Haraway, Donna: *Companion Species Manifesto,* 72; and ethics of relating, 70–73; legacy of, 74–76; *When Species Meet,* ix
health: vs. animal welfare concerns, 123; vegetarianism/veganism and, 101, 119, 137, 142
Hearne, Vicki, 72
heart and head, idiom of, 222–24; in legislative process, 221–22, 224–25, 229, 231, 239–45, 250; role of interest in, 234
Heidegger, Martin, 24
Hinduism: as animal-friendly form of "spirituality," 36; fascist neo/imperialist strand of, 280n14; and vegetarianism/veganism, 282n1
Hirschman, Albert, 234
horsemanship, alternative, 75–76
human(s)/humanity: animal cruelty and concern about state of, 188–89, 212–13; aspirations for, animal protection tied to, 55, 187, 188–89; dominion of, biblically informed assertion of, 60; and ethical choice, 132–33; personalities of, reflections on, 150, 153–54
humanitarianism, and animal protection, 57, 279n13, 282n3, 284n3
hunting: call for ban on, 256; evil in, 208; indigenous peoples and, 35. *See also* foxhunting; shooting estates
Hunt Saboteur Association, 56
hypergood(s), 34

Iain (IT officer; pseudonym), xvi; date balls of, 62, 134; and ethos of positivity, 196; on Girls' Day Out event, 109; on hidden motivations behind ethical choices, 133–35; moral biography of, 12, 128, 129–38; on new inclusive vision, 137; office of, 48; personality of, 151–52; relationship with Caley (cat), 138–39, 145, 148; relationship with Euan, 153; role of, 129, 134; and sci-fi hoax film, 1–6, 148; sense of humor, 134; vegetarianism/veganism of, 130–32, 135–38 image(s): affective powers of, 130–31, 240–45, 248; in annual reports of Edinburgh animal group, 40*f*, 66*f*, 174*f*, 190*f*; ethos of positivity and, 98–99, 240; and fellow feeling, generation of, 106, 107; as tool of public messaging, 54, 98; undercover investigations and, 39, 40*f*, 157–59, 241, 243

immanent ethics: Davé on, 261–62, 285n4; and wins in legislative process, 262–63

inconsistency. *See* contradiction

India, animal activism in: colonizing project and, 277n1, 278n4; critical encounters with animal suffering and, 23; genealogies behind, 280n14, 282n1; term used for, 278n2. *See also* Davé, Naisargi

indifference, of mainstream public: alienation from natural connection and, 60; animal people's puzzlement over, 10, 11, 13, 144, 211; in Britain, 207; Edinburgh group's project to combat, 86; Oxfam's "be humankind" campaign targeting, 103, 281n2

Indifference (Davé), ix–x, 24, 281n4, 285n4; inspiration drawn from, xi, xii

indigenous peoples, attitudes toward, 35–36, 143, 266, 268

industrialized animal husbandry, critiques of, 5, 88

interest(s): balancing in legislative process, 231–39; as dominant language of government, 234

interspecies being, Davé on, x, 281n4

Israel, animal activism in, 279n13

Jainism, 36, 279n14

Jones McVey, Rosie, 75–76

journalist(s)/journalism: animal protection groups perceived as "antis" by, 89, 90; and balancing of interests, 232; communications officer's work with, 81, 114, 196–97; idiom of heart and head in, 223; tabloid report on animal cruelty (Cat Bin Lady story), 191–98, 192*f*

Kalahari Meerkat Project, 173–74, 175

Kant, Immanuel, 24, 129

Karlsson, Mikael M., 20

Keane, Webb, 125, 126, 128, 279n11

Kelly, Therese, 284n1

Kelly, Tobias, 245, 247

kindness, acts of: Edinburgh animal group and emphasis on, 44, 100–104; popularity of websites promoting, 103–4

laboratories: animal abuse in, 176–78, 182; contradictions in, 184–85; practices of habituation in, 173; self-justification for animal abuse in, 183; undercover investigations in, 55, 164–65, 172, 176–79; violent protests against, 56; welfare protocols in, 282n1. *See also* animal testing

Laidlaw, James: on freedom of moral subjects, 129; on other-interested action, 133; on religious traditions and animal liberation movement, 279n14; on responsibility, 37–38

Lambek, Michael, 118

language: ethos of positivity and, 97–98; limitations of, in expression of moral patiency, 29–30, 32, 99

Laura (fundraising assistant; pseudonym), 83, 108–9

law/lawmaking: animal movement's views on, 39; balancing of interests in, 231–39, 245; concern for animal welfare in, 219, 284n3; conscience vote in, 246–51; Edinburgh animal group and, 39, 40*f*, 228–29, 233, 240–50, 262–63; ethics of heart and head in, 221–22, 224–25, 229, 231, 239–45; moral patiency in, 38, 46–47; professional classes in, 222, 228; transplantation of legislation from elsewhere, 58; undercover investigations and, 169–70; values of moderateness in, 46, 228–29, 231; views on animal protection groups and, 89; "wins" in, immanent ethics and, 262–63

League against Cruel Sports, 52

Lévi-Strauss, Claude, 149, 155

Lienhardt, Godfrey, 28–29

Linzey, Andrew, 281n1

Liu, Xin, 30, 87, 279n10

live animal transportation, 64, 162

lobbyist(s). *See* policy director/lobbyist

Lynch, Michel E., 175, 182, 184–85

The Machine Question (Gunkel), 270
Maggie (policy director and lobbyist;
    pseudonym), xvii, 43; on animal
    sentience, 92; on animal suffering, 221;
    BBC documentary and, 221, 224–25;
    Burns night suppers hosted by, 61–62;
    charges of collusion against, 252–56,
    258, 263, 284n1; companion animal of,
    49; expertise of, 46, 219, 228, 229, 231,
    236–39, 240–41, 248–50, 264; at fringe
    meeting for SNP delegates, 243–44;
    gradualist approach of, 256–57; and
    idiom of heart and head, 221–22, 224–25,
    229, 231, 239–45, 250; inconsistencies
    of, 266; interests apart from animal
    protection, 264–66; on legislative process,
    58, 262–63; moderate stance of, 219, 228–
    29, 231, 256; moral biography of, 229–31;
    on moral patiency in lawmaking process,
    46–47; as Morningside Lady, 226–28;
    office of, 49; personality of, 151; on
    policy handbook, 50, 54; and undercover
    investigator, xvi–xvii, 49, 159–60, 241,
    243; vegetarianism of, 229–30
Mahmood, Saba, 26, 278n7
mainstream public(s): and accusations of
    inconsistency, 260; vs. animal people,
    10–11, 12–13, 22, 41; author as
    representative of, 43; depictions of, lack
    of diversity in, 5, 35, 277n2; indifference
    of, 10, 11, 13, 60, 86, 144; metaphor of
    addiction used to describe, 113, 124; and
    moral patiency, 40, 274; and self-ethics,
    121–22, 124
mainstream public(s), engaging: anxieties
    regarding, 47, 87, 108–10, 119–23, 127;
    confronting evil as challenge for, 205;
    Edinburgh animal group's project of,
    4–5, 41, 44, 54, 78, 79–87, 89, 91, 93–94,
    108–10, 113, 119–20, 189–90, 209–10,
    260–61, 269; ethos of positivity and, 91,
    93–94, 97; family relations as model for,
    137; moderate disposition and, 217–18;
    movements of ethical reform and, 87–88;
    social media and, 1, 85, 101–2, 134
Mairi (pseudonym), xi, 151
Majorca, feral cats in, 140
Make Poverty History Movement, 85
Marine Scotland, 253–54
Matthews, Bernard, 197
Mattingly, Cheryl, 118
Mazzarella, William, 24–25, 26, 27, 37
medicines, ethical choice regarding, 123

Members of the European Parliament (MEPs),
    lobbying of, 10, 63
Mexico, animal activism in, 278n2, 283n5
Minding Animals conference, 71
Mittermaier, Amira, 27–28
moderate animal activist(s)/activism, x, xii,
    217–18; appreciation of contradictions
    by, 264; charges against, 252–56,
    258–59, 261, 263; dilemma of, x–xi, 47;
    Edinburgh group as, x, 8, 41, 46, 53–54,
    122–23, 217, 219–20, 252; involvement
    in legislative process, 219; moral patiency
    of, 240; National Anti-Vivisection
    Society (NAVS) as, 55, 56; vs. politics of
    prefiguration, 218
moderateness: element of performativity in,
    220; values of, in lawmaking process, 46,
    228–29, 231; virtue of, 218–19
modern culture, moral sources in, 34
moral absolutism, 261
moral agent(s)/agency: animal others as,
    resistance to idea of, 19; of animal people,
    21, 41, 46; conditionalities attached to,
    128; conscience and, 275; definitions
    of, 15–16, 18, 111; and ethical choice,
    111–12, 132–33; ethos of positivity and,
    99–100; internal tension and, 133; and
    moral patiency, co-construction of, 32,
    279n11; and moral patiency, dynamic
    shifts between, 47, 274–75, 285n6;
    vs. moral patiency, 14–20, 269, 270;
    personality and, 147, 153, 275–76; as
    radical inaction, 113; responsibility as
    defining marker of, 15–16, 38, 132–33;
    and vegetarianism, 44
moral community, 105–6; Edinburgh animal
    group and vision of, 107; value-feelings as
    basis of, 105, 106, 107
moral enlightenment: critical witness
    encounters with animal suffering
    and, 17–18, 21, 23–24; direct encounters
    with nonhuman animals and, 5–6, 9–10,
    11, 12–13, 17–18; gradual evolution
    of, 8–9
moral patient(s)/patiency: affective status
    of, 21; after-the-fact explanations for,
    22, 25, 31, 125; and animal cruelty,
    274; and animal love, 112; of animal
    people, 21, 25–26, 27, 29, 46, 170, 273;
    before-the-fact explanations for, 22, 25,
    71, 268, 269; being good as form of, 46;
    children and, 16, 18; and conscience,
    247–48, 275; Dave's work and notion

of, 23–24, 26–27; definitions of, 16, 18; derivative status of, 274–75; encounters with nonhuman animals and, 27, 86; and ethical choice, 112, 124; ethos of surrender and, 24, 26–27; Gunkel on, 272–73, 274–75; heuristic of, test of value of, 32–33; and inconsistencies, 268; in lawmaking process, 38, 46–47; Lienhardt's work and notion of, 28–29; linguistically inhibited expression of, 29–30, 32, 99; literature on, 23–29; machines and, 19, 270; mainstream publics and, 40, 274; Mazzarella's work and notion of, 24–25, 26, 27, 37; mechanistic metaphors of, 272, 285n5; and moral agency, co-construction of, 32, 279n11; and moral agency, dynamic shifts between, 47, 274–75, 285n6; vs. moral agency, 14–20, 269, 270; nonhuman animals and, 16, 20–21, 273; personality and, 153; as philosophical category, 274; and policy work of moderate animal activists, 240; project to engage mainstream and, 44; Regan on, 16–17, 25–26, 31, 125, 248, 269; religion and, 26, 27–29; respect as natural expression of, 269; secular-liberal imaginary and, 37–38; undercover investigator and, 170, 179

moral sources: for Edinburgh animal group, 34–37; in modern age, 34

moral subject(s): animal people as, x, xii; biographic sense of, xi; companion animals as, 273, 285n6; definition of, 18, 270; expansion of category of, 270, 272–74; freedom of, 129; inconsistencies of, 263–69, 285n6; machines as, 19, 270–71; in secular-liberal imaginary, 33

Morningside Lady, 226–28, 227f

mountain hares: Burns's poem about, 60–61; campaign for protection of, xv, 61

multiform sentient world, idea of, 5

multispecies ethnography, ix, 70–76

National Anti-Vivisection Society (NAVS), 9, 55–56; Edinburgh group as breakaway offshoot of, 55, 56, 63

New Age beliefs, and animal protection, 57

Obama, Barack, 124

Oxfam, "be humankind" campaign, 103, 281n2

Oxford Group, 56, 281n1

personality: animal science of, 150; animal sentience and, 45, 147, 149, 153; vs. character, 276; of companion animals, 152; as evaluation tool, 151–53; expert positions and, 147, 154–55; human, reflections on, 150, 153–54; and moderate animal activism, 219; and moral agency, 147, 153, 275–76; and moral patiency, 153; and office culture, 147, 153; performance of, undercover investigator and, 154, 157, 159, 180, 181, 265; as totem, 149, 155, 180–81; variable, recognition of, 150, 152–53, 181

personhood: companion animals and, 148; strategic valence of, 146, 147, 148–49

PETA (People for the Ethical Treatment of Animals), x; campaign to ban bearskin use in British Army, 12; cruelty-free lifestyle guides of, 126; Edinburgh group compared to, 53–54, 57; expansion of, 4, 53–54; expertise in publicity, 54, 98

Pigeon Trouble (Song), ix, x; inspiration drawn from, xi, xii

Pleasurable Kingdom (Balcombe), 93

policy director/lobbyist: and idiom of heart and head, 221–22, 224–25, 229, 250; moderateness of, virtue of, 218–19, 228–29; personality of, 155; role/expertise of, 46, 219, 228, 229, 231, 236–39, 240–41, 248–50, 264; and undercover investigator, cooperation between, xvi–xvii, 49, 159–60, 241, 243

positivity, ethos of: Edinburgh group and, 44, 91–104, 95f, 124, 189, 196, 198, 240; and kind acts, 44, 100–104; and moral agency, 99–100; Obama campaign as example of, 124; resistance to, 110

posthuman utopian visions, 142–45

pragmatism: Edinburgh animal group and, 14, 53, 73, 88, 122, 144, 231; moderate animal activism and, 218

predator control: legislative debates on, 222; on shooting estates, 166–68, 220, 221; shooting lobby on, 225

primates, nonhuman: practices of habituation in study of, 173–74; use as laboratory animals, 164–65, 174f, 176–78

professionalized animal group(s), 7, 42, 58, 281n2; Edinburgh animal group as, 57–58; vs. radical activists, 253 radical animal activists: charges of collusion by, 252–56, 258, 284n1; charges of inconsistency by, 258–59, 261; charges of naivete/laziness against, 256–57; rejection of compromise by, 264

reason, as moral source, 34

Regan, Tom: on animal rights, 72; *The Case for Animal Rights*, 15–17, 25–26, 284n1; exclusions contained in moral arguments of, 271; and idiom of heart and head, 284n1; influence of, 56, 280n1; on moral agency, 15–17, 111, 269; on moral patiency, 16–17, 25–26, 31, 125, 248, 269; on oscillation as necessary part of moral experience, 285n6

relating, ethics of, 70–73

religion: and animal-friendly forms of "spirituality," 36, 279n14; animal people's suspicion of, 34–35, 37, 215; anthropology of, 26, 27–29; and meat eating, 282n1; and moral patiency position, 26, 27–29

respect, moral patiency and, 269

responsibility: claim of exceptionalism and, 248; consciousness-raising and increasing sense of, 126; dynamic and distributed status of, 37–38; enacted, animal movement and, 112; in encounters with nonhuman animals, 140; Haraway's reconceptualization of, 73, 74; and moral agency, after-the-fact explanations of, 15–16, 38, 132–33; secular-liberal imaginaries and, 38–39, 40

robots/machines: Edinburgh animal group's attitude toward, 271–72; moral status of, 18–19, 270–71, 272

Rogers, Crystal, 280n14

Royal Society for the Prevention of Cruelty to Animals (RSPCA), 53, 162, 192

Sandoval-Cervantes, Iván, 278n2, 283n5

Sarah (assistant fundraiser; pseudonym), xvi, 48; companion animal of, 48; at Girls' Day Out, 83, 108–9; inspiration poster of, 108; rescue work of, 77

Saunders, George, 285n3

Scheler, Max, 105–6, 107

science and research manager: role of, 90–91. *See also* Craig

sci-fi hoax film, 1–6, 71; on alien status of nonhuman animals, 5; and ethos of positivity, 94; moral implications of, 5–6; moral patiency in, 27; and narrative of

making contact, 1–6, 148, 188; political alignment in, 35; social media campaign preceding release of, 1, *2f*

Scotland: animal movement's legislative wins in, xv, 263; and Edinburgh animal group's identity, 3, 58–65; independence referendum in, idiom of heart and head in, 223–24. *See also* shooting estates

Scottish National Party (SNP): animal protection issue and, 235; annual conference of, Edinburgh animal group and fringe meeting at, 242–45, 247–48

Scottish Parliament: Cross-Party Group for Animal Welfare in, 43; devolved powers over animal legislation, 64; Public Petitions Committee (PPC) of, 232; and Wildlife and Natural Environment Bill (WANE bill), xvii, 235–39, 249, 250

Scottish Society for the Prevention of Cruelty to Animals (SSPCA), 53, 162

seal culls: animal groups' positions on, 253–54; legislative wins regarding, xv, 263

secular-liberal imaginary: and animal protection, 33, 35; and moral patiency, 37–38; moral subjects in, 33; and redistribution of responsibility, 38–39, 40; and vegetarianism/veganism, 282n1

self-control: moderate activism and, 219; as moral source, 34

self-ethics, 121–22, 124

self-expression, as moral source, 34

self-interest, being good and, 133–35

self-sacrifice: animal rescue and, 213–14; undercover investigations and, 159, 160–61, 169–72, 175–76

sentience: focus on, ethos of positivity and, 91–93, 94, 95*f*, 99, 102, 149; multiform, sci-fi hoax film on, 5; negative vs. positive dimensions of, 92; personality and, 45, 147, 149, 153

Shelia (office manager; pseudonym), xvi, 48, 151

shooting estates: BBC documentary on, 220–21, 224–25, 232; economic contribution of, debate on, 221, 225; predator control on, 166–68, 220, 221; as target of animal protection, 88; undercover investigations of, 43, 165–69, 172

Singer, Peter: and accusations of complicity and collusion, 252–53, 254; *Animal Liberation*, 9, 15, 112, 252–53; exclusions contained in moral arguments of, 271; influence of, 56, 148, 234, 280n1

slaughterhouses, CCTV cameras in, xv
Sloterdijk, Peter, 24
Smith, Mick, 105–6, 107
snare(s)/snaring: ban on, as conscience matter, 246; campaign against, 52, 64*f*, 165, 236, 240–45, 248–50; legislative debates on, 222; legislative wins regarding, xv, 263; of predators, on shooting estates, 166–68; rabbit caught in, childhood memory of, 11; welfare regulations regarding, 168
social media: campaign preceding sci-fi hoax film, 1, 2*f*; emphasis on kind acts in, 101–2; and ethos of positivity, efforts to promote, 196; Girls' Day Out event and engagements on, 83; idiom of heart and head in, 223; and mainstream, Edinburgh group's project to engage, 1, 85, 101–2, 134; PETA's deployment of, 54; and transnational animal protection, 4
social media strategist, 134. *See also* David
Song, Hoon: on animal cruelty, 189; on early history of animal protection, 55; on "gloved love," 75; inspiration drawn from, xi, xii; *Pigeon Trouble*, ix, x; on undercover investigator, 158, 161
specialist animal groups, 52
SSPCA (Scottish Society for the Prevention of Cruelty to Animals), 53, 162
stakeholders, in legislative process, 232–33; insider, 236–37
statistical analysis, and redistribution of responsibility, 38–39, 40
Strathern, Marilyn, 31, 70
subjectiveless selfhood, moral state of, 180
suffering. *See* animal suffering
surrender, ethos of, 24, 26–27

Taylor, Charles, 34
technology, ethics and, 18–20, 22, 270–71
Tennyson, Alfred, 59
Ticktin, Miriam, 278n3, 282n3, 284n3
"To a Mouse" (Burns), 59, 60
transnational network(s): of animal protection groups, 4, 65; for animal rescue, xvii, 9–10, 191, 200, 203, 205–7

undercover investigator/investigation, 156–86; on animal cruelty, 179, 181–86, 188–89; and animal love, inability to express, 170–71; in circuses, 156–57, 162, 175; and evidence of animal abuse, 39, 40*f*, 157–59, 241, 243; expertise of, 45, 156,

158–59, 160, 161–62, 165–69; history of, 55; in laboratories, 55, 164–65, 172, 176–79; and legal success, rate of, 169–70; moral authority of, 158–59, 179; and moral patiency, 170, 179; personalities cultivated by, 154, 157, 159, 180, 181, 265; and policy work, xvi–xvii, 49, 159–60, 241; practices of habituation in, 172–79; process of becoming, 163–64, 168; relationships formed during, 175–79; and self-sacrifice, 159, 160–61, 169–72, 175–76, 186, 265; on shooting estates, 43, 165–69, 172; as solitary pursuit, 162–63, 169, 170, 186; in UK, 162; unique intermediary position of, 159, 160; as witness to animal abuse, 45, 157–59, 171–72, 179, 181–86; in zoos, 164, 172
United States: animal protection movement, 4, 56; Labor Day pigeon shoot in, ix, 75
University College London, 55
utopian visions, posthuman, 142–45

value-feelings, as basis for moral community, 105, 106, 107
vegetarianism/veganism: Buddhism and, 142–43; challenges associated with, awareness of, 136; concern for consistency and, 123–24, 259–60; and do-no-harm principle, 113; Edinburgh group's project to engage mainstream and, 119–20; ethical choice and, 44, 112–13, 117–18, 120, 122, 123, 127, 130–32, 135–38, 139, 141–42; exemplary status of, impact on other choices, 44–45; and family relations, 135–38; and health, 101, 119, 137, 142; Hinduism and, 280n14, 282n1; in history of animal movement, 112; informed concern and, 229–30; levels of commitment in, 120–21; nonchalant, 114–19, 122, 127–28; reluctant, 210; strong childhood feelings and, 8; utopian vision of, 144; witnessing of animal suffering and, 130–31, 142
vicarious responsibility, 39, 238, 249
virtue ethics, 34

website, of Edinburgh animal group, 81, 83, 94, 98, 103, 134; Amazing Animal Facts section of, 94–95, 99, 134, 150, 151
Weiss, Erica, 279n13
When Species Meet (Haraway), ix
Wildlife and Natural Environment Bill (WANE bill), xvii, 235–39, 249, 250

Williams, Bernard, 37
women: aristocratic and bourgeois, in early animal protection movement, 55, 65; engaging, Edinburgh animal group's effort toward, 79–83, 86; in National Anti-Vivisection Society (NAVS), 56
World War II, conscientious objectors during, 245–46, 247

"The Wounded Hare" (Burns), 60–61
Wrenn, Corey Lee, 281n2

zoo(s): closure of, utopian vision of, 145; practices of habituation in, 173; as target of animal protection, 88, 90; undercover investigations in, 164, 172; working in, experience of, 163
Zoo Check, 164